Current Developments in Mathematics

2012

edited by

David Jerison
Mark Kisin
Tomasz Mrowka
Richard Stanley
Horng-Tzer Yau
Shing-Tung Yau

International Press
www.intlpress.com

Current Developments in Mathematics, 2012

Editorial board:

Mark Kisin
Horng-Tzer Yau
Shing-Tung Yau

David Jerison
Tomasz Mrowka
Richard Stanley

Harvard University
Cambridge, Massachusetts, U.S.A.

Massachusetts Institute of Technology
Cambridge, Massachusetts, U.S.A.

ISBN 978-1-57146-240-4

Typeset using the LaTeX system.

Contents

Classifying automorphic representations

James Arthur

ABSTRACT. We shall describe a recent classification of the automorphic representations of orthogonal and symplectic groups. In order to place it in perspective, we shall also review Langlands' general principle of functoriality and related matters. Finally, we shall add a few remarks on the explicit application of the results to those cases of functoriality on which they bear.

CONTENTS

Preface

This article is an introduction to the monograph [**ECR**], the purpose of which was to classify the automorphic representations of a family of classical groups. The groups are quasisplit, special orthogonal and symplectic groups G. Their representations are classified in terms of those of general linear groups $GL(N)$. The monograph is based on the stabilization of the trace formula for G, established for any connected group in [**A1**]. It also depends on the stabilization of the twisted trace formula for $GL(N)$, which represents work in progress by Moeglin and Waldspurger [**W5**]–[**W7**], [**MW2**]. Until it has been completed, the classification will remain conditional.

There are already two short surveys [**A4**], [**A5**] of some of the main results of [**ECR**]. This article is somewhat different. I have tried to write it as a longer

2010 *Mathematics Subject Classification.* Primary 11F70, 11F66.

report that might be suitable for the broader readership of Current Developments in Mathematics. The monograph [**ECR**] is long, and often quite complex. It draws on techniques from many diverse sides of the subject, which would be hard to present in any detail here. Moreover, what has been included in this report might still be difficult in places. I hope, however, that I have given enough motivation to offer some perspective on the modern theory of automorphic forms, as well as the actual contents of [**ECR**].

For the most part, we confine ourselves to the orthogonal and symplectic groups G whose representations we classify. We recall these groups in §1, and their relations with Langlands' principle of functoriality. In §2, we discuss the automorphic Langlands group L_F. Its existence is far from known, but its expected properties offer much guidance. In §3, we describe how Langlands parameters and their generalizations, which in their global form would be defined on the hypothetical group L_F, suggest how to relate representations of G with those of general linear groups $GL(N)$. This simple exercise in linear algebra also provides an entry into the theory of endoscopy, which underlies the statements (and proofs) of our theorems.

The representation theory of $GL(N)$ is relatively simple, and quite well understood now, thanks to the work of a number of mathematicians over the past forty years. We shall review some of it in §4, taking the opportunity also to review the theory of arithmetic and automorphic L-functions. We shall use the automorphic representations of $GL(N)$ in §5 as a foundation for ad hoc global parameters for G that do not depend on the hypothetical group L_F. The construction requires two "seed" theorems (Theorems 5.1 and 5.2), which we state but (like everything else) do not prove.

In §6, we review the theory of endoscopy, whose spectral roots we encountered in §3. We use it to state a critical local result (Theorem 6.1), which is the starting point for the local classification. We will then be in a position to state the main theorems (Theorems 7.1, 7.2 and 7.3) in §7. In the final §8, we will add some supplementary comments on how the theorems relate to the two fundamental cases of functoriality discussed in §1. These observations do not appear in [**ECR**].

We follow the discussion from [**ECR**] closely in some places, and reorder it in others. We are also including supplementary background material from the theory of automorphic forms and the Langlands program. We have not tried to cross-reference statements here with those of [**ECR**], but a reader will have no difficulty seeing how they correspond. This report takes us through Sections 1.1–2.1 (and the beginning of §2.2) of [**ECR**], the part of the monograph given to the statements of the main theorems. Most of the rest of the monograph, specifically Sections 2.3–8.2 (and the remaining part of §2.2), is devoted to the proofs. The argument is long and complex, but it has a certain unity. It comes with several layers of induction, of which we will see hints in §5. We will add nothing further to this, except to note that the heart of the argument is an endoscopic comparison of trace formulas.

Acknowledgments. The author was supported in part by NSERC Grant A3483.

1. Classical groups and functoriality

The groups we consider will be attached to the four infinite families of complex simple Lie algebras. These are represented by the following four infinite families of Coxeter-Dynkin diagrams, for which I am indebted to W. Casselman.

Type $\mathbf{A_n}$

Type $\mathbf{B_n}$

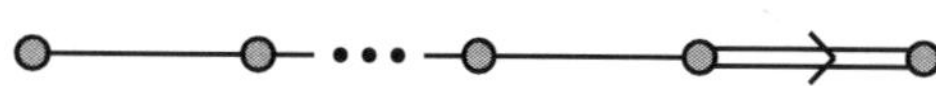

Type $\mathbf{C_n}$

Type $\mathbf{D_n}$

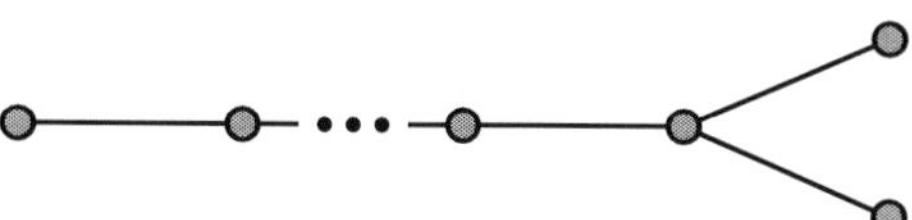

For corresponding complex groups, we could take the special linear groups $SL(n+1,\mathbb{C})$, the odd orthogonal groups $SO(2n+1,\mathbb{C})$, the symplectic groups $Sp(2n,\mathbb{C})$ and the even orthogonal groups $SO(2n,\mathbb{C})$. The family $\mathbf{A}_n$ will be our starting point. Since the theory here is simplest for general linear groups, we will take the reductive groups $GL(N,\mathbb{C})$, $N = n+1$, as the complex representatives for this family.

We actually want to take these groups over a number field F (such as the rational numbers $\mathbb{Q}$) or one of its completions (such as the real number field $\mathbb{R}$ or a p-adic field $\mathbb{Q}_p$). We therefore take F to be any local or global field of characteristic 0.

We recall the fundamental theorem of Chevalley, which implies that any of these complex groups has a canonical F-structure. In other words, for each of the diagrams, there is a canonical classical group that is defined over F. It is the *split* group attached to the given diagram (and centre). Our interest is actually in quasisplit groups. They represent a broader class of groups, obtained by twisting any given split group by a Galois action on the diagram. The symmetry group of a diagram is the group of bijections of the set of vertices that preserve all of the edges and arrows. It is isomorphic to $\mathbb{Z}/2\mathbb{Z}$ in type $\mathbf{A}_n$, trivial in types $\mathbf{B}_n$ and $\mathbf{C}_n$, and equal[1] to $\mathbb{Z}/2\mathbb{Z}$ in type $\mathbf{D}_n$. A quasisplit group is determined by a homomorphism from the Galois group

$$\Gamma_F = \Gamma_{\overline{F}/F} = \mathrm{Gal}(\overline{F}/F)$$

---

[1] If $n = 4$, this group is actually isomorphic to S_3, but we agree to consider only the standard symmetries that interchange the two right hand vertices in the diagram.

of an algebraic closure $\overline{F}$ over F to the symmetry group of the diagram. The monograph [**ECR**] does not treat nonsplit, quasisplit groups of type $\mathbf{A}_n$. (These are unitary groups, for which we refer the reader to [**Mok**].) Since a quasisplit group of type $\mathbf{B}_n$ or $\mathbf{C}_n$ is split, we have only to consider type $\mathbf{D}_n$. In this case, a quasisplit group is determined by a quotient of Γ_F of order 1 or 2, or in other words, a Galois extension E/F of degree 1 or 2.

One of the remarkable discoveries of Langlands has been the fundamental role played by a certain dual group. We take G to be the group over F we are working with, either a general linear group $GL(N)$ or a quasisplit special orthogonal or symplectic group. The dual group $\widehat{G}$ is a complex classical group, which is attached to the dual diagram obtained by reversing the directions of any arrows in the diagram of G. Chevalley's theorem is based on an identification of the symmetry group of the diagram with the group of outer automorphisms of the split group (or rather, a group of F-automorphisms that represent those outer automorphisms). This transfers to a dual Galois action of Γ_F by complex analytic automorphisms of $\widehat{G}$, which factors through the quotient

$$\Gamma_{E/F} = \mathrm{Gal}(E/F)$$

of Γ_F. Langlands built this action into the dual group by forming the semidirect product

$$^L G = \widehat{G} \rtimes \Gamma_F$$

that is now known as the L-group.

The L-group is actually a more concrete object than might be suggested by the large Galois factor. For many purposes, one can replace Γ_F by any quotient $\Gamma_{E/F}$ through which the Galois action factors, and in particular, by the minimal such quotient above. If G is split, for example, one can often take $E = F$ and $^L G = \widehat{G}$. Equipped with this minimal form $^L G = {}^L G_{E/F}$ of the L-group, our four families of groups are then as follows.

Type $\mathbf{A_n}$: $G = GL(N)$ is split, and $\widehat{G} = GL(N, \mathbb{C}) = {}^L G$, where $N = n + 1$.

Type $\mathbf{B_n}$: $G = SO(2n + 1)$ is split, and $\widehat{G} = Sp(2n, \mathbb{C}) = {}^L G$.

Type $\mathbf{C_n}$: $G = Sp(2n)$ is split, and $\widehat{G} = SO(2n + 1, \mathbb{C}) = {}^L G$.

Type $\mathbf{D_n}$: $G = SO(2n)$ is quasisplit, and $\widehat{G} = SO(2n, \mathbb{C})$;
$\qquad\qquad {}^L G = SO(2n, \mathbb{C}) \rtimes \Gamma_{E/F}$, where $\deg(E/F) \in \{1, 2\}$.

To keep the scope of the article within bounds, we will generally confine our discussion to groups from this list. Unless otherwise stated, groups G, G', etc. will be assumed to be taken from our four families of quasisplit classical groups, or possibly to be direct products of such groups. In fact, we will often restrict G to be of type $\mathbf{B}_n$, $\mathbf{C}_n$ or $\mathbf{D}_n$, and explicitly write $GL(N)$ for any one of our groups of type $\mathbf{A}_n$.

We are interested in the representation theory of G. If F is local, our concern will be the set $\Pi(G)$ of equivalence classes of irreducible representations of $G(F)$, together with its subsets

$$\Pi_{\mathrm{temp}}(G) \subset \Pi_{\mathrm{unit}}(G) \subset \Pi(G)$$

of representations that are respectively *tempered* and *unitary*. (A tempered representation can be described informally as an irreducible representation π such that the tensor product

$$\pi \otimes \pi^{\vee}, \qquad \pi^{\vee}(x) = {}^{t}\pi(x)^{-1},\ x \in G(F),$$

occurs in the decomposition of the regular representation

$$\big(R(y_1, y_2)\phi\big)(x) = \phi(y_1^{-1}xy_2), \qquad x, y_1, y_2 \in G(F),\ \phi \in L^2\big(G(F)\big),$$

of $G(F) \times G(F)$ on the Hilbert space $L^2\big(G(F)\big)$. It is automatically unitary.) If F is global, we are interested in the set $\Pi(G)$ of automorphic representations of G. These are irreducible representations of the adelic group $G(\mathbb{A})$, which are of a very special sort.

We recall that the adeles $\mathbb{A} = \mathbb{A}_F$ of F form a locally compact ring, in which F embeds as a discrete subring. The adelic group is then a restricted direct product

$$(1.1) \qquad\qquad G(\mathbb{A}) = \widetilde{\prod_{v}} G(F_v),$$

taken over the valuations v on F. For any v, F_v is the locally compact field obtained by completing F with respect to v. It is modeled on the standard case of the completion $F_v = \mathbb{R}$ of $F = \mathbb{Q}$ with respect to the usual absolute value $|\cdot|_v = |\cdot|$. We recall that the complementary valuations for $F = \mathbb{Q}$ are the nonnegative functions

$$|u|_p = \begin{cases} p^{-r}, & \text{if } u = \dfrac{a}{b}p^r,\ \text{for } a, b, r \in \mathbb{Z},\ (a, p) = (b, p) = 1, \\ 0, & \text{if } u = 0, \end{cases}$$

on $\mathbb{Q}$, parametrized by prime numbers p. In general, the restricted direct product is the group of elements

$$x = \prod_{v} x_v, \qquad x_v \in G(F_v),$$

in the direct product such that for almost all valuations v, x_v lies in the maximal compact subgroup $G(\mathfrak{o}_v)$ of points in $G(F_v)$ with values in the compact subring

$$\mathfrak{o}_v = \{u_v \in F_v : |u_v|_v \leq 1\}$$

of integers in F_v. It becomes a locally compact group under the appropriate direct limit topology. The group $G(F)$ embeds in $G(F_v)$ (as a dense subgroup). The diagonal embedding of $G(F)$ into $G(\mathbb{A})$ exists (because an element in $G(F)$ is integral at almost all valuations v), and is easily seen to have discrete image.

Since $G(F)$ is discrete in $G(\mathbb{A})$, the quotient $G(F)\backslash G(\mathbb{A})$ is a reasonable object. It comes with a right invariant measure, which is determined up to a positive multiplicative constant. One can therefore form the associated space $L^2\big(G(F)\backslash G(\mathbb{A})\big)$ of square-integrable functions. It is a Hilbert space, equipped with the unitary representation

$$\big(R(y)\phi\big)(x) = \phi(xy), \qquad x, y \in G(\mathbb{A}),\ \phi \in L^2\big(G(F)\backslash G(\mathbb{A})\big),$$

of $G(\mathbb{A})$ by right translation. An *automorphic representation* is an irreducible representation of $G(\mathbb{A})$ that occurs in the spectral decomposition of R.

The description of an automorphic representation just given is more of an informal characterization than a definition. It can be made precise, but it is also more restrictive than the formal definition[2] in [**L3**]. The representation R has a subrepresentation R_{disc} that decomposes discretely[3] into a direct sum of irreducible representations (like the 1-dimensional representations

$$x \longrightarrow e^{2\pi i n x}, \qquad x \in \mathbb{R},\ n \in \mathbb{Z},$$

from the theory of Fourier series), and a complementary subrepresentation R_{cont} that decomposes continuously into a direct integral of irreducible representations (like the 1-dimensional representations

$$x \longrightarrow e^{\lambda x}, \qquad x \in \mathbb{R},\ \lambda \in i\mathbb{R},$$

from the theory of Fourier transforms). The point is that an automorphic representation can occur in either the discrete or the continuous spectrum. In another sense, however, the analogy with classical Fourier analysis is misleading. This is because an automorphic representation π of $G(\mathbb{A})$ has much more structure than these familiar 1-dimensional representations. According to [**F**], any such π can be written as a restricted tensor product

$$(1.2) \qquad\qquad \pi = \overset{\sim}{\bigotimes_{v}} \pi_v, \qquad \pi_v \in \Pi_{\text{unit}}(G_v),$$

of irreducible unitary representations π_v for the local groups G_v over F_v. It is one thing to be able to put together an irreducible representation π of $G(\mathbb{A})$ explicitly as a product of this sort. We would require some knowledge of the sets $\Pi_{\text{unit}}(G_v)$, an interesting problem to be sure, and one that has not been solved in general, but nothing further. It is quite another matter to determine which such products are automorphic. This new constraint imposes profound relations among the different constituents π_v of π.

In fact, it is pretty clear that we will never be able to write down all automorphic representations explicitly. Langlands' point of view was more subtle. He discovered unexpected reciprocity laws among the automorphic representations for different groups. Thus, although an explicit construction of all automorphic representations will not be available, there will still be hidden ties that bind different automorphic representations. Langlands formulated them as a fundamental principle [**L1**], which later became known as functoriality. It pertains to any analytic homomorphism ρ between two L-groups $^LG'$ and ^LG, taken as groups over the full Galois group Γ_F, that commutes with the two projections onto Γ_F. (A homomorphism with these properties is called an *L-homomorphism*.)

PRINCIPLE 1.1 (Langlands Functoriality). *Suppose that G and G' are general quasisplit groups over a number field F. Then any L-homomorphism*

$$\rho : {}^LG' \longrightarrow {}^LG$$

between their L-groups determines a natural correspondence

$$\pi' \longrightarrow \pi$$

---

[2]This refers to the definition on p. 203 of [**L3**] that was shown in Proposition 2 of [**L3**] to be equivalent to the formal definition of an automorphic representation in [**BJ**].

[3] To be correct, we should really be speaking of the *relative discrete spectrum*. That is, R_{disc} is the representation of $G(\mathbb{A})$ on the invariant subspace $L^2_{\text{disc}}(G(F)\backslash G(\mathbb{A}))$ of $L^2(G(F)\backslash G(\mathbb{A}))$ that decomposes discretely modulo the centre $Z = Z(G)$ of G. This distinction is only relevant to the reductive group $GL(N)$, in which $Z(F)\backslash Z(\mathbb{A})$ is noncompact.

of their automorphic representations, which is compatible with their local decomposi-tions (1.1), and depends only on the orbit of ρ under the action of $\widehat{G}$ by conjugation on $^L G$.

By correspondence we mean a *relation* rather than a function, and for this assertion, automorphic representations are to be understood in the inclusive sense of the precise definition in [**L3**], which we have not given, rather than the restrictive characterization we have given. Langlands also stated a local form of the principle of functoriality, which applies to any local field F of characteristic 0. The compatibility assertion for global F above is that if

$$\pi' \longrightarrow \pi,$$

then

$$\pi'_v \longrightarrow \pi_v,$$

for any completion F_v of F. In other words, if π' and π are automorphic representa-tions of G' and G that correspond under global functoriality for ρ, then their local components π'_v and π_v correspond under local functoriality for the completion

$$\rho_v : \ ^L G'_v \longrightarrow \ ^L G_v$$

of ρ.

This is the informal version of functoriality stated by Langlands in his original paper [**L1**]. It leaves unspecified the nature of the correspondence. Nevertheless, it still gives a sense of the depth of what was a completely new phenomenon. What-ever might be the internal relations among the constituents π_v of the automorphic representation π of G, they will be reflections of the corresponding internal relations among the constituents π'_v of the automorphic representation π' of G'. This will become clearer when we describe more precise versions of functoriality in the final section of the paper.

The principle of functoriality is one of the great problems of mathematics. It has been established in a significant number of cases. However, these pale in comparison with the cases that remain unknown. Langlands has introduced some striking ideas for attacking the general case through a completely new application of the trace formula. (See [**FLN**], [**L6**] and [**A3**, Afterword]. It is not known, however, how far these ideas will ultimately take us. Even if they work in principle, there will still be many years of effort required by many mathematicians before they can be fully realized.

One of the goals of the monograph [**ECR**] was to establish the principle of functoriality in two basic cases. We shall describe them in turn.

The first case arises from the natural embedding of a complex classical group into a complex general linear group. Suppose that G belongs to one of our families $\mathbf{B}_n$, $\mathbf{C}_n$ or $\mathbf{D}_n$. There is then a canonical embedding of the dual group $\widehat{G}$ into a general linear group $GL(N, \mathbb{C})$, for N equal to $2n$, $2n + 1$ and $2n$ respectively. If G is split over F, this extends trivially to a canonical embedding

$$^L G = \widehat{G} \times \Gamma_F \longrightarrow \ ^L\big(GL(N)\big) = GL(N, \mathbb{C}) \times \Gamma_F$$

of the full L-group of G to that of $GL(N)$. In the special case of type $\mathbf{C}_n$, we also obtain a nonstandard embedding of $^L G$ into $^L\big(GL(N)\big)$ for any quadratic extension E/F, by mapping the quotient $\Gamma_{E/F} = \Gamma_F/\Gamma_E$ isomorphically into the central subgroup $\{\pm 1\}$ of the image of $O(2n+1, \mathbb{C})$ in $GL(N, \mathbb{C})$. If G is not split over F, it is of type $\mathbf{D}_n$. The associated quadratic quotient $\Gamma_{E/F}$ then acts on $\widehat{G} = SO(2n, \mathbb{C})$ through the nonidentity connected component of the complex group $O(2n, \mathbb{C})$. This leads again to a canonical embedding

$$^L G = \widehat{G} \rtimes \Gamma_F \ \longrightarrow \ {}^L(GL(N)) = GL(N, \mathbb{C}) \times \Gamma_F$$

of L-groups. We will discuss the various L-embeddings for this basic case of functoriality from a different perspective in Section 3.

In the second case, G is above. This time, however, we take a product

$$G' = G_1' \times G_2'$$

of smaller such groups. We require that the dual group

$$\widehat{G}' = \widehat{G}_1' \times \widehat{G}_2'$$

come with a natural embedding into $\widehat{G}$. This means that

$$\widehat{G}' = Sp(2m, \mathbb{C}) \times Sp(2n - 2m, \mathbb{C}) \subset Sp(2n, \mathbb{C}) = \widehat{G},$$

$$\widehat{G}' = SO(2m, \mathbb{C}) \times SO(2n + 1 - 2m, \mathbb{C}) \subset SO(2n + 1, \mathbb{C}) = \widehat{G},$$

and

$$\widehat{G}' = SO(2m, \mathbb{C}) \times SO(2n - 2m, \mathbb{C}) \subset SO(2n, \mathbb{C}) = \widehat{G},$$

for integers $0 \leq m \leq n$, where G is of type $\mathbf{B}_n$, $\mathbf{C}_n$ and $\mathbf{D}_n$ respectively. If G is of type $\mathbf{B}_n$, G' is split, and the embedding of $^L G'$ into $^L G$ extends trivially to an L-embedding of $^L G'$ into $^L G$. If G is of type $\mathbf{C}_n$, G and G_2' are split, but G_1' can be a quasisplit group defined by a quadratic extension E_1 of F. In this case, we obtain an L-embedding

$$^L G' = (\widehat{G}_1' \times \widehat{G}_2') \rtimes \Gamma_F \ \longrightarrow \ {}^L G = \widehat{G} \times \Gamma_F$$

from the nonstandard embedding of the second factor $^L G_2'$ attached to the quadratic extension E_1/F. Finally, if G is of type $\mathbf{D}_n$, it is the quasisplit group defined by an extension $E = F(\sqrt{d})$ of degree 1 or 2. We can then take G_1' and G_2' to be quasisplit groups of types D_m and D_{n-m} defined by any quadratic extensions $E_1 = F(\sqrt{d_1})$ and $E_2 = F(\sqrt{d_2})$ such that $d_1 d_2$ equals d. It is then easy to see that there is a canonical embedding

$$^L G' = (\widehat{G}_1' \times \widehat{G}_2') \rtimes \Gamma_F \ \longrightarrow \ {}^L G = \widehat{G} \rtimes \Gamma_F$$

of L-groups.

We thus obtain two basic cases of the principle of functoriality by taking the L-homomorphism ρ of Principle 1.1 to be any one of the L-embeddings we have just described. The first is at the heart of the classification of representations of G (both local and global) in terms of those of $GL(N)$. The second provides the foundation for an understanding of the precise functorial correspondence from G to $GL(N)$.

2. The automorphic Langlands group L_F

About ten years after formulating the principle of functoriality [**L1**], Langlands introduced something just as surprising [**L4**]. It is a hypothetical group that would be universal, in the sense that it could be used to characterize the automorphic representations of any group (connected and reductive, but not necessarily belonging to one of our four families) over a number field F. The existence of the Langlands group is closely related to the principle of functoriality, even somewhat deeper, but its hypothetical properties are still very useful for guidance and motivation.

In the earlier paper [**L1**], Langlands had already predicted a major role for a smaller group, the variant of the absolute Galois group Γ_F introduced by Weil. The Weil group W_F is a locally compact group that is defined if F is either local or global. It comes with a continuous homomorphism

$$\phi: W_F \longrightarrow \Gamma_F,$$

with dense image and connected kernel, under which the preimage of any subgroup Γ_E of Γ_F of finite index equals the Weil group W_E of the associated finite field extension E of F. For each such E, W_F is equipped also with a topological isomorphism

$$r_E: C_E \longrightarrow W_E^{\mathrm{ab}}$$

where $W_E^{\mathrm{ab}} = W_E/W_E^c$ is the abelianization of W_E, and

$$C_E = \begin{cases} E^*, \text{ the multiplicative group of } E, \text{ if } E \text{ is local,} \\ \mathbb{A}_E^*/E^*, \text{ the idèle class group of } E, \text{ if } E \text{ is global.} \end{cases}$$

The triplet $(W_F, \phi, \{r_E\})$, which one usually denotes simply by W_F, is subject to four natural conditions [**T**, §1.1, (W_1)–(W_4)]. With these conditions, W_F is defined and uniquely determined up to an isomorphism, which itself is uniquely determined up to conjugation by an element in the kernel of ϕ. (See [**T**].)

If $F = \mathbb{C}$, for example, W_F equals the multiplicative group $\mathbb{C}^*$. If $F = \mathbb{R}$, W_F is a nontrivial central extension

$$1 \longrightarrow W_\mathbb{C} \longrightarrow W_\mathbb{R} \longrightarrow \mathbb{Z}/2\mathbb{Z} \longrightarrow 1$$

of $W_\mathbb{C}$. It can be identified with the explicit group generated by $\mathbb{C}^*$ and a symbol $\sigma_\mathbb{R}$ with $\sigma_\mathbb{R}^2 = -1$, subject to the conjugation relation

$$\sigma_\mathbb{R} z \sigma_\mathbb{R} = \overline{z}, \qquad z \in \mathbb{C}^*.$$

If F is a local field that is nonarchimedean (which is to say that it is not equal to $\mathbb{C}$ or $\mathbb{R}$), W_F is an extension

$$1 \longrightarrow I_F \longrightarrow W_F \longrightarrow \langle \mathrm{Frob} \rangle \longrightarrow 1$$

of an infinite cyclic group by a compact subgroup I_F (the inertia group) of Γ_F. The element Frob represents the Frobenius automorphism in the absolute Galois group

$$\mathrm{Gal}(\overline{k}/k) \cong \Gamma_F/I_F$$

of the residue field k of F, a compact, totally disconnected group, in which the cyclic group generated by Frob is dense. In this case, W_F is still a reasonably explicit object. If F is global, W_F is more complicated. But as in the local case, the computations with W_F that arise in representation theory can often be made quite explicit.

The global Weil group has some extra structure. It comes with embeddings of the local Weil groups W_{F_v}, which are determined up to conjugacy, and are compatible with the associated embeddings of Galois groups. In other words, for every valuation v on F, we have a commutative diagram

$$(2.1) \qquad \begin{array}{ccc} W_{F_v} & \longrightarrow & \Gamma_{F_v} \\ \cup & & \cup \\ \downarrow & & \downarrow \\ W_F & \longrightarrow & \Gamma_F \end{array}$$

in which the vertical embeddings are determined up to conjugacy in W_F and Γ_F.

Suppose that G is one of our groups over F. Since

$$W_F/W_E \cong \Gamma_F/\Gamma_E,$$

for any finite extension E of F, the Weil group acts on $\widehat{G}$. We can therefore take the Weil form

$$^L G = \widehat{G} \rtimes W_F$$

of the L-group. This seems more cumbersome than the Galois form, especially if we take the minimal Galois form with Γ_F replaced by the finite group $\Gamma_{E/F}$. However, it is ultimately the best version of the L-group to work with. We can still talk about L-homomorphisms in this context, and we are free to formulate the principle of functoriality as in §1. We shall use the Weil form of the L-group in the rest of this section, as well as in the next.

Langlands conjectured the existence of a natural mapping $\phi \to \Pi_\phi$, from L-homomorphisms

$$\phi \colon W_F \longrightarrow {}^L G,$$

taken up to conjugacy of $^L G$ by its subgroup $\widehat{G}$, to packets Π_ϕ of irreducible representations. The proposed sets Π_ϕ later become known as L-packets, since they were conjectured to consist of representations with the same L-functions and ε-factors. For local F, they should be finite subsets of $\Pi(G)$. For global F, they should be compatible with the localizations $\phi \to \phi_v$ defined by the embeddings (2.1). More precisely, the global L-packet of a global parameter ϕ would simply be defined as the set

$$(2.2) \qquad \Pi_\phi = \left\{ \pi = \overset{\sim}{\bigotimes_v} \pi_v \; : \; \pi_v \in \Pi_{\phi_v}, \; \pi_v \text{ unramified}^4 \text{ for almost all } v \right\}$$

of irreducible representations of $G(\mathbb{A})$. The conjectural part of this global definition can be taken as an assertion that the global L-functions attached to representations $\pi \in \Pi_\phi$ have analytic continuation and functional equation. This conjecture, local and global, became known as the Langlands correspondence. Notice that W_F can be regarded as the Weil form of the L-group $^L G'$, for the trivial group $G' = \{1\}$ over F. The Langlands correspondence can therefore be regarded as a very special case of principle of functoriality. It is an illustration of the depth of functoriality, as a conjecture that governs the very foundations of the subject in spite of its seemingly innocuous statement.

If F equals $\mathbb{C}$ or $\mathbb{R}$, Langlands established a correspondence $\phi \to \Pi_\phi$ for any G over F. Part of the problem was of course to formulate the definitions, which necessarily go to the roots of Harish-Chandra's monumental contributions to harmonic analysis on real groups. Langlands then showed that $\Pi(G)$ is a disjoint union of the packets Π_ϕ, taken over the ($\widehat{G}$-orbits) of L-homomorphisms ϕ. His results can be regarded as a classification of the representations of a real group.

---

^{4} We will leave this notion undefined for the present, and return to it in §4.

If F is local nonarchimedean, much less is known, in part because there is no p-adic analogue of Harish-Chandra's classification of the discrete series. It was also clear from early examples that the L-packets Π_ϕ would not exhaust the representations in $\Pi(G)$. If F is global, the problem not surprisingly is much deeper. In particular, it is known that the automorphic representations $\pi \in \Pi(G)$ that happen to lie in some Weil packet Π_ϕ are really quite sparse. It was with this understanding that Langlands was led to introduce a larger group that would replace W_F in his conjectural correspondence.

Assume that F is global. In Langlands' original article [**L4**], the universal automorphic Galois group was to be an object in the category of complex, reductive pro-algebraic groups. It was formulated as an extension of the absolute Galois group Γ_F by a connected, complex, reductive pro-algebraic group. Kottwitz [**K**] later pointed out that the group would be simpler if it were taken in the category of locally compact topological groups, like the Weil group itself. In this formulation, the global Langlands group L_F should be an extension of W_F by a connected, compact group. It would thus take its place in a sequence

$$L_F \longrightarrow W_F \longrightarrow \Gamma_F$$

of three locally compact groups, all having fundamental ties to the arithmetic of F. The group L_F should also have a local analogue L_{F_v}, for any valuation v on F, that embeds into an extension

$$(2.3) \qquad \begin{array}{ccc} L_{F_v} & \longrightarrow W_{F_v} & \longrightarrow \Gamma_{F_v} \\ \cap \quad & \cap \quad & \cap \quad \\ \downarrow \quad & \downarrow \quad & \downarrow \quad \\ L_F & \longrightarrow W_F & \longrightarrow \Gamma_F \end{array}$$

of the commutative diagram (2.1). In particular, L_F should be equipped with a canonical embedding of L_{F_v}, which as in the cases of W_F and Γ_F, is determined only up to conjugacy.

Suppose that F is local. The local Langlands group L_F is known. It is defined as

$$(2.4) \qquad L_F = \begin{cases} W_F, & \text{if } F \text{ is archimedean,} \\ W_F \times SU(2), & \text{if } F \text{ is nonarchimedean.} \end{cases}$$

Given G over F, we write $\Phi(G)$ for the set of L-homomorphisms

$$(2.5) \qquad \phi : L_F \longrightarrow {}^L G,$$

taken up to $\widehat{G}$-conjugacy. The conjectural Langlands correspondence for G again takes the form of a mapping

$$\phi \longrightarrow \Pi_\phi, \qquad \phi \in \Phi(G),$$

from $\Phi(G)$ to finite subsets Π_ϕ of $\Pi(G)$. However, this time it should have the property

$$\Pi(G) = \coprod_{\phi \in \Phi(G)} \Pi_\phi$$

of exhaustion. In particular, the extra unitary factor $SU(2)$ in the definition (2.4) would be what is needed to obtain a full classification in case F is p-adic. It is equivalent to the supplementary data that had been used earlier to construct the Weil-Deligne group. (See [**T**], [**L4**].)

Suppose again that F is global. The global Langlands group L_F is far from known. It would be much larger than W_F, for the reason that "most" automorphic representations contain information that cannot be reduced to something as simple as the Weil group. In the article [**A2**], we described a conjectural construction of L_F. It is given by an extension

$$(2.6) \qquad\qquad 1 \longrightarrow K_F \longrightarrow L_F \longrightarrow W_F \longrightarrow 1$$

of W_F by an infinite product K_F of compact, connected, *simply connected* groups. The factors of K_F are parametrized by certain very basic automorphic representations of G (which we called *primitive* in [**A2**]), as G ranges over *all* simply connected, quasisplit groups. The group L_F would then be somewhat self-referential, in that it is supposed to classify automorphic representations, and yet at the same time, is built out of certain automorphic representations. This does not make its existence any easier to establish. One requires the principle of functoriality for all groups G, and more, to define the primitive automorphic representations at the core of L_F.

In any case, the global Langlands group is supposed to be characterized in terms of automorphic representations of general linear groups. It was predicated by Langlands on the assumption that for any N, there is a bijective correspondence between irreducible N-dimensional representations of L_F, and cuspidal automorphic representations of the group $GL(N)$. We recall that the cuspidal automorphic representations are the fundamental building blocks for general automorphic representations, and always occur in the discrete spectrum. (See [**BJ**, p. 191–197] and [**L3**, Proposition 2].) More generally, given one of our groups G over F, and assuming the existence of the global group L_F, we define the associated set of L-homomorphisms $\Phi(G)$ as in the local case (2.5). The elements $\phi \in \Phi(G)$ would again parametrize global L-packets Π_ϕ, defined in terms of their local components by (2.2). As in the special case of a Weil parameter, we could state the conjectural part of this global definition by saying that global L-functions attached to representations $\pi \in \Pi_\phi$ should have analytic continuation and functional equation.

In terms of representation theory, the real global conjecture concerns automorphic representations in a packet Π_ϕ. It is convenient to write $\Phi_{\mathrm{bdd}}(G)$ (in both the local and global cases) for the subset of parameters in $\Phi(G)$ whose image in $^L G$ projects onto a bounded (that is, relatively compact) subset of $\widehat{G}$. If $G = GL(N)$, for example, elements in $\Phi_{\mathrm{bdd}}(G)$ project to N-dimensional representations of L_F that are unitary. With F being global, there will be no automorphic representations π in the packet Π_ϕ of any global parameter in the complement of $\Phi_{\mathrm{bdd}}(G)$. This is only because we have taken the restricted definition of automorphic representations, as elements $\pi \in \Pi(G)$ that occur in the automorphic spectral decomposition, rather than including the supplementary representations obtained by analytic continuation into the complex domain, as in [**L3**]. By the same token, if F is local, there will be no tempered representations π in the packet of a local parameter ϕ taken from the complement of $\Phi_{\mathrm{bdd}}(G)$.

If F is global, the packets Π_ϕ of parameters $\phi \in \Phi_{\mathrm{bdd}}(G)$ should contain many automorphic representations π. However, they will still not exhaust the set $\Pi(G)$ of all automorphic representations. The situation here is like that of a nonarchimedean local field F, where we had to define local parameters on the product $W_F \times SU(2)$ rather than just W_F. The final step for the global field F is likewise to add a factor $SU(2)$, and form a product $L_F \times SU(2)$. The localization of this group at v, which will be needed to construct the local components of representations attached to the resulting global parameters, will be a new product $L_{F_v} \times SU(2)$. In particular, if F_v is nonarchimedean, we will be dealing with a local product

$$W_{F_v} \times SU(2) \times SU(2)$$

with two factors $SU(2)$.

If F is local or global, we write $\Psi(G)$ for the set of L-homomorphisms

$$\psi : \ L_F \times SU(2) \ \longrightarrow \ {}^L G,$$

taken up to $\widehat{G}$-conjugacy, but with the property that the restriction of ψ to L_F lies in the subset $\Phi_{\mathrm{bdd}}(G)$ of $\Phi(G)$. If F is global, the diagram (2.3) gives a localization mapping[5]

$$\psi \ \longrightarrow \ \psi_v$$

from $\Psi(G)$ to $\Psi(G_v)$. If F is local, we conjecture that there is again a natural mapping $\psi \to \Pi_\psi$, this time from parameters $\psi \in \Psi(G)$ to finite subsets Π_ψ of $\Pi_{\mathrm{unit}}(G)$. If F is global, we will then be able to attach a global packet

$$(2.7) \qquad \Pi_\psi = \left\{ \pi = \overset{\sim}{\bigotimes_v} \pi_v : \ \pi_v \in \Pi_{\pi_v}, \ \pi_v \text{ unramified for almost all } v \right\}$$

of irreducible unitary representations of $G(\mathbb{A})$ to parameters $\psi \in \Psi(G)$. Many of the representations in these packets should be automorphic. Moreover, they ought now to exhaust the set of all automorphic representations.

The set $\Psi(G)$, defined for F local or global, is where the process ends. It contains $\Phi_{\mathrm{bdd}}(G)$, as the set of parameters on $L_F \times SU(2)$ that vanish on the second factor. The set $\Phi_{\mathrm{bdd}}(G)$ in turn contains the set $\Phi_{\mathrm{bdd}}^W(G)$ of bounded parameters on W_F, as the parameters on L_F that are pullbacks from W_F. We obtain embeddings

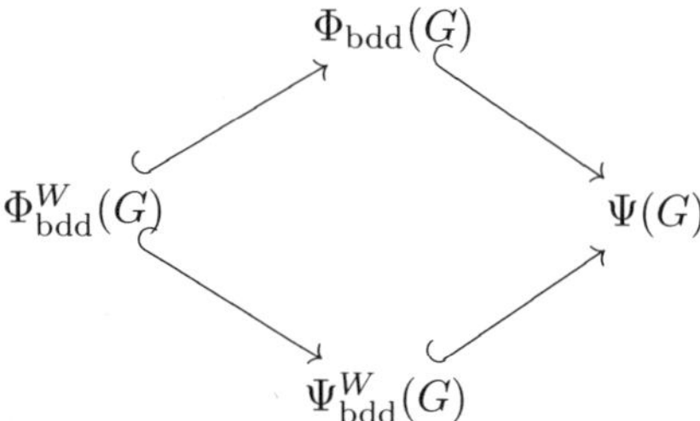

where $\Phi_{\mathrm{bdd}}^W(G)$ is the subset of parameters in $\Psi(G)$ whose restrictions to L_F lie in the subset $\Phi_{\mathrm{bdd}}^W(G)$. The global family $\Psi(G)$ appears to be the right set of parameters for classifying automorphic representations of G. Its localizations are then forced on us if we want to describe the local constituents of automorphic representations.

There is still something more to say. We would of course like to know how to construct the local packets Π_ψ. We would also like to characterize which representations in a global packet Π_ψ are actually automorphic. The answers to these questions are closely related to a certain finite group attached to ψ.

---

[5] We are assuming in this heuristic discussion that the generalized Ramanujan conjecture is valid for $GL(N)$. Since this is not known, one must make a minor adjustment in practice. See the discussion of [**ECR**, (1.3.10)] and of (4.12) in §4 here.

For any $\psi \in \Psi(G)$, with F being either local or global, we have the centralizer

$$(2.8) \qquad\qquad S_\psi = \mathrm{Cent}\big(\mathrm{im}(\psi), \widehat{G}\big)$$

of the image of ψ in $\widehat{G}$, a complex, reductive subgroup of $\widehat{G}$. We also have its finite group

$$(2.9) \qquad\qquad \mathcal{S}_\psi = \overline{S}_\psi / \overline{S}_\psi^0 = S_\psi / S_\psi^0 Z(\widehat{G})^{\Gamma_F}$$

of connected components, or rather, connected components in the quotient

$$(2.10) \qquad\qquad \overline{S}_\psi = S_\psi / Z(\widehat{G})^{\Gamma_F}$$

of S_ψ by the group of elements in the centre $Z(\widehat{G})$ of $\widehat{G}$ that are invariant under the action of the Galois group Γ_F. (The group $Z(\widehat{G})^{\Gamma_F}$ equals $Z(\widehat{G})$ unless G is a nonsplit group $SO(2)$, and in particular, of type D_n with $n = 1$.)

With our standing agreement that G belongs to one of the families $\mathbf{A}_n$–$\mathbf{D}_n$, the quotient $\mathcal{S}_\psi$ is an abelian 2-group. If F is local, the representations in the local packet Π_ψ are closely tied to the group of (linear) characters on $\mathcal{S}_\psi$. If F is global, the group S_ψ is contained in the centralizer S_{ψ_v} of any localization ψ_v. It follows that there is a canonical mapping from $\mathcal{S}_\psi$ into $\mathcal{S}_{\psi_v}$. The representations in a global packet Π_ψ that are automorphic are then tied to characters on the product

$$\mathcal{S}_{\psi,\mathbb{A}} = \prod_v \mathcal{S}_{\psi_v}$$

whose restrictions to the diagonal image of $\mathcal{S}_\psi$ equal a certain character ε_ψ on $\mathcal{S}_\psi$ that is defined explicitly in terms of arithmetic invariants attached to ψ. The case that ε_ψ is nontrivial will obviously be interesting, as also will be the case that the mapping from $\mathcal{S}_\psi$ to $\mathcal{S}_{\psi,\mathbb{A}}$ is not injective. But if neither of these exceptional conditions hold for ψ, we see a particularly clear analogy between the local groups $G(F_v)$ and global quotient $G(F) \backslash G(\mathbb{A})$ introduced in §1 and the "miniature models" $\mathcal{S}_{\psi_v}$ and $\mathcal{S}_\psi \backslash \mathcal{S}_{\psi,\mathbb{A}}$ we have just defined.

3. Self-dual, finite dimensional representations

The last section contained a good deal of information, some of which is perhaps difficult to absorb at the early stages of our presentation. Moreover, it seems to have been built on the shaky foundation of a hypothetical group L_F. Some such discussion, however, is a necessary part of the exposition. It helps us to see what we should be looking for.

We are trying to classify representations of groups G of type $\mathbf{B}_n$, $\mathbf{C}_n$ and $\mathbf{D}_n$ in terms of those of general linear groups $GL(N)$. If representations are indeed attached to parameters on a group $L_F \times SU(2)$, it makes sense to try to compare parameters for G with those of $GL(N)$. We shall do so in this section. The calculation will be an elementary exercise in linear algebra, which follows the discussion of §1.2 of [**ECR**].

We are assuming that F is a local or global field of characteristic 0. Suppose that Λ_F is any locally compact group over W_F with connected kernel. We are of course thinking of the case that Λ_F equals the product $L_F \times SU(2)$ (which is hypothetical if F is global), but the arguments will be the same for any Λ_F. We have already noted the obvious fact that an L-homomorphism from Λ_F to the L-group of $GL(N)$

projects onto the dual group $GL(N, \mathbb{C})$ of $GL(N)$, and can therefore be identified with an N-dimensional (continuous, complex) representation

$$r : \Lambda_F \longrightarrow GL(N, \mathbb{C})$$

of Λ_F.

We say that r is *self-dual* if it is equivalent to its contragredient representation

$$r^\vee(\lambda) = {}^t r(\lambda)^{-1}, \qquad \lambda \in \Lambda_F,$$

where $x \to {}^t x$ is the usual transpose mapping. In other words, the equivalence class of r is invariant under the usual automorphism

$$\theta(x) = x^\vee = {}^t x^{-1}, \qquad x \in GL(N),$$

of $GL(N)$. This condition depends only on the inner class of θ. It remains the same if θ is replaced by any conjugate

$$\theta_g(x) = g^{-1}\theta(x)g, \qquad g \in GL(N).$$

We shall analyze the self-dual representations r in terms of orthogonal and symplectic subgroups of $GL(N, \mathbb{C})$.

We decompose a given representation r into a direct sum

(3.1) $$r = \ell_1 r_1 \oplus \cdots \oplus \ell_r r_r,$$

for inequivalent representations

$$r_k : \Lambda_F \longrightarrow GL(N_k, \mathbb{C}), \qquad 1 \le k \le r,$$

and multiplicities ℓ_k with

$$N = \ell_1 N_1 + \cdots + \ell_r N_r.$$

The representation is self-dual if and only if there is an involution $k \to k^\vee$ on the indices such that for any k, $r_k^\vee$ is equivalent to $r_{k^\vee}$ and $\ell_k = \ell_{k^\vee}$. We say that r is *elliptic* if it satisfies the further constraint that for each k, $k^\vee = k$ and $\ell_k = 1$. We shall concentrate on this case.

Assume that r is elliptic. Then

(3.2) $$r = r_1 \oplus \cdots \oplus r_r,$$

for distinct, irreducible, self-dual representations r_i of Λ_F of degree N_i. If i is any index, we can write

$$r_i^\vee(\lambda) = A_i r_i(\lambda) A_i^{-1}, \qquad \lambda \in \Lambda_F,$$

for a fixed element $A_i \in GL(N_i, \mathbb{C})$. Applying the automorphism θ to each side of this equation, we see that

$$r_i(\lambda) = A_i^\vee r_i^\vee(\lambda)(A_i^\vee)^{-1} = (A_i^\vee A_i) r_i(\lambda)(A_i^\vee A_i)^{-1}.$$

Since r_i is irreducible, the product $A_i^\vee A_i$ is a scalar matrix. We can therefore write

$${}^t A_i = c_i A_i, \qquad c_i \in \mathbb{C}^*.$$

If we take the transpose of each side of this equation, we see further that $c_i^2 = 1$. Thus, c_i equals $+1$ or -1, and the nonsingular matrix A_i is either symmetric or skew-symmetric. The mapping

$$x_i \longrightarrow (A_i^{-1}) {}^t x_i A_i, \qquad x_i \in GL(N_i),$$

represents the adjoint relative to the bilinear form defined by A_i. Therefore $r_i(\lambda)$ belongs to the corresponding orthogonal group $O(A_i, \mathbb{C})$ or symplectic group $Sp(A_i, \mathbb{C})$, according to whether c_i equals $+1$ or -1.

Let us write I_O and I_S for the set of indices i such that c_i equals $+1$ and -1 respectively. We then write

$$r_\varepsilon(\lambda) = \bigoplus_{i \in I_\varepsilon} r_i(\lambda), \qquad \lambda \in \Lambda_F,$$

$$A_\varepsilon = \bigoplus_{i \in I_\varepsilon} A_i,$$

and

$$N_\varepsilon = \sum_{i \in I_\varepsilon} N_i,$$

for a symbol ε that can be either O or S. Thus A_O is a symmetric matrix in $GL(N_O, \mathbb{C})$, A_S is a skew-symmetric matrix in $GL(N_S, \mathbb{C})$, and r_O and r_S are representations of Λ_F that take values in the respective groups $O(A_O, \mathbb{C})$ and $Sp(A_S, \mathbb{C})$. We have established a canonical decomposition

$$r = r_O \oplus r_S$$

of the self-dual representation r into orthogonal and symplectic components.

It is only the equivalence class of r that is relevant. We can therefore replace $r(\lambda)$ with its conjugate by any matrix $B \in GL(N, \mathbb{C})$. This has the effect of replacing the matrix

$$A = A_O \oplus A_S$$

in $GL(N, \mathbb{C})$ by ${}^t\!BAB$. In particular, we can take A_O to be any symmetric matrix in $GL(N_O, \mathbb{C})$ and A_S to be any skew-symmetric matrix in $GL(N_S, \mathbb{C})$. We can therefore put the orthogonal and symplectic groups that contain the images of r_O and r_S into some standard form.

It is generally most convenient to choose the standard form for these groups so that the subgroup of diagonal matrices is a maximal torus. For our standard orthogonal group in $GL(N)$, we therefore take

$$O(N) = O(J),$$

where

$$J = J(N) = \begin{pmatrix} 0 & & 1 \\ & \reflectbox{$\ddots$} & \\ 1 & & 0 \end{pmatrix}$$

is the second diagonal in $GL(N)$. It is a group with two connected components, whose identity component is the special orthogonal group

$$SO(N) = \{x \in O(N) : \det(x) = 1\}.$$

As the standard symplectic group in $GL(N)$, defined for N even, we can take

$$Sp(N) = Sp(\widetilde{J}),$$

where

$$\widetilde{J} = \widetilde{J}(N) = \begin{pmatrix} & & & 0 & & 1 \\ & & & & -1 & \\ & & \iddots & & & \\ (-1)^{N+1} & & & 0 & & \end{pmatrix}$$

for any N. Notice that $Sp(N)$ is the group of fixed points of the automorphism

$$\widetilde{\theta}(N) = \mathrm{Int}(\widetilde{J}) \circ \theta : \quad x \longrightarrow \widetilde{J}\theta(x)\widetilde{J}^{-1},$$

while $O(N)$ is the group of fixed points of the automorphism

$$\mathrm{Int}(\widetilde{I}) \circ \widetilde{\theta}(N)$$

in the inner class of $\widetilde{\theta}(N)$ (and θ), where

$$\widetilde{I} = \widetilde{I}(N) = J\widetilde{J}^{-1} = \begin{pmatrix} 1 & & & & 0 \\ & -1 & & & \\ & & \ddots & & \\ 0 & & & (-1)^{N+1} \end{pmatrix}.$$

It is customary to treat $\widetilde{\theta}(N)$ as the standard automorphism in the inner class of θ, since it stabilizes the standard splitting [**K**, (1.3)] of $GL(N)$.

Returning to our discussion, we can arrange that A equals $J(N_O) \oplus \widetilde{J}(N_S)$. In the interest of symmetry, we actually take A to be the matrix

$$J_{O,S} = J(N_O, N_S) = \begin{pmatrix} 0 & & \widetilde{J}(N_S') \\ & J(N_O) & \\ \widetilde{J}(N_S') & & 0 \end{pmatrix}, \qquad N_S = 2N_S',$$

obtained from the obvious embedding of $J(N_O) \oplus \widetilde{J}(N_S)$ into $GL(N,\mathbb{C})$. The associated representation r from the given equivalence class then maps Λ_F to the corresponding subgroup of $GL(N,\mathbb{C})$, namely the subgroup

$$O(N_O,\mathbb{C}) \times Sp(N_S,\mathbb{C})$$

defined by the embedding

$$(x,y) \longrightarrow \begin{pmatrix} y_{11} & 0 & y_{12} \\ 0 & x & 0 \\ y_{21} & 0 & y_{22} \end{pmatrix},$$

where y_{ij} are the four $(N_S' \times N_S')$-block components of the matrix $y \in Sp(N_S,\mathbb{C})$.

The subrepresentations r_O and r_S of r can be analyzed separately. The symplectic factor r_S is the simpler of the two. Its image is contained in the *connected* complex group

$$\widehat{G}_S = Sp(N_S,\mathbb{C}).$$

This in turn is the dual group of the split special orthogonal group

$$G_S = SO(N_S + 1).$$

We need say nothing more in this case.

The orthogonal factor r_O is complicated by the fact that its image is contained in the disconnected group $O(N_O,\mathbb{C})$. Its projection

$$\Lambda_F \longrightarrow O(N_O,\mathbb{C})/SO(N_O,\mathbb{C}) = \mathbb{Z}/2\mathbb{Z}$$

onto the corresponding group of connected components is a character η of Λ_F of order 1 or 2. Since the kernels of the mappings $\Lambda_F \to W_F$ and $W_F \to \Gamma_F$ are connected, η can be identified with a character on the Galois group Γ_F of order 1 or 2. This determines a field extension E of F of degree 1 or 2.

Suppose first that N_O is odd. In this case, the matrix $(-I)$ in $O(N_O)$ belongs to the nonidentity component, and the orthogonal group is a direct product

$$O(N_O, \mathbb{C}) = SO(N_O, \mathbb{C}) \times \mathbb{Z}/2\mathbb{Z}.$$

We write

$$SO(N_O, \mathbb{C}) = \widehat{G}_O,$$

where G_O is the split group $Sp(N_O - 1)$ over F. We then use η to identify the direct product

$$^L G_{O,E/F} = \widehat{G}_O \times \Gamma_{E/F}$$

with a subgroup of $O(N_O, \mathbb{C})$, namely $SO(N_O, \mathbb{C})$ or $O(N_O, \mathbb{C})$, according to whether η has order 1 or 2. We thus have an embedding of a restricted form $^L G_{O,E/F}$ of the L-group of G_O into $GL(N_O, \mathbb{C})$.

Assume next that N_O is even. In this case, the nonidentity component in $O(N_O)$ acts by an outer automorphism on $SO(N_O)$. We write

$$SO(N_O, \mathbb{C}) = \widehat{G}_O,$$

where G_O is now the corresponding quasisplit orthogonal group $SO(N_O, \eta)$ over F defined by η. In other words, G_O is the split group $SO(N_O)$ if η is trivial, and the nonsplit group obtained by twisting $SO(N_O)$ over E by the given outer automorphism if η is nontrivial. Let $\widetilde{\omega}(N_O)$ be the permutation matrix in $GL(N_O)$ that interchanges the middle two co-ordinates, and leaves the other co-ordinates invariant. We take this element as a representative of the nonidentity component of $O(N_O, \mathbb{C})$. We then use η to identify the semidirect product

$$^L G_{O,E/F} = \widehat{G} \rtimes \Gamma_{E/F}$$

with a subgroup of $O(N_O, \mathbb{C})$, namely $SO(N_O, \mathbb{C})$ or $O(N_O, \mathbb{C})$ as before. We again obtain an embedding of a restricted (in this case minimal) L-group of G_O into $GL(N_O, \mathbb{C})$.

We have shown that the elliptic self-dual representation r factors through the embedded subgroup

$$^L G_{E/F} = {}^L (G_O \times G_S)_{E/F}$$

of $GL(N, \mathbb{C})$ attached to the quasisplit group

$$G = G_O \times G_S$$

over F. The group G is called a *twisted endoscopic group* for $GL(N)$. It is determined by r, and in fact, by the decomposition $N = N_O + N_S$ and the character $\eta = \eta_G$ attached to r. The same is true of the L-embedding

$$\xi = \xi_{O,S,\eta} : \ ^L G = \widehat{G} \rtimes \Gamma_F \ \hookrightarrow \ ^L(GL(N)) = GL(N, \mathbb{C}) \times \Gamma_F,$$

obtained by inflating the embedding above to the full L-groups. A third object we can associate to the decomposition $N = N_O + N_S$, and hence to r, is the product

$$s = s_{O,S} = J_{O,S}^{-1} \rtimes \theta.$$

It is a semisimple element in the coset

$$(3.3) \qquad\qquad \widehat{\widetilde{G}}(N) = GL(N, \mathbb{C}) \rtimes \theta$$

of θ in the semidirect product $\widetilde{\widehat{G}}(N)^+$ of $GL(N,\mathbb{C})$ with the group θ^+ of order 2 generated by θ. The complex group $\widehat{G} = \widehat{G}_O \times \widehat{G}_S$ is then the connected centralizer of s in the subgroup

$$GL(N,\mathbb{C}) = GL(N,\mathbb{C}) \rtimes 1.$$

The triplet (G, s, ξ) is called a *twisted endoscopic datum* for $GL(N)$.

The triplet (G, s, ξ) belongs to a special class, called elliptic twisted endoscopic data. This is a consequence of our condition that the original self-dual representation is elliptic. A general (nonelliptic) twisted endoscopic datum for $GL(N)$ is again a triplet (G, s, ξ), where G is a quasisplit group over F, s is a semisimple element in $GL(N,\mathbb{C}) \rtimes \theta$ whose connected centralizer in $GL(N,\mathbb{C})$ equals $\widehat{G}$, and ξ is an L-embedding of ^LG into $^LGL(N)$. We require that ξ equal the identity on $\widehat{G}$, and that the projection onto $GL(N,\mathbb{C})$ of the image of ξ lie in the full centralizer of s. The twisted endoscopic group G (or datum) is said to be *elliptic* if its subgroup $Z(\widehat{G})^{\Gamma_F}$ is finite.

We have been following a special case of the general terminology of [**KS**, p. 16]. The notion of *isomorphism* between general endoscopic data is defined in [**KS**, p. 18]. In the case at hand, it is given by an element g in the dual group $GL(N,\mathbb{C})$ whose action by conjugation is compatible (in a natural sense we do not spell out, but which is easy to imagine) with the two twisted endoscopic data. We write $\widetilde{\mathrm{Aut}}_N(G)$ for the group of isomorphisms of the twisted endoscopic datum G to itself. The main role for this group is in its image

$$\widetilde{\mathrm{Out}}_N(G) = \widetilde{\mathrm{Aut}}_N(G)/\widetilde{\mathrm{Int}}_N(G)$$

in the group of outer automorphisms of the group G over F. (Following standard practice, we often let the twisted endoscopic group G represent a full datum (G, s, ξ), or even an isomorphism class of data.) If G represents one of the elliptic data above, $\widetilde{\mathrm{Out}}_N(G)$ is trivial if the integer N_O is odd or zero. In the remaining case that N_O is even and positive, $\widetilde{\mathrm{Out}}_N(G)$ is a group of order 2, the nontrivial element being the outer automorphism induced by the nontrivial component of $O(N_O,\mathbb{C})$, which we have represented by the permutation matrix $\widetilde{\omega}(N_O)$ above.

We write[6]

$$\widetilde{\mathcal{E}}(N) = \mathcal{E}\big(\widetilde{G}(N)\big)$$

for the set of isomorphism classes of twisted endoscopic data for $GL(N)$, and

$$\widetilde{\mathcal{E}}_{\mathrm{ell}}(N) = \mathcal{E}_{\mathrm{ell}}\big(\widetilde{G}(N)\big)$$

for the subset of classes that are elliptic. The data (G, s, ξ), attached to equivalence classes of elliptic, self-dual representations r as above, form a set of representative of $\widetilde{\mathcal{E}}_{\mathrm{ell}}(N)$. The set $\widetilde{\mathcal{E}}_{\mathrm{ell}}(N)$ is thus parametrized by triplets (N_O, N_S, η), where $N_O + N_S = N$ is a decomposition of N into nonnegative integers with N_S even, and $\eta = \eta_G$ is a character of Γ_F of order 1 or 2 such that $\eta = 1$ if $N_O = 0$ and $\eta \neq 1$ if $N_O = 2$. (The last constraint is required in order that the datum be elliptic.)

Our general goal is to describe the representations of a quasisplit, special orthogonal or symplectic group in terms of those of general linear groups. The twisted endoscopic groups G are obviously relevant to the problem. Indeed, they are just the

---

[6] It is understood that $\widetilde{G}(N) = GL(N) \rtimes \theta$ is the nonidentity component of the reductive, nonconnected group $\widetilde{G}(N)^+ = GL(N) \rtimes \theta^+$ over F, and that the complex variety $\widetilde{\widehat{G}}(N) = GL(N,\mathbb{C}) \rtimes \theta$ above is the associated "dual set".

groups we want to study. But they are also part of the broader theory of endoscopy, which can therefore be brought to bear on the task. Since the general arguments are often inductive, the case that $\widehat{G}$ is either purely orthogonal or purely symplectic will have a special role. Accordingly, we write

$$\widetilde{\mathcal{E}}_{\mathrm{sim}}(N) = \mathcal{E}_{\mathrm{sim}}\big(\widetilde{G}(N)\big)$$

for the subset of elements in $\widetilde{\mathcal{E}}_{\mathrm{ell}}(N)$ that are *simple*, in the sense that one of the integers N_O or N_S vanishes. We then have a chain of sets

$$(3.4) \qquad\qquad \widetilde{\mathcal{E}}_{\mathrm{sim}}(N) \subset \widetilde{\mathcal{E}}_{\mathrm{ell}}(N) \subset \widetilde{\mathcal{E}}(N),$$

which are all finite if F is local, and all infinite if F is global.

The objects we have been discussing concern the L-embeddings for the first case of functoriality at the end of §1. We have provided some motivation for them. We have also used them as a way to introduce the theory of twisted endoscopy for $GL(N)$. We shall now consider the L-embeddings of the second case of functoriality from the end of §1.

To this end, we note that in addition to the twisted endoscopic data for $GL(N)$, one has also to work with ordinary (untwisted) endoscopic data. These are attached to our quasisplit, special orthogonal or symplectic groups, which is to say, the groups $G \in \widetilde{\mathcal{E}}_{\mathrm{sim}}(N)$. An endoscopic datum for G is similar to what we have described for $GL(N)$ above. It is a triplet (G', s', ξ'), where G' is a (connected) quasisplit group over F, s' is a semisimple element in $\widehat{G}$ of which $\widehat{G}'$ is the connected centralizer, and ξ' is an L-embedding of $^L G'$ into $^L G$. We again require that ξ' equal the identity on $\widehat{G}'$, and that its image lie in the centralizer of s' in $^L G$. (See [**LS**, (1.2)], a special case of the general definition in [**KS**], which we have specialized further to the case at hand.) There is again the notion of isomorphism of endoscopic data, which allows us to form the associated finite group

$$\mathrm{Out}_G(G') = \mathrm{Aut}_G(G')/\mathrm{Int}_G(G')$$

of any given G'. (We again often let the endoscopic group G' stand for an endoscopic datum (G', s', ξ'), or an isomorphism class of data.) We write $\mathcal{E}(G)$ for the set of isomorphism classes of endoscopic data G' for G, and $\mathcal{E}_{\mathrm{ell}}(G)$ for the subset of data that are elliptic, in the sense that $Z(\widehat{G}')^{\Gamma_F}$ is finite. We write $\mathcal{E}_{\mathrm{sim}}(G)$ for the subset of data $G' \in \mathcal{E}_{\mathrm{ell}}(G)$ such that $Z(\widehat{G}')^{\Gamma_F}$ equals the minimal group $Z(\widehat{G})^{\Gamma_F}$. It consists of the simple group G alone. We then have a second chain of sets

$$(3.5) \qquad\qquad \mathcal{E}_{\mathrm{sim}}(G) \subset \mathcal{E}_{\mathrm{ell}}(G) \subset \mathcal{E}(G),$$

which is parallel to (3.4). Similar definitions apply to groups G that represent more general data in $\widetilde{\mathcal{E}}(N)$.

The explicit description of elements $G' \in \mathcal{E}_{\mathrm{ell}}(G)$ is similar to our analysis of the set $\widetilde{\mathcal{E}}_{\mathrm{ell}}(N)$ above. If $G \in \widetilde{\mathcal{E}}_{\mathrm{sim}}(N)$ is of type $\mathbf{B}_n$, we have

$$\widehat{G}' = Sp(N_1', \mathbb{C}) \times Sp(N_2', \mathbb{C}) \subset Sp(N, \mathbb{C}) = \widehat{G},$$

for a decomposition $N = N_1' + N_2'$ of $N = 2n$ into even integers with $0 \leq N_1' \leq N_2'$. If G is of type $\mathbf{C}_n$, we have

$$\widehat{G}' = SO(N_1', \mathbb{C}) \times SO(N_2' + 1, \mathbb{C}) \subset SO(N, \mathbb{C}) = \widehat{G},$$

for a decomposition $N - 1 = N_1' + N_2'$ of $N - 1 = 2n$ into nonnegative, even integers. If G is of type $\mathbf{D}_n$, we have

$$\widehat{G}' = SO(N_1', \mathbb{C}) \times SO(N_2', \mathbb{C}) \subset SO(N, \mathbb{C}) = \widehat{G},$$

for a decomposition $N = N_1' + N_2'$ of $N = 2n$ into even integers with $0 \leq N_1' \leq N_2'$. It is clear that the products are all connected centralizers of diagonal matrices $s' \in \widehat{G}'$ with entries ± 1. The reader will also be able to construct all endoscopic groups $G' = G_1' \times G_2'$ with these dual groups, referring as needed to the end of §1, from quasisplit twists of the even orthogonal factors G_i'. The endoscopic groups G' determine the endoscopic data (G', s', ξ'), unlike what was the case for twisted endoscopic data for $GL(N)$. (See the end of [**ECR**, §1.3] for more remarks.)

We have completed our brief examination of elliptic, self-dual representations r. We are regarding these objects as parameters for $GL(N)$, in the spirit of §2. We have seen that any such parameter factors into a product of two parameters for two different quasisplit classical groups. These products are governed by twisted endoscopic data $G \in \widetilde{\mathcal{E}}_{\mathrm{ell}}(N)$. They can be refined further according to ordinary endoscopic data $G' \in \widetilde{\mathcal{E}}_{\mathrm{ell}}(G)$. Thus, while the parameters will not be available in the global case of ultimate concern (for lack of a global Langlands group L_F), the endoscopic data that control many of their properties will be. Before we can examine the ramifications of this, however, we have first to formulate a makeshift substitute for global parameters attached to our quasisplit special orthogonal and symplectic groups. We shall do so in §5, after a discussion in §4 of the automorphic representations of $GL(N)$ that will serve as the global parameters for this group.

We have considered only the self-dual representations r that are elliptic, since it is these objects that pertain directly to our global theorems. We might ask what happens if r is not elliptic. With a little reflection, one sees that any such r factors through subgroups of $GL(N, \mathbb{C})$ attached to *several* data $G \in \widetilde{\mathcal{E}}_{\mathrm{ell}}(N)$, in contrast to what we have seen in the elliptic case. It also factors through subgroups attached to data $G \in \widetilde{\mathcal{E}}(N)$ in the complement of $\widetilde{\mathcal{E}}_{\mathrm{ell}}(N)$. These matters are best formulated in terms of the centralizers

$$\widetilde{S}_r(N) = S_r\big(\widetilde{G}(N)\big) = \mathrm{Cent}\big(\mathrm{im}(r), \widehat{\widetilde{G}}(N)\big)$$

and

$$S_r = S_r(G) = \mathrm{Cent}\big(\mathrm{im}(r), \widehat{G}\big)$$

of the images of r. Their analysis is closely tied to an elementary but important bijective correspondence (5.11) we will describe later.

4. The case of $GL(N)$

In §2, we introduced the kind of parameters that ought to characterize representations of a given group G. In §3, we examined how the parameters for an orthogonal or symplectic group G are related to those for a general linear group $GL(N)$. This was in support of our goal, to classify representations of G in terms of those of $GL(N)$. What makes the goal worthwhile is the fact that much of the representation theory of $GL(N)$ is well understood and relatively simple. In this section, we shall review what we need of the theory, following [**ECR**, §1.3] and [**A5**, §1].

Assume for the time being that F is global. If the global Langlands group L_F existed, its corresponding set $\Psi\big(GL(N)\big)$ of global parameters would be identified with the set of equivalence classes of unitary, N-dimensional representations of the locally compact group $L_F \times SU(2)$. Following notation from [**ECR**], we could then write $\Psi_{\mathrm{sim}}\big(GL(N)\big)$ for the associated subset of irreducible representations. It would correspond to the set of pairs (μ, ν), where μ is an irreducible unitary representation of L_F of dimension m, and ν is the irreducible representation of $SU(2)$ of dimension n, for some decomposition $N = mn$ into a product of positive integers.

We do not, of course, have the group L_F at our disposal. However, its irreducible N-dimensional representations are supposed to correspond to something we do have, cuspidal automorphic representations of $GL(N)$. We shall therefore take $\Psi_{\mathrm{sim}}(N) = \Psi_{\mathrm{sim}}\big(GL(N)\big)$ to be the set of pairs (μ, ν) as above, with μ now being a unitary, cuspidal automorphic representation of $GL(m)$ instead of an irreducible, unitary m-dimensional representation of the hypothetical group L_F.

THEOREM 4.1 (Moeglin-Waldspurger [**MW1**]). *There is a canonical bijection*

$$\psi \longrightarrow \pi_\psi, \qquad \psi \in \Psi_{\mathrm{sim}}(N),$$

from $\Psi_{\mathrm{sim}}(N)$ *onto the set of irreducible unitary representations of* $GL(N, \mathbb{A})$ *that occur in the automorphic, relative discrete spectrum* $L^2_{\mathrm{disc}}\big(GL(N, F)\backslash GL(N, \mathbb{A})\big)$ *of* $GL(N)$. *Moreover, for any* ψ, π_ψ *occurs in the relative discrete spectrum with multiplicity one.*

More generally, we let $\Psi(N) = \Psi\big(GL(N)\big)$ denote the set of formal, unordered sums

$$(4.1) \qquad\qquad \psi = \psi_1 \boxplus \cdots \boxplus \psi_r, \qquad \psi_i \in \Psi_{\mathrm{sim}}(N_i),$$

for some partition

$$N = N_1 + \cdots + N_r$$

of N. It would correspond to the set

$$\psi_1 \oplus \cdots \oplus \psi_r$$

of N-dimensional direct sums of irreducible unitary representations ψ_i of the hypothetical group $L_F \times SU(2)$, or in other words, the set of all unitary, N-dimensional representations of $L_F \times SU(2)$.

COROLLARY 4.2. *There is a canonical bijection*

$$\psi \longrightarrow \pi_\psi, \qquad \psi \in \Psi(N),$$

from $\Psi(N)$ *onto the set of irreducible constituents of the full automorphic spectrum* $L^2\big(GL(N, F)\backslash GL(N, \mathbb{A})\big)$ *of* $GL(N)$.

The irreducible constituents of the automorphic spectrum are just the automorphic representations, according to the informal definition from §1 we are working with. The corollary tells us that there is a natural bijection from $\Psi(N)$ onto the set of automorphic representations of $GL(N)$. This is a consequence of Langlands' theory of Eisenstein series [**L2**], which provides an explicit construction of the automorphic continuous spectrum of any group in terms of (relative) discrete spectra for smaller groups.

If $\psi = (\mu, \nu)$ belongs to $\Psi_{\mathrm{sim}}(N)$, Moeglin and Waldspurger construct the representation π_ψ as a global Langlands quotient. It is by definition the unique irreducible quotient of the representation of $GL(n, \mathbb{A})$ obtained by parabolic induction from the nonunitary representation

$$x \longrightarrow \mu(x_1)|\det x_1|^{\frac{n-1}{2}} \otimes \mu(x_2)|\det x_2|^{\frac{n-3}{2}} \otimes \cdots \otimes \mu(x_n)|\det x_n|^{-\frac{(n-1)}{2}}$$

of the standard Levi subgroup

$$M_P(\mathbb{A}) = \big\{ x = (x_1, \ldots, x_n) : \ x_i \in GL(m, \mathbb{A}) \big\}.$$

More generally, if ψ is a general element in $\Psi(N)$ of the form above, π_ψ is the irreducible representation of $GL(N, \mathbb{A})$ obtained by parabolic induction from the unitary representation

$$\pi_{\psi_1}(x_1) \otimes \cdots \otimes \pi_{\psi_r}(x_r)$$

of the standard Levi subgroup

$$M_P(\mathbb{A}) = \big\{ x = (x_1, \ldots, x_r) : \ x_i \in GL(N_i, \mathbb{A}) \big\}.$$

In both cases, Eisenstein series provide functions that represent intertwining operators from the representations π_ψ to the associated constituents of the automorphic spectrum.

Suppose that π is an irreducible representation of $GL(N, \mathbb{A})$. We assume implicitly that π is *admissible* [**F**, p. 182], a broad condition that is always valid if π is automorphic. The condition includes a property of weak continuity, which in view of the direct limit topology on the restricted direct product (1.1), implies that π is unramified for almost all valuations v of F. We alluded to unramified representations in our heuristic description of the general global packets (2.2) and (2.7). For the group $G = GL(N)$, an irreducible representation π_v of $G(F_v)$ is *unramified* by definition if F_v is nonarchimedean, and the restriction of π_v to the open compact subgroup $G(\mathfrak{o}_v)$ of $G(F_v)$ contains the trivial, 1-dimensional representation. Unramified representations have a very simple classification. A well known integral transform, introduced into p-adic harmonic analysis by Satake, leads to a canonical bijection

$$(4.2) \qquad\qquad \pi_v \longrightarrow c(\pi_v),$$

from the set of irreducible unramified representations π_v of $GL(N, F_v)$ onto the set of semisimple conjugacy classes c_v in the dual group $GL(N, \mathbb{C})$ of $GL(N)$. The given global representation π thus gives rise to a family of conjugacy classes

$$(4.3) \qquad\qquad c^S(\pi) = \big\{ c_v(\pi) = c(\pi_v) : \ v \notin S \big\},$$

parametrized by a cofinite set of valuations of F. In order to remove its dependence on S, some finite set that contains the archimedean valuations, we write $c(\pi)$ for the equivalence class of $c^S(\pi)$ with respect to the relation defined by setting $c^S \sim (c')^{S'}$ if $c_v = c'_v$ for almost all v.

We call $c(\pi)$ (or any of its representatives $c^S(\pi)$ within the equivalence class) a *Hecke eigenfamily*. It represents a set of simultaneous eigenvalues for the action of the factors of the restricted tensor product

$$\mathcal{H}^S_{\mathrm{un}}(N) = \overset{\sim}{\bigotimes_{v \notin S}} \mathcal{H}_{v,\mathrm{un}}(N)$$

of local unramified Hecke algebras

$$\mathcal{H}_{v,\mathrm{un}}(N) = C_c^\infty\big(GL(N, \mathfrak{o}_v) \backslash GL(N, F_v) / GL(N, \mathfrak{o}_v)\big)$$

under convolution, relative to the hyperspecial maximal compact subgroup

$$GL(N, \mathfrak{o}^S) = \prod_{v \notin S} GL(N, \mathfrak{o}_v)$$

of $GL(N, \mathbb{A}^S)$, on the space of $GL(N, \mathfrak{o}^S)$-invariant vectors of π^S. Our interest of course is the case that π is automorphic.

Suppose that ψ is an element (4.1) in the set $\Psi(N)$. We then obtain a Hecke eigenfamily

$$(4.4) \qquad c^S(\psi) = c^S(\pi_\psi) = \{c_v(\psi) = c(\pi_{\psi,v}) : \ v \notin S\},$$

with equivalence class $c(\psi) = c(\pi_\psi)$, from the irreducible representation π_ψ of $GL(N, \mathbb{A})$. It is to be regarded as a concrete datum, which is attached to the formal object ψ through the automorphic representation π_ψ. According to the remarks following the statement of Corollary 4.2, $c(\psi_i)$ will be given explicitly in terms of Hecke eigenfamilies $c(\mu_i)$, represented by sets

$$c^S(\mu_i) = \{c_v(\mu_i) : \ v \notin S\}, \qquad 1 \le i \le r,$$

which we extract from the cuspidal components μ_i of constituents ψ_i of ψ. More precisely, if $\psi = (\mu, \nu)$ belongs to $\Psi_{\mathrm{sim}}(N)$ then

$$(4.5) \qquad c_v(\psi) = c_v(\mu) \otimes c_v(\nu) = c_v(\mu) q_v^{\frac{n-1}{2}} \oplus \cdots \oplus c_v(\mu) q_v^{-\frac{n-1}{2}}, \qquad v \notin S,$$

while if ψ is a general element (4.1) in $\Psi(N)$, we have

$$(4.6) \qquad c_v(\psi) = c_v(\psi_1) \oplus \cdots \oplus c_v(\psi_r), \qquad v \notin S.$$

These elements are to be regarded as diagonal matrices in $GL(N, \mathbb{C})$, which of course represent semisimple conjugacy classes.

We write

$$(4.7) \qquad \mathcal{C}(N) = \{c(\psi) : \ \psi \in \Psi(N)\}$$

for the set of Hecke eigenfamilies attached to elements in $\Psi(N)$.

THEOREM 4.3 (Jacquet-Shalika [**JS**]). *The mapping*

$$\psi \longrightarrow c(\psi), \qquad \psi \in \Psi(N),$$

is a bijection from $\Psi(N)$ *to* $\mathcal{C}(N)$.

As we noted in [**A5**, §1], which we have been following closely here, Theorem 4.3 predates Theorem 1.1. As stated in [**JS**], it applied to a class of automorphic representations Langlands introduced in [**L4**], and called *isobaric*. At the time, it was not known whether this class included the constituents of the automorphic (relative) discrete spectrum. Theorem 4.1 implies that such constituents are distinct and isobaric. It therefore yields the interpretation we have stated above for the original theorem of Jacquet and Shalika.

Theorem 4.3 is a striking result. It implies that any information that might be contained in a constituent π_ψ of the automorphic spectrum of $GL(N)$ will be captured in the corresponding Hecke eigenfamily $c(\psi)$. Since $c(\psi)$ appears to contain less information, and since it is just a concrete set of complex parameters, the assertion is indeed remarkable. It is the more so for our expressed goal of classifying automorphic representations of the other groups G in terms of those of $GL(N)$, and thus in terms of Hecke eigenfamilies $c(\psi)$.

Before turning to the other groups, we shall discuss the local theory for $GL(N)$. The local Langlands correspondence is known in this case. In order to review it, we shall first say something about local L-functions and ε-factors.

Assume now that F is local. There are two kinds of local L-functions, arithmetic and representation theoretic. The former are attached to finite dimensional representations of the local Langlands group L_F, the latter to irreducible representations of general linear groups over F.

Suppose that

$$\phi : \; L_F \; \longrightarrow \; GL(N, \mathbb{C})$$

is an N-dimensional (semisimple, continuous) representation of L_F. The associated arithmetic L-function $L(s, \phi)$ is a meromorphic function of $s \in \mathbb{C}$. One can also form the local arithmetic ε-factor $\varepsilon(s, \phi, \psi_F)$, a monomial of the form ab^{-s}, which depends on a nontrivial additive character ψ_F of F. The definition of the coefficient a for nonarchimedean F is by far the most subtle part of the process. It was constructed canonically by Deligne, following ideas of Artin and Langlands, by a global argument. If F is archimedean, we refer the reader to the definition in [**T**, §3]. If F is nonarchimedean, one extends ϕ analytically to a representation to the product of W_F with the complexification $SL(2, \mathbb{C})$ of the compact unitary subgroup $SU(2)$ of L_F. This gives a representation

$$\mu_\phi(w) = \phi\left(w, \begin{pmatrix} |w|^{\frac{1}{2}} & 0 \\ 0 & |w|^{-\frac{1}{2}} \end{pmatrix}\right), \qquad w \in W_F,$$

of W_F, in which $|w|$ is the absolute value on W_F, and a nilpotent matrix

$$N_\phi = \log\left(\phi\left(1, \begin{pmatrix} 1 & 1 \\ 0 & 1 \end{pmatrix}\right)\right).$$

The pair $V_\phi = (\mu_\phi, N_\phi)$ provides representation of what is known as the Weil-Deligne group, from which one can define an L-function

$$L(s, \phi) = Z(V_\phi, q_F^{-s})$$

and ε-factor

$$\varepsilon(s, \phi, \psi_F) = \varepsilon(V_\phi, q_F^{-s}),$$

with q_F being the order of the residue field of F, following notation in [**T**, §4]. Of particular interest are the tensor product L-function

$$L(s, \phi_1 \times \phi_2) = L(s, \phi_1 \otimes \phi_2)$$

and ε-factor

$$\varepsilon(s, \phi_1 \times \phi_2, \psi_F) = \varepsilon(s, \phi_1 \otimes \phi_2, \psi_F)$$

attached to any pair of representations ϕ_1 and ϕ_2 of L_F.

Representation theoretic L-functions $L(s, \pi, r)$ and ε-factors $\varepsilon(s, \pi, r, \psi_F)$ ought to be attached naturally to irreducible representations π of $G(F)$ and finite dimensional representations r of $^L G$, for *any* reductive group G over F. For general G, this can be done in only relatively simple cases. However, if G is a product $GL(N_1) \times GL(N_2)$ of general linear groups, there is a broader theory [**JPS**]. It applies to any representation $\pi = \pi_1 \otimes \pi_2$, with r being the standard representation

$$r(g_1, g_2) : \; X \; \longrightarrow \; g_1 X {}^t g_2, \qquad g \in GL(N_i, \mathbb{C}),$$

of

$$\widehat{G} = GL(N_1, \mathbb{C}) \times GL(N_2, \mathbb{C})$$

on the space of complex $(N_1 \times N_2)$-matrices X. The theory yields functions

$$L(s, \pi_1 \times \pi_2) = L(s, \pi, r)$$

and

$$\varepsilon(s, \pi_1 \times \pi_2, \psi_F) = \varepsilon(s, \pi, r, \psi_F)$$

known as Rankin-Selberg products.

We write $\Phi(N) = \Phi\big(GL(N)\big)$, $\Phi_{\mathrm{bdd}}(N) = \Phi_{\mathrm{bdd}}\big(GL(N)\big)$ and $\Psi(N) = \Psi\big(GL(N)\big)$ for the specialization to $GL(N)$ of the local parameter sets from §2, and $\Pi(N) = \Pi\big(GL(N)\big)$, $\Pi_{\mathrm{temp}}(N) = \Pi_{\mathrm{temp}}\big(GL(N)\big)$ and $\Pi_{\mathrm{unit}}(N) = \Pi_{\mathrm{unit}}\big(GL(N)\big)$ for the associated sets of irreducible representations of $GL(N, F)$. The local Langlands correspondence applies to the first of these. It is characterized essentially by its compatibility with Rankin-Selberg products.

THEOREM 4.4 (Langlands [**L5**], Harris-Taylor [**HT**], Henniart [**H**], Scholze [**Sch**]). *There is a canonical bijective correspondence*

$$\phi \longrightarrow \pi_\phi$$

from $\Phi(N)$ to $\Pi(N)$ such that

$$(4.8) \qquad L(s, \pi_{\phi_1} \times \pi_{\phi_2}) = L(s, \phi_1 \times \phi_2)$$

and

$$(4.9) \qquad \varepsilon(s, \pi_{\phi_1} \times \pi_{\phi_2}, \psi_F) = \varepsilon(s, \phi_1 \times \phi_2, \psi_F),$$

for any pair of local parameters $\phi_1 \in \Phi(N_1)$ and $\phi_2 \in \Phi(N_2)$.

The correspondence also satisfies other natural conditions. These include compatibility with the automorphism $\widetilde{\theta}(N)$ of $GL(N)$, with tensor products by 1-dimensional representations, and with the one-dimensional correspondence given by class field theory (as it relates to determinants and central characters). If we append these three supplementary conditions to its compatibility (4.8) and (4.9) with Rankin-Selberg products, the correspondence becomes unique. It is in this sense that it is canonical.

The local correspondence has other properties as well. It restricts to a bijection from the subset $\Phi_{\mathrm{bdd}}(N)$ of $\Phi(N)$ to the subset $\Pi_{\mathrm{temp}}(N)$ of $\Pi(N)$. It restricts further to a bijection from the subset $\Phi_{\mathrm{sim,bdd}}(N)$ of irreducible representations in $\Phi_{\mathrm{bdd}}(N)$ to the subset $\Pi_{2,\mathrm{temp}}(N)$ of irreducible representations of $GL(N, F)$ that occur in the relative discrete spectrum of $L^2\big(GL(N, F)\big)$. In the case that F is nonarchimedean, it also takes the subset of representations ϕ in $\Phi_{\mathrm{sim,temp}}(N)$ that are trivial on the factor $SU(2)$ of L_F onto the subset of representations in $\Pi_{\mathrm{sim,temp}}(N)$ that are supercuspidal.

The local Langlands correspondence also bears on the local set $\Psi(N)$. For any $\psi \in \Psi(N)$, we write

$$\phi_\psi(w) = \psi\left(w, \begin{pmatrix} |w|^{\frac{1}{2}} & 0 \\ 0 & |w|^{-\frac{1}{2}} \end{pmatrix}\right), \qquad w \in L_F,$$

where $|w|$ is now the pullback to L_F of the absolute value on W_F. We obtain a mapping

$$(4.10) \qquad \psi \longrightarrow \phi_\psi, \qquad \psi \in \Psi(N),$$

from $\Psi(N)$ to $\Phi(N)$, which is in fact injective, as one sees easily from the fact that restriction of ψ to L_F lies in $\Phi_{\mathrm{bdd}}(N)$. The representation

$$\pi_\psi = \pi_{\phi_\psi}, \qquad \psi \in \Psi(N),$$

defined by the local correspondence is called a *Speh* representation. It is known to be unitary. We thus have an injective mapping

(4.11) $$\psi \longrightarrow \pi_\psi$$

from $\Psi(N)$ to $\Pi_{\mathrm{unit}}(N)$. This is the local analogue of the global bijection of Corollary 4.1.2. In contrast to the global setting, however, it does not have a natural interpretation in terms of local harmonic analysis. It is also not surjective.

Suppose that ψ is a general local parameter in $\Psi(N)$. The centralizer S_ψ defined (2.8) in §2 is the group of invertible intertwining operators of the N-dimensional representation given by ψ. It is a product of complex general linear groups, embedded diagonally in $GL(N, \mathbb{C})$, and is therefore connected. The associated group of connected components $\mathcal{S}_\psi$ is consequently trivial. Our heuristic discussion at the end of §2 suggests that the packet Π_ψ ought then to consist of only the one representation π_ψ. This in fact matches the formal definition we will discuss in §6. It is one more indication that the representation theory of $GL(N)$ is simpler than that of other groups.

Suppose again that F is global. If $\psi = (\mu, \nu)$ belongs to the global set $\Psi_{\mathrm{sim}}(N)$, any local constituent $\pi_{\psi,v}$ of the automorphic representation of Theorem 4.1 equals the Speh representation π_{ψ_v} attached to the local parameter[7] $\psi_v = \mu_v \otimes \nu$. This is not hard to establish from the definitions of π_ψ and π_{ψ_v} as Langlands quotients. The generalized Ramanujan conjecture asserts that as a local constituent of μ, μ_v is a *tempered* representation of $GL(m, F_v)$, or equivalently, that it is a *unitary* m-dimensional representation of L_{F_v}. Since we do not know that this conjecture is valid, however, we can assume only that the local parameter ψ_v lies in the set $\Psi_v^+(N)$ of N-dimensional, not necessarily unitary representations of $L_{F_v} \times SU(2)$. More generally, suppose that

$$\psi = \psi_1 \boxplus \cdots \boxplus \psi_r, \qquad \psi_k \in \Psi_{\mathrm{sim}}(N_k),$$

is a general element in $\Psi(N)$. As in the case above that ψ is simple, one sees that

$$\pi_{\psi,v} = \pi_{\psi_v},$$

for the local parameter

$$\psi_v = \psi_{1,v} \oplus \cdots \oplus \psi_{r,v}$$

in $\Psi_v^+(N)$. We thus obtain a natural localization mapping

(4.12) $$\psi \longrightarrow \psi_v, \qquad \psi \in \Psi(N),$$

from $\Psi(N)$ to $\Psi_v^+(N)$. We also see that the global packet Π_ψ, defined heuristically (2.7) in terms of L_F, should consist of the one representation

$$\pi_\psi = \overset{\sim}{\bigotimes_v} \pi_{\psi_v}.$$

---

[7] We are writing μ_v for both the local constituent of the cuspidal automorphic representation μ of $GL(m)$, and the m-dimensional representation of L_{F_v} to which it corresponds. This is natural if we consider that μ should really represent an irreducible, m-dimensional representation of the global group L_F.

Once again, the representation theory of $GL(N)$ seems to be about as simple as it can be.

As we noted in §2, global packets should have implications for global L-functions. Arithmetic global L-functions and ε-factors are attached to N-dimensional representations ϕ of W_F (or more generally, of the hypothetical group L_F). They are defined as products

$$L(s, \phi) = \prod_v L(s, \phi_v)$$

and

$$\varepsilon(s, \phi) = \prod_v \varepsilon(s, \phi_v, \psi_{F_v})$$

of their local analogues, where ψ_F is a nontrivial additive character on the quotient $\mathbb{A}/F$. Automorphic (representation theoretic) versions are attached to admissible representations π of $G(\mathbb{A})$ and finite dimensional representations r of $^L G$, for *any* reductive group G over F. They are again defined as products

$$(4.13) \qquad L(s, \pi, r) = \prod_v L(s, \pi_v, r_v)$$

and

$$(4.14) \qquad \varepsilon(s, \pi, r) = \prod_v \varepsilon(s, \pi_v, r_v, \psi_{F_v})$$

of their localizations. The products of ε-factors can be taken over a finite set. The products of L-functions converge for the real part of s in some right half plane.[8] They should have analytic continuation[9] to meromorphic functions of s, which satisfy the functional equations

$$L(s, \phi) = \varepsilon(s, \phi) L(1 - s, \phi^{\vee})$$

and

$$(4.15) \qquad L(s, \pi, r) = \varepsilon(s, \pi, r) L(1 - s, \pi, r^{\vee}),$$

where $\phi^{\vee}$ and $r^{\vee}$ denote the contragredients of the finite dimensional representations ϕ and r. The two kinds of objects should ultimately be related by identities

$$(4.16) \qquad L(s, \pi, r) = L(s, r \circ \phi)$$

and

$$(4.17) \qquad \varepsilon(s, \pi, r) = \varepsilon(s, r \circ \phi),$$

for any representation π in the global packet of ϕ. However, these "reciprocity laws" are still far away in general, and as in the local case, one needs to study the arithmetic and automorphic objects independently.

The correspondence $\psi \to \phi_\psi$ of (4.10) will be completely general. As in the local case discussed for $GL(N)$ above, it will map general parameters ψ on $L_F \times SU(2)$ injectively to parameters on the group L_F. One then defines arithmetic L-functions and ε-factors for any ψ and r by

$$L(s, r \circ \psi) = L(s, r \circ \phi_\psi)$$

---

[8] In the representation theoretic case, one must impose a property of weak growth on π, which always holds if π is automorphic.

[9] In the representation theoretic case, π should now be automorphic.

and

$$\varepsilon(s, r \circ \psi) = L(s, r \circ \phi_\psi).$$

They should be equal to representation theoretic functions $L(s, \pi_\psi, r)$ and $\varepsilon(s, \pi_\psi, r)$ respectively, for any representation π_ψ in the global packet Π_{ϕ_ψ}.

A reader unfamiliar with these notions might be overwhelmed now with the details of what has taken the world of number theory many years to absorb, while at the same time, an expert might be quite tired of my repetition. At the risk of exhausting the patience of both parties, let me say a word on the concrete roots of global L-functions.

An irreducible admissible representation π of $G(\mathbb{A})$ still has unramified local constituents π_v at almost all v. Moreover, the Satake transform (4.2) remains valid. It is a bijection from the set of unramified representations π_v of $G(F_v)$ onto the set of semisimple conjugacy classes $c(\pi_v)$ in the dual group whose image in the factor W_{F_v} of $^L G_v$ is a Frobenius element. The given global representation π then gives rise to a family

$$c^S(\pi) = \left\{ c_v(\pi) = c(\pi_v) : \ v \notin S \right\}$$

of semisimple conjugacy classes in $^L G$, and a corresponding equivalence class $c(\pi)$ that is independent of S. For any unramified representation r_v of $^L G_v$, the local L-function of π_v and r_v is defined

$$L(s, \pi_v, r_v) = \det \left(1 - r_v\big(c(\pi_v)\big) q_v^{-s} \right)^{-1}, \qquad q_v = q_{F_v},$$

in terms of the characteristic polynomial of the conjugacy class $r_v\big(c(\pi_v)\big)$. If r is allowed to range over representations of $^L G$ that are unramified outside S (a set we can allow to vary), the partial global L-functions

$$L^S(s, \pi, r) = \prod_{v \notin S} L(s, \pi_v, r_v)$$

give an elegant way to package the concrete data from $c(\pi)$. If π is automorphic, each associated partial L-function is supposed to have analytic continuation to a meromorphic function of s, which satisfies a functional equation that is now quite complex. The object is then to define supplementary factors $L(s, \pi_v, r_v)$ (which should be relatively straightforward) for $v \in S$, and $\varepsilon(s, \pi_v, r_v, \psi_v)$ (which are deep if $v \in S$, but equal to 1 if $v \notin S$) so that the products (4.13) and (4.14) satisfy the simple functional equation (4.15). The Langlands-Shahidi method [**Sha**] has been the most successful technique for studying these questions, but as in the local situation, it applies only to relatively simple cases.

Global Rankin-Selberg products are well understood, thanks again to [**JPS**]. They apply to the representation r of the group $G = GL(N_1) \times GL(N_2)$ on the space of $(N_1 \times N_2)$-matrices. Suppose that $\psi_1 \in \Psi(N_1)$ and $\psi_2 \in \Psi(N_2)$ are elements in the sets we have introduced as substitutes for general global parameters. The corresponding representations

$$\pi_i = \pi_{\psi_i} = \pi_{\phi_{\psi_i}}, \qquad i = 1, 2,$$

of $GL(N_i, \mathbb{A})$ are automorphic, according to Corollary 4.2. We write

(4.18) $$L(s, \psi_1 \times \psi_2) = L(s, \pi_1 \times \pi_2)$$

and

(4.19) $$\varepsilon(s, \psi_1 \times \psi_2) = \varepsilon(s, \pi_1 \times \pi_2)$$

for the associated global L-functions and ε-factors (4.13) and (4.14). It follows from [**JPS**] that they have analytic continuation and functional equation (4.15). The definitions (4.18) and (4.19) are suggestive. They look like examples of the fundamental reciprocity laws between arithmetic and automorphic L-functions. But because we have had to define the global sets $\Psi(N)$ in terms of cuspidal automorphic representations instead of representations of the Langlands group L_F, they remain just definitions. However, in a formal sense, they confirm the expectation that global L-functions attached to representations in global packets (2.2) and (2.7) should have analytic continuation and functional equation. They also represent formal global versions of the compatibility conditions (4.8) and (4.9) that characterize the local Langlands correspondence.

5. Global parameters for G

In the last section we introduced a set of global "parameters" $\Psi(N)$ for $GL(N)$. It is our substitute for the actual set of parameters, which would consist of unitary representations of the hypothetical global group $L_F \times SU(2)$. We also saw how to describe the automorphic representation theory of $GL(N)$ in terms of $\Psi(N)$. In this section, we turn to our classical groups G. Motivated by the heuristic discussion of §3, we will try to identify the subset $\Psi(G)$ of $\Psi(N)$ that will serve as a set of global parameters for G. Among other things, we will have to attach a subgroup S_ψ of $\widehat{G}$ to any $\psi \in \Psi(G)$, since its group of connected components governs global multiplicities. We follow the discussion in [**ECR**, §1.4].

We assume in this section that the field F is global. The notation (4.1) for a general element $\psi \in \Psi(N)$ should not be confused with the similar way we denoted an elliptic representation (3.2) of the abstract group Λ_F. This is because the simple factors ψ_i in (4.1) are not required to be distinct. In future, we shall prefer the analogue of the notation (3.1). That is, we denote a general element in $\Psi(N)$ as a formal, unordered direct sum

$$(5.1) \qquad \psi = \ell_1 \psi_1 \boxplus \cdots \boxplus \ell_r \psi_r,$$

for positive integers ℓ_k and *distinct* elements $\psi_k = (\mu_k, \nu_k)$ in $\Psi_{\mathrm{sim}}(N_k)$. The ranks are positive integers $N_k = m_k n_k$ such that

$$N = \ell_1 N_1 + \cdots + \ell_r N_r = \ell_1 m_1 n_1 + \cdots + \ell_r m_r n_r.$$

We will also often denote the simple components ψ_k of ψ as formal tensor products

$$\psi_k = (\mu_k, \nu_k) = \mu_k \boxtimes \nu_k,$$

since they replace tensor products of irreducible representations of the two groups L_F and $SU(2)$.

We are looking to §3 for guidance. The essential objects there were the representations of Λ_F, an abstract group over W_F, that were self-dual. The corresponding duality operator on $\Psi(N)$ is given by the outer automorphism $\theta \colon x \to x^\vee$ of $GL(N)$. It transforms a general parameter (5.1) to its contragredient

$$\psi^\vee = \ell_1 \psi_1^\vee \boxplus \cdots \boxplus \ell_r \psi_r^\vee$$
$$= \ell_1 (\mu_1^\vee \boxtimes \nu_1) \boxplus \cdots \boxplus \ell_r (\mu_r^\vee \boxtimes \nu_r),$$

since

$$\psi_k^\vee = (\mu_k \boxtimes \nu_k)^\vee = \mu_k^\vee \boxtimes \nu_k^\vee = \mu_k^\vee \boxtimes \nu_k.$$

We are writing

$$\mu_k^\vee(x) = \mu_k(\,^t x^{-1}) \cong \,^t\mu_k(x)^{-1}, \qquad x \in GL(m_k, \mathbb{A}),$$

for the contragredient of the cuspidal automorphic representation μ_k of $GL(m_k)$, and we have written $\nu_k^\vee = \nu_k$, since any representation of $SU(2)$ is self-dual. The contragredient $\pi_\psi^\vee$ of the associated automorphic representation π_ψ of $GL(N)$ then equals $\pi_{\psi^\vee}$, as follows from the various definitions. Since the local correspondence of Theorem 4.4 commutes with duality, according to one of its supplementary conditions, the mapping $\psi \to \psi^\vee$ is the analogue for $\Psi(N)$ of duality for N-dimensional representations of $L_F \times SU(2)$.

We write

$$\widetilde{G}(N) = \widetilde{G}(N)^0 \rtimes \widetilde{\theta}(N) = GL(N) \rtimes \theta$$

as in §3 for the nonidentity connected component of the semidirect product

$$\widetilde{G}(N)^+ = \widetilde{G}(N)^0 \rtimes \widetilde{\theta}(N)^+ = GL(N) \rtimes \theta^+$$

of $GL(N)$ with the group of order 2 generated by either of the automorphisms $\widetilde{\theta}(N)$ or θ. We also write

$$(5.2) \qquad \widetilde{\Psi}(N) = \Psi\big(\widetilde{G}(N)\big) = \{\psi \in \Psi(N) : \ \psi^\vee = \psi\}$$

for the set of self-dual elements in $\Psi(N)$. This subset is indeed associated with the component $\widetilde{G}(N)$. It consists of the elements $\psi \in \Psi(N)$ such that the automorphic representation $\pi_\psi \cong \pi_\psi^\vee$ of $GL(N) = \widetilde{G}(N)^0$ has an extension to the group $\widetilde{G}(N, \mathbb{A})^+$ generated by $\widetilde{G}(N, \mathbb{A})$. We observe also that there is pointwise action

$$c = \{c_v\} \ \longrightarrow \ c^\vee = \{c_v^\vee\}$$

of θ on the set of (equivalence classes of) Hecke eigenfamilies $\mathcal{C}(N)$. The mapping of Theorem 4.3 restricts to a bijection from $\widetilde{\Psi}(N)$ onto the subset

$$\widetilde{\mathcal{C}}(N) = \{c \in \mathcal{C}(N) : \ c^\vee = c\}$$

of self-dual elements in $\mathcal{C}(N)$.

We are particularly interested in the subset

$$\widetilde{\Psi}_{\mathrm{sim}}(N) = \widetilde{\Psi}(N) \cap \Psi_{\mathrm{sim}}(N)$$

of $\widetilde{\Psi}(N)$. It consists of the simple parameters $\psi = \mu \boxtimes \nu$ in $\widetilde{\Psi}(N)$. Among these, we have the smaller subset

$$\widetilde{\Phi}_{\mathrm{sim}}(N) = \widetilde{\Psi}_{\mathrm{cusp}}(N)$$

of parameters that are *generic*, in the sense that ν is trivial. A simple generic parameter is therefore a self-dual, cuspidal automorphic representation of $GL(N)$.

The term *simple* was also applied in §1.2 to endoscopic data. It was used to designate the subset $\widetilde{\mathcal{E}}_{\mathrm{sim}}(N)$ of elliptic twisted endoscopic data $G \in \widetilde{\mathcal{E}}_{\mathrm{ell}}(N)$ that are not composite. An element in $\widetilde{\mathcal{E}}_{\mathrm{sim}}(N)$ is therefore one of our quasisplit, special orthogonal or symplectic groups G over F, but armed with some extra structure (namely, a choice of endoscopic embedding) in case $G = Sp(2n)$ and $\widehat{G} = SO(2n + 1, \mathbb{C})$. For any such G, we need to introduce the subset $\widetilde{\Psi}(G)$ of $\widetilde{\Psi}(N)$ that will serve as global parameters for G. Its construction will be based on the following fundamental case.

THEOREM 5.1. *Suppose that $\phi \in \widetilde{\Phi}_{\mathrm{sim}}(N)$ is a simple generic global parameter. Then there is a unique $G_\phi \in \widetilde{\mathcal{E}}_{\mathrm{ell}}(N)$, regarded as an isomorphism class of twisted endoscopic data $(G_\phi, s_\phi, \xi_\phi)$ for $GL(N)$, such that*

$$c(\phi) = \xi_\phi\big(c(\pi)\big),$$

for some automorphic representation π of G. Moreover, G_ϕ is simple, and π occurs in the automorphic discrete spectrum $L^2_{\mathrm{disc}}\big(G(F)\backslash G(\mathbb{A})\big)$.

The theorem asserts that among all twisted elliptic endoscopic data $G \in \widetilde{\mathcal{E}}_{\mathrm{ell}}(N)$, there is exactly one that contains the automorphic source for the Hecke eigenfamily $c(\phi)$ of ϕ. The result serves as a "seed theorem" for our construction of $\widetilde{\Psi}(G)$. However, it has to be proved at the same time as broader theorems that pertain to the set $\widetilde{\Psi}(G)$ under construction. In this process, Theorem 5.1 becomes an induction hypothesis in the general treatment of parameters in $\widetilde{\Psi}(G)$. We shall assume it in what follows.

Suppose that ψ is a fixed element in $\widetilde{\Psi}(N)$. It is convenient to write K_ψ for the set $\{1, \ldots, r\}$ that indexes the simple constituents ψ_k of ψ. Since ψ is self-dual, there is an involution $k \to k^\vee$ on K_ψ such that $\psi_{k^\vee} = \psi_k^\vee$ and $\ell_{k^\vee} = \ell_k$. The indexing set is then a disjoint union

$$K_\psi = I_\psi \amalg J_\psi \amalg J_\psi^\vee, \qquad J_\psi^\vee = \{j^\vee : j \in J_\psi\},$$

where I_ψ is the set of fixed points of the involution, and J_ψ is some complementary set of representatives of the orbits of order 2. With this notation, we write

$$\psi = \Big(\underset{i \in I_\psi}{\boxplus} \ell_i \psi_i \Big) \boxplus \Big(\underset{j \in J_\psi}{\boxplus} \ell_j (\psi_j \boxplus \psi_{j^\vee}) \Big).$$

If i belongs to I_ψ, we apply Theorem 5.1 to the simple generic factor $\mu_i \in \widetilde{\Phi}_{\mathrm{sim}}(m_i)$ of $\psi_i = \mu_i \boxtimes \nu_i$. This gives an endoscopic datum $(G_{\mu_i}, s_{\mu_i}, \xi_{\mu_i})$ in $\widetilde{\mathcal{E}}_{\mathrm{sim}}(m_i)$, which we denote by H_i. If j belongs to J_ψ, we set $H_j = GL(m_j)$. We thus obtain a connected, reductive group H_k over F for any index k in either I_ψ or J_ψ, which is to say, in the set

$$\{K_\psi\} \cong I_\psi \amalg J_\psi$$

of orbits of the involution on K_ψ. Let $^L H_k$ be the Galois form of its L-group. We can then form the fibre product

$$(5.3) \qquad \mathcal{L}_\psi = \prod_{k \in \{K_\psi\}} (\,^L H_k \longrightarrow \Gamma_F)$$

of these groups over Γ_F.

We will use the group $\mathcal{L}_\psi$ as a substitute for the global Langlands group L_F in our study of automorphic representations attached to ψ. To make matters slightly more transparent, we have formulated it in algebraic form, as an extension of the pro-finite (and hence pro-algebraic) group Γ_F by a complex reductive group, rather than an extension of the locally compact group W_F by a compact topological group. For this reason we will work from now on with the Galois forms of L-groups rather than their Weil forms.

If $k = i$ is any index in I_ψ, we have the L-embedding

$$\widetilde{\mu}_i = \xi_{\mu_i} : \ ^L H_i \longrightarrow \ ^L\big(GL(m_i)\big) = GL(m_i, \mathbb{C}) \times \Gamma_F$$

that comes with H_i as a (twisted) endoscopic datum for

$$\widetilde{G}(m_i) = GL(m_i) \rtimes \widetilde{\theta}(m_i).$$

If $k = j$ belongs to J_ψ, we define an L-embedding

$$\widetilde{\mu}_j : {}^L H_j \longrightarrow {}^L\big(GL(2m_j)\big) = GL(2m_j, \mathbb{C}) \times \Gamma_F$$

by setting

$$\widetilde{\mu}_j(h_j \times \sigma) = \big(h_j \oplus \widetilde{\theta}_j(h_j)\big) \times \sigma, \qquad h_j \in \widehat{H}_j = GL(m_j, \mathbb{C}), \ \sigma \in \Gamma_F,$$

where

$$\widetilde{\theta}_j(h_j) = \widetilde{\theta}(m_j)(h_j) = \widetilde{J}(m_j) \cdot {}^t h_j^{-1} \cdot \widetilde{J}(m_j)^{-1}.$$

We then define an L-embedding

$$\widetilde{\psi} : \mathcal{L}_\psi \times SL(2, \mathbb{C}) \longrightarrow {}^L\big(GL(N)\big) = GL(N, \mathbb{C}) \times \Gamma_F$$

as the direct sum

$$(5.4) \qquad \widetilde{\psi} = \Big(\bigoplus_{i \in I_\psi} \ell_i(\widetilde{\mu}_i \otimes \nu_i) \Big) \oplus \Big(\bigoplus_{j \in J_\psi} \ell_j(\widetilde{\mu}_j \otimes \nu_j) \Big).$$

Our use of $SL(2, \mathbb{C})$ here in place of $SU(2)$ is purely notational, and is in keeping with our construction of $\mathcal{L}_\psi$ as a complex pro-algebraic group. We are of course free to interpret the embedding $\widetilde{\psi}$ also as an N-dimensional representation of $\mathcal{L}_\psi \times SL(2, \mathbb{C})$. With either interpretation, we shall be primarily interested in the equivalence class of $\widetilde{\psi}$ as a $GL(N, \mathbb{C})$-conjugacy class of homomorphisms from $\mathcal{L}_\psi \times SL(2, \mathbb{C})$ to either $GL(N, \mathbb{C})$ or $GL(N, \mathbb{C}) \times \Gamma_F$.

We can now define the set of global parameters attached to any of our classical groups. Suppose that $G \in \widetilde{\mathcal{E}}_{\mathrm{sim}}(N)$. We write $\widetilde{\Psi}(G)$ for the set of elements $\psi \in \widetilde{\Psi}(N)$ such that $\widetilde{\psi}$ factors through ${}^L G$. By this we mean that there exists an L-homomorphism

$$\widetilde{\psi}_G : \mathcal{L}_\psi \times SL(2, \mathbb{C}) \longrightarrow {}^L G$$

such that

$$(5.5) \qquad \xi \circ \widetilde{\psi}_G = \widetilde{\psi},$$

where ξ is the L-embedding of ${}^L G$ into ${}^L\big(GL(N)\big)$ that is part of the twisted endoscopic datum represented by G. Since $\widetilde{\psi}$ and ξ are to be regarded as $GL(N, \mathbb{C})$-conjugacy classes of homomorphisms, $\widetilde{\psi}_G$ is determined up to the stabilizer in $GL(N, \mathbb{C})$ of its image, a group that contains $\widehat{G}$. The quotient of this stabilizer by $\widehat{G}$ equals the group $\widetilde{\mathrm{Out}}_N(G)$ of outer automorphisms of $\widehat{G}$ described in §2. It is trivial unless G is of type $\mathbf{D}_n$, the case of an even orthogonal group in which $\widetilde{\mathrm{Out}}_N(G)$ equals $\mathbb{Z}/2\mathbb{Z}$. This case complicates our study of automorphic representations in a number of ways, all stemming from the fact that there can be two $\widehat{G}$-orbits of homomorphisms $\widetilde{\psi}_G$ in the larger class of $\widetilde{\psi}_G$. It is why we write $\widetilde{\Psi}(G)$ in place of the more familiar symbol $\Psi(G)$.

More generally, suppose that G belongs to the larger set $\widetilde{\mathcal{E}}_{\mathrm{ell}}(N)$ of elliptic data, or even the full set $\widetilde{\mathcal{E}}(N)$ of (twisted) endoscopic data for $\widetilde{G}(N)$. As a group over F, G equals a direct product

$$G = \prod_\alpha G_\alpha$$

of groups G_α that range over (quasisplit) special orthogonal and symplectic groups and (split) general linear groups. We define the set of global parameters for G as the product

$$\widetilde{\Psi}(G) = \prod_\alpha \Psi_\alpha(G_\alpha),$$

where $\Psi_\alpha(G_\alpha)$ equals $\widetilde{\Psi}(G_\alpha)$ if G_α is special orthogonal or symplectic, and equals $\Psi(G_\alpha)$ if G_α is a general linear group. It is not hard to see that an element in $\widetilde{\Psi}(G)$ can be identified with a pair $(\psi, \widetilde{\psi}_G)$, for a parameter $\psi \in \widetilde{\Psi}(N)$ and an L-homomorphism

$$\widetilde{\psi}_G : \mathcal{L}_\psi \times SL(2, \mathbb{C}) \longrightarrow {}^L G$$

that satisfies (5.5), and is defined as a $\widehat{G}$-orbit only up to the action of $\widetilde{\mathrm{Out}}_N(G)$. The projection

$$(\psi, \widetilde{\psi}_G) \longrightarrow \psi$$

is not generally injective, in contrast to the injective embedding of $\widetilde{\Psi}(G)$ into $\widetilde{\Psi}(N)$ for simple G that is an implicit part of our original definition. However, we still sometimes denote elements in the more general sets $\widetilde{\Psi}(G)$ by ψ when there is no danger of confusion. In the global situation at hand, one is usually concerned only with the case that G is elliptic, but the mapping is still not injective if G is not simple.

Suppose that ψ belongs to the set $\widetilde{\Psi}(G)$ we have just defined for any endoscopic datum $G \in \widetilde{\mathcal{E}}(N)$. We can then define the group

$$S_\psi = S_\psi(G) = \mathrm{Cent}\big(\mathrm{im}(\widetilde{\psi}_G), \widehat{G}\big)$$

we have been looking for as the centralizer of the image of $\widetilde{\psi}_G$. It is a reductive subgroup of $\widehat{G}$, whose quotient

$$\mathcal{S}_\psi = \mathcal{S}_\psi(G) = S_\psi / S_\psi^0 Z(\widehat{G})^{\Gamma_F}$$

is a finite abelian 2-group. Notice that there is a canonical element

$$(5.6) \qquad s_\psi = \widetilde{\psi}_G\left(1, \begin{pmatrix} -1 & 0 \\ 0 & -1 \end{pmatrix}\right)$$

in S_ψ. Its image in $\mathcal{S}_\psi$ plays a role in the classification of nontempered automorphic representations of G. We can also assign a twisted centralizer

$$\widetilde{S}_\psi(N) = S_\psi\big(\widetilde{G}(N)\big) = \mathrm{Cent}\big(\mathrm{im}(\widetilde{\psi}), GL(N, \mathbb{C}) \rtimes \theta\big)$$

to ψ, as well as its untwisted analogue

$$\widetilde{S}_\psi^0(N) = S_\psi\big(\widetilde{G}^0(N)\big) = \mathrm{Cent}\big(\mathrm{im}(\widetilde{\psi}), GL(N, \mathbb{C})\big).$$

Then $\widetilde{S}_\psi^0(N)$ is a product of complex general linear groups, embedded diagonally in $GL(N, \mathbb{C})$, as we noted in §4, and acts simply transitively by right or left translation on $\widetilde{S}_\psi(N)$. Since $\widetilde{S}_\psi(N)$ and $\widetilde{S}_\psi^0(N)$ are both connected, they do not complicate the automorphic representation theory of $GL(N)$.

Following §3, we write

$$\widetilde{\Psi}_{\mathrm{ell}}(N) = \Psi_{\mathrm{ell}}\big(\widetilde{G}(N)\big)$$

for the subset of parameters $\psi \in \widetilde{\Psi}(N)$ such that the indexing set J_ψ is empty, and such that $\ell_i = 1$ for each $i \in I_\psi$. These objects are analogous to the self-dual representations r we called elliptic in §2. Using the group $\mathcal{L}_\psi \times SL(2, \mathbb{C})$ in place of Λ_F, we can carry out the discussion of §3 without change here. Among other things, it tells us that any $\psi \in \widetilde{\Psi}_{\mathrm{ell}}(N)$ has a unique source in one of the sets

$$\widetilde{\Psi}(G), \qquad G \in \widetilde{\mathcal{E}}_{\mathrm{ell}}(N).$$

To be more precise, let $\widetilde{\Psi}_2(G)$ be the preimage of $\widetilde{\Psi}_{\mathrm{ell}}(N)$ in $\widetilde{\Psi}(G)$, for any $G \in \widetilde{\mathcal{E}}_{\mathrm{ell}}(N)$. The mapping from $\widetilde{\Psi}(G)$ to $\widetilde{\Psi}(N)$ then takes $\widetilde{\Psi}_2(G)$ injectively[10] onto a subset of $\widetilde{\Psi}_{\mathrm{ell}}(N)$, which we identify with $\widetilde{\Psi}_2(G)$. Moreover, $\widetilde{\Psi}_{\mathrm{ell}}(N)$ is a disjoint union

$$\widetilde{\Psi}_{\mathrm{ell}}(N) = \coprod_{G \in \widetilde{\mathcal{E}}_{\mathrm{ell}}(N)} \widetilde{\Psi}_2(G)$$

of these subsets. We thus have parallel chains of parameter sets

$$\widetilde{\Psi}_{\mathrm{sim}}(N) \subset \widetilde{\Psi}_{\mathrm{ell}}(N) \subset \widetilde{\Psi}(N)$$

and

$$(5.7) \qquad \widetilde{\Psi}_{\mathrm{sim}}(G) \subset \widetilde{\Psi}_2(G) \subset \widetilde{\Psi}(G), \qquad G \in \widetilde{\mathcal{E}}_{\mathrm{ell}}(N),$$

where $\widetilde{\Psi}_{\mathrm{sim}}(G)$ denotes the intersection of $\widetilde{\Psi}_{\mathrm{sim}}(N)$ with $\widetilde{\Psi}_2(G)$. Observe that $\widetilde{\Psi}_2(G)$ is the subset of parameters $\psi \in \widetilde{\Psi}(G)$ such that the centralizer S_ψ is finite, while $\widetilde{\Psi}_{\mathrm{sim}}(G)$ consists of those ψ such that S_ψ equals the minimal group $Z(\widehat{G})^{\Gamma_F}$.

The group $\mathcal{L}_\psi$ now seems quite promising. As we have just seen, it leads to the kind of objects we would attach to parameters defined on the product $L_F \times SU(2)$. In particular, we now have L-homomorphisms $\widetilde{\psi}$ and $\widetilde{\psi}_G$ from $\mathcal{L}_\psi \times SL(2, \mathbb{C})$ to groups $^L\big(GL(N)\big)$ and $^L G$, with the corresponding centralizers $\widetilde{S}_\psi(N)$ and S_ψ we will need. There is no denying that the process is pretty crude, starting with the ad hoc definition (5.3) that requires Theorem 5.1 as a long term induction hypothesis. It is also not appealing that $\mathcal{L}_\psi$ depends on the parameter ψ. Nevertheless, the group $\mathcal{L}_\psi$ does in the end serve our purpose. It is a kind of endoscopic hull of what would be the image of the Langlands group L_F under a parameter ψ.

Much of our discussion here has been taken directly from §1.4 of [**ECR**]. The reader can refer to this section of [**ECR**] for some further discussion, having to do with the following natural questions:

(i) given $\psi \in \widetilde{\Psi}_{\mathrm{ell}}(N)$, determine the unique $G \in \widetilde{\mathcal{E}}_{\mathrm{ell}}(N)$ such that ψ belongs to $\widetilde{\Psi}_2(G)$;

(ii) more generally, given any $\psi \in \widetilde{\Psi}(N)$, determine all $G \in \widetilde{\mathcal{E}}_{\mathrm{ell}}(N)$ such that ψ lies in the image of $\widetilde{\Psi}(G)$;

(iii) given $G \in \widetilde{\mathcal{E}}_{\mathrm{sim}}(N)$ and $\psi \in \widetilde{\Psi}(G)$, determine the centralizer $S_\psi = S_\psi(G)$ explicitly.

However, there is another point from [**ECR**, §1.4] that we do need to mention here. It concerns a bijective correspondence, which is elementary, but which is also in some sense the theoretical center of the theory of endoscopy.

In §3, we described endoscopic data $G' \in \mathcal{E}(G)$ for G. Since any such G' is again a direct product

$$G' = \prod_{\alpha'} G'_{\alpha'}$$

---

[10] so in particular, the mapping from $\widetilde{\Psi}(G)$ to $\widetilde{\Psi}(N)$ that we agreed above was *not* injective if G lies in the complement of $\widetilde{\mathcal{E}}_{\mathrm{sim}}(N)$ in $\widetilde{\mathcal{E}}_{\mathrm{ell}}(N)$, *is* injective upon restriction to the subset $\widetilde{\Psi}_2(G)$ of $\widetilde{\Psi}(G)$.

of groups of the kind we have studied, we can define the set of parameters

$$\widetilde{\Psi}(G') = \prod_{\alpha'} \Psi(G'_{\alpha'}),$$

as above. We can also form the centralizer group $S_{\psi'} = S_{\psi'}(G')$, for any ψ' in this set.

Consider a pair

$$(5.8) \qquad\qquad (G', \psi'), \qquad G' \in \mathcal{E}(G),\ \psi' \in \widetilde{\Psi}(G').$$

We recall that G' really represents an isomorphism class of triplets (G', s', ξ'), where s' is a semisimple element in $\widehat{G}'$ and ξ' is an L-embedding of ${}^L G'$ into ${}^L G$. The parameter $\psi' = \psi_{G'}$ can be identified with a pair $(\psi, \widetilde{\psi}')$, for a parameter $\psi \in \widetilde{\Psi}(G)$ and an L-embedding

$$\widetilde{\psi}' = \widetilde{\psi}_{G'} :\ \mathcal{L}_\psi \times SL(2, \mathbb{C}) \ \longrightarrow\ {}^L G'$$

such that

$$(5.9) \qquad\qquad\qquad \xi' \circ \widetilde{\psi}' = \widetilde{\psi}_G,$$

in the notation above, where $\widetilde{\psi}'$ is defined as a $\widehat{G}'$-orbit only up to the action of the finite group

$$\widetilde{\mathrm{Out}}_N(G') = \mathrm{Out}_G(G') \times \widetilde{\mathrm{Out}}_N(G).$$

The pair (G', ψ') gives rise to a second pair

$$(5.10) \qquad\qquad (\psi, s), \qquad \psi \in \widetilde{\Psi}(G),\ s \in \mathcal{E}(\overline{S}_\psi),$$

where $\mathcal{E}(\overline{S}_\psi) = \mathcal{E}(\overline{S}_{\psi,\mathrm{ss}})$ denotes the set of semisimple conjugacy classes in the complex reductive group $\overline{S}_\psi = S_\psi / Z(\widehat{G})^{\Gamma_F}$. Indeed $\psi = \psi_G$ is attached to ψ' as above, while s is just the image $\xi'(s')$ of s' in $\widehat{G}$.

Conversely, suppose that it is a pair (ψ, s) of the second sort (5.10) that we are given. With this information, we set $\widehat{G}'$ equal to the connected centralizer of s in $\widehat{G}$, and s' equal (somewhat superfluously) to the preimage of s in $\widehat{G}'$. The product

$$\mathcal{G}' = \widehat{G}' \cdot \widetilde{\psi}_G\big(\mathcal{L}_\psi \times SL(2, \mathbb{C})\big)$$

of $\widehat{G}'$ with the image of $\widetilde{\psi}_G$ can be identified with an L-subgroup of ${}^L G$, for which the identity embedding ξ' is an L-homomorphism. We define G' to be a quasisplit group for which $\widehat{G}'$, equipped with the L-action induced by $\mathcal{G}'$, is a dual group. The triplet (G', s', ξ') represented by G' is then an endoscopic datum for G, as defined in §3. Since s lies in the centralizer of the image of the L-embedding $\widetilde{\psi}_G$ attached to ψ, $\widetilde{\psi}_G$ factors through ${}^L G'$. We obtain an L-embedding $\widetilde{\psi}'$ of $\mathcal{L}_\psi \times SL(2, \mathbb{C})$ into ${}^L G'$ that satisfies (5.9), and hence, an element $\psi' \in \widetilde{\Psi}(G')$. The pair (ψ, s) thus leads in the other direction to a pair (G', ψ') of the first sort (5.8).

The bijective correspondence

$$(5.11) \qquad\qquad\qquad (G', \psi') \ \longrightarrow\ (\psi, s)$$

is a general phenomenon. It applies to arbitrary endoscopic data, twisted or otherwise, and corresponding spectral parameters. In particular, it has a natural variant in the case that (G, G') is replaced by $\big(\widetilde{G}(N), G\big)$. It also applies without change if F is replaced by a local field. In all cases, the correspondence transforms questions on the transfer of characters to questions on the groups S_ψ.

Finally, we will need to know how to localize global parameters for G. For any v, we have the local parameter set $\Psi(G_v)$ defined in §4. We write $\Psi^+(G_v)$ for the larger set of parameters on $L_{F_v} \times SU(2)$, in which the image in $\widehat{G}$ is not required to be bounded. To match our global convention, we will also write $\widetilde{\Psi}^+(G_v)$ for the quotient of $\Psi^+(G_v)$ by the group $\widetilde{\mathrm{Out}}_N(G_v)$. We would like to show that for any $G \in \widetilde{\mathcal{E}}_{\mathrm{sim}}(N)$, the mapping (4.12) from $\Psi(N)$ to $\Psi_v^+(N)$ takes the subset $\widetilde{\Psi}(G)$ of $\Psi(N)$ into the subset $\widetilde{\Psi}^+(G_v)$ of $\Psi_v^+(N)$. This property is not elementary. It is a consequence of a second "seed theorem", which we state as a complement to Theorem 5.1, but which like Theorem 5.1, has in the end to be proved at the same time as broader theorems.

THEOREM 5.2. *Suppose that $\phi \in \widetilde{\Phi}_{\mathrm{sim}}(N)$ is simple generic, as in Theorem 5.1. Then the localization ϕ_v of ϕ at any v, a priori an element in the subset $\widetilde{\Phi}_v(N)$ of local generic parameters in $\widetilde{\Psi}_v^+(N)$, lies in the subset $\widetilde{\Phi}(G_{\phi,v})$ of $\widetilde{\Phi}_v(N)$ attached to the localization $G_{\phi,v}$ of the global datum $G_\phi \in \widetilde{\mathcal{E}}_{\mathrm{sim}}(N)$ of Theorem 5.1.*

The theorem asserts that the N-dimensional representation ϕ_v of L_{F_v}, which is attached by the local Langlands correspondence to the cuspidal automorphic representation of $GL(N)$ given by ϕ, factors through the local endoscopic embedding

$$\xi_{\phi,v} : {}^L G_{\phi,v} \longrightarrow {}^L\big(GL(N)_v\big).$$

It allows us to identify ϕ_v with a local L-homomorphism from L_{F_v} to ${}^L G_{\phi,v}$. Like its global companion Theorem 5.1, Theorem 5.2 is proved by a long induction argument that includes the proof of broader theorems. Since we will not be able to present the general argument here, we will just assume both Theorems 5.1 and 5.2 in what follows.

We apply Theorem 5.2 to each of the orthogonal and symplectic factors ${}^L H_k$ in the fibre product (5.3). Putting them together, and applying only the local correspondence of Theorem 4.4 in case $H_k = GL(m_k)$, we obtain a conjugacy class of L-homomorphisms

$$(5.12) \qquad \begin{array}{ccc} L_{F_v} & \longrightarrow & \Gamma_{F_v} \\ \downarrow & & \downarrow \\ \mathcal{L}_\psi & \longrightarrow & \Gamma_F \end{array}$$

It is determined up the action of the group

$$\widetilde{\mathrm{Out}}(\mathcal{L}_\psi) = \prod_k \widetilde{\mathrm{Out}}_{m_k}(H_k),$$

where $\widetilde{\mathrm{Out}}_{m_k}(H_k) = 1$ in case $H_k = GL(m_k)$. This is the analogue of (2.3) for our makeshift substitute $\mathcal{L}_\psi$ for L_F.

Suppose that ψ belongs to $\widetilde{\Psi}(G)$ for some $G \in \widetilde{\mathcal{E}}_{\mathrm{sim}}(N)$, or indeed, for any G in the general set $\widetilde{\mathcal{E}}(N)$. It then follows from this discussion that we can identify the localization of ψ, as an element in $\widetilde{\Psi}_v^+(N)$, with an L-homomorphism

$$\psi_v : L_{F_v} \times SU(2) \longrightarrow {}^L G_v.$$

This fits into a larger commutative diagram of L-homomorphisms

$$(5.13)$$

$$
\begin{array}{ccccc}
L_{F_v} \times SU(2) & \xrightarrow{\psi_v} & {}^L G_v & \longrightarrow & \Gamma_{F_v} \\
\downarrow & & \downarrow & & \downarrow \\
\mathcal{L}_\psi \times SL(2,\mathbb{C}) & \xrightarrow{\tilde{\psi}_G} & {}^L G & \longrightarrow & \Gamma_F
\end{array}
$$

where the left hand vertical arrow is given by the mapping of L_{F_v} into $\mathcal{L}_\psi$ in (5.12), and the embedding of $SU(2)$ into $SL(2,\mathbb{C})$. In particular, we obtain a localization mapping of the form (4.12) that takes $\tilde{\Psi}(G)$ into the set $\tilde{\Psi}^+(G_v)$. Moreover, since ψ_v is essentially the restriction of the global embedding $\tilde{\psi}_G$ to the image of $L_{F_v} \times SU(2)$, the global centralizer S_ψ is contained in S_{ψ_v}. From this, it follows that there is a canonical mapping

$$x \longrightarrow x_v, \qquad x \in \mathcal{S}_\psi,$$

of $\mathcal{S}_\psi$ into $\mathcal{S}_{\psi_v}$.

6. Transfer and the fundamental lemma

The field F will be local in this section unless stated otherwise. The essential problem is to establish the local Langlands correspondence for a special orthogonal or symplectic group $G \in \tilde{\mathcal{E}}_{\mathrm{sim}}(G)$. It is closely related to local functoriality, specifically the second of two cases described at the end of §1, in which $G' \in \tilde{\mathcal{E}}(G)$ is an endoscopic group for G. Within the general principle of functoriality, this case is distinguished by being also a part of the separate theory of endoscopy. As such, it should come with a characterization of the image of the functorial correspondence

$$\pi' \longrightarrow \pi, \qquad \pi' \in \Pi(G'),$$

of Principle 1.1. The reason for this is that the functorial correspondence of representations will be dual to a transfer of functions from $G(F)$ to $G'(F)$.

The transfer of functions is based on harmonic analysis. Its domain is a space of test functions on $G(F)$, such as the space $C_c^\infty(G)$ of functions on $G(F)$ that are *smooth* (which means infinitely differentiable if F is archimedean, and locally constant if F in nonarchimedean) and compactly supported. Following [**ECR**], we will instead take the Hecke algebra $\mathcal{H}(G)$, a convolution algebra that equals $C_c^\infty(G)$ if F is nonarchimedean, but that is the proper subalgebra of functions $f \in C_c^\infty(G)$ that satisfy a supplementary finiteness condition under left and right translation of f by elements in a fixed maximal compact subgroup of $G(F)$, if F is archimedean.

An element $\gamma \in G(F)$ is called *strongly regular* if its centralizer

$$G_\gamma = \mathrm{Cent}(\gamma, G) = \{x \in G : x^{-1}\gamma x = \gamma\}$$

is a (maximal) torus in G. For any such γ, we have the associated *invariant orbital integral*

$$f_G(\gamma) = |D(\gamma)|^{\frac{1}{2}} \int_{G_\gamma(F)\backslash G(F)} f(x^{-1}\gamma x)\,dx$$

of any test function $f \in \mathcal{H}(G)$, where dx is a fixed, right invariant measure on the coset space $G_\gamma(F)\backslash G(F)$. We have normalized $f_G(\gamma)$ by the Weyl discriminant

$$D(\gamma) = \det\left((1 - \mathrm{Ad}(\gamma))_{\mathfrak{g}/\mathfrak{g}_\gamma}\right),$$

where $\mathfrak{g}$ and $\mathfrak{g}_\gamma$ are the Lie algebras of G and G_γ. We regard f_G as a function on the set of strongly regular points γ, and write

$$(6.1) \qquad \mathcal{I}(G) = \{f_G : f \in \mathcal{H}(G)\}$$

for the image of $\mathcal{H}(G)$ under this transform.

The functions in $\mathcal{I}(G)$ also have a spectral interpretation. Any representation $\pi \in \Pi(G)$ has a character, which can be identified with the linear form

$$\operatorname{tr}\big(\pi(f)\big) = \operatorname{tr}\Big(\int_{G(F)} f(x)\pi(x)dx \Big), \qquad f \in \mathcal{H}(G),$$

on $\mathcal{H}(G)$. We set

$$f_G(\pi) = \operatorname{tr}\big(\pi(f)\big),$$

in analogy with the notation we have used for the dual orbital integrals. It can then be shown that either of the two functions $\{f_G(\gamma)\}$ or $\{f_G(\pi)\}$ attached to f determines the other. We can therefore regard any element f_G in $\mathcal{I}(G)$ as a function of either γ or π. It is *invariant,* in the sense that it depends only on the conjugacy class of γ or the equivalence class of π. It also remains invariant if its preimage $f \in \mathcal{H}(G)$ is replaced by any conjugate

$$f^y(x) = f(yxy^{-1}), \qquad x, y \in G(F).$$

The theory of endoscopy is founded on the fact that conjugacy in $G(F)$ is finer than geometric conjugacy. Two strongly regular elements in $G(F)$ are said to be *stably conjugate* if they are conjugate as elements in the group $G(\overline{F})$ of geometric points in G. For the local field F, it is easy to show that there are only finitely many $G(F)$-conjugacy classes γ in any (strongly regular) stable conjugacy class δ. The corresponding sum

$$f^G(\delta) = \sum_\gamma f_G(\gamma), \qquad f \in \mathcal{H}(G),$$

of orbital integrals is called the *stable orbital integral* of the given function f at δ. We write

$$(6.2) \qquad \mathcal{S}(G) = \{f^G : f \in \mathcal{H}(G)\}$$

for the space of functions of δ obtained in this way. As we shall see, L-packets arise when we try to find a spectral interpretation for the functions f^G in $\mathcal{S}(G)$ analogous to the values $f_G(\pi)$ of functions f_G in $\mathcal{I}(G)$. In general, a distribution on $G(F)$, or more correctly a continuous linear form on $\mathcal{H}(G)$, is said to be *stable* if its value at any f depends only on f^G. If this is so, we can identify S with the linear form

$$(6.3) \qquad \widehat{S}(f^G) = S(f), \qquad f \in \mathcal{H}(G),$$

on $\mathcal{S}(G)$. The spectral question above is then to attach stable distributions to representations π.

Suppose that G' is an endoscopic datum for G. Langlands and Shelstad [**LS**] define a strongly regular element δ' in $G'(F)$ to be *strongly G-regular* if its image in $G(F)$ (under any admissible embedding [**LS**, (1.3)] of its centralizer $G'_{\delta'}$ into G) is strongly regular for G. The space of strongly G-regular elements remains open and dense in $G'(F)$, so functions in the space $\mathcal{S}(G')$ are determined by their values on strongly G-regular, stable conjugacy classes in $G'(F)$. The point of the article [**LS**] was to introduce an explicit function $\Delta(\delta', \gamma)$ of a strongly G-regular stable conjugacy class δ' in $G'(F)$ and a strongly regular conjugacy class γ in $G(F)$,

40   J. ARTHUR

which they called a *transfer factor* for G and G'. By construction, this function
vanishes unless γ belongs to the stable conjugacy class of the image of δ' in $G'(F)$.
It therefore has finite support in either of the variables when the complementary
variable is fixed. The role of $\Delta(\delta', \gamma)$ is as the kernel function for the transfer
mapping that sends a function $f \in \mathcal{H}(G)$ to the function

$$(6.4) \qquad f'(\delta') = f_\Delta^{G'}(\delta') = \sum_\gamma \Delta(\delta', \gamma) f_G(\gamma)$$

of δ'. Langlands and Shelstad conjectured that the function $f'(\delta')$ belongs to the
space $\mathcal{S}(G')$.

The Langlands-Shelstad transfer conjecture remained a fundamental problem
for twenty years. It had been established earlier for archimedean F (and ad hoc
transfer factors) by Shelstad [**She1**]. But for nonarchimdean F, it was closely tied
to the fundamental lemma. The fundamental lemma is a related conjecture, posed
originally by Langlands, which became precise with the introduction of the transfer factors of [**LS**]. It applies to the case that the quasisplit groups G and G' are
unramified, which means that they are both split over an unramified extension of
the nonarchimedean field F. With this condition, $G(F)$ has a hyperspecial maximal compact subgroup K_F, an important object determined uniquely up to the
appropriate analogue of stable conjugacy. The fundamental lemma asserts that if f
is the characteristic function of K_F (a function in $\mathcal{H}(G)$ since $K_F \subset G(F)$ is open),
then f' equals the image in $\mathcal{S}(G')$ of the characteristic function of any hyperspecial
maximal compact subgroup K'_F of $G'(F)$. It thus represents a more precise version
of the transfer conjecture in a special case.

Kottwitz and Shelstad [**KS**] later extended the results of [**LS**] to twisted endoscopic data. They had a number of new problems to deal with, but for our setting
here, we need include only a small modification of the discussion above. To do so,
we replace the group G by the component $\widetilde{G}(N) = GL(N) \rtimes \widetilde{\theta}(N)$, and its endoscopic datum G' by a twisted endoscopic datum $G \in \widetilde{\mathcal{E}}(N)$. Endoscopic transfer in
this setting is again tied to local functoriality, this time to the first of the two cases
introduced at the end of §1. From what we have just said, it is clear that this case
is also distinguished by being part of the separate theory of endoscopy.

One defines the Hecke module $\widetilde{\mathcal{H}}(N) = \mathcal{H}\big(\widetilde{G}(N)\big)$ of functions $\widetilde{f}$ on $\widetilde{G}(N, F)$,
and the notion of a strongly regular element $\widetilde{\gamma}$ in $\widetilde{G}(N, F)$. One then defines the
twisted orbital integral $\widetilde{f}_N(\widetilde{\gamma})$ of $\widetilde{f}$ over the orbit of $\widetilde{\gamma}$ under the group $\widetilde{G}^0(N, F) =$
$GL(N, F)$ acting by conjugation on $\widetilde{G}(N, F)$. The twisted transfer factor from [**KS**]
is an explicit, but sophisticated function $\widetilde{\Delta}(\delta, \widetilde{\gamma})$ of a strongly $\widetilde{G}(N)$-regular, stable
conjugacy class δ in $G(F)$ and a strongly regular, twisted conjugacy class $\widetilde{\gamma}$ in
$\widetilde{G}(N, F)$. It serves as the kernel function for the transfer mapping that sends a
function $\widetilde{f} \in \widetilde{\mathcal{H}}(N)$ to the function

$$(6.5) \qquad \widetilde{f}^G(\delta) = \widetilde{f}_\Delta^G(\delta) = \sum_{\widetilde{\gamma}} \widetilde{\Delta}(\delta, \widetilde{\gamma}) \widetilde{f}_N(\widetilde{\gamma})$$

of δ. For reasons similar to those above, the sum can again be taken over a finite
set that depends only on δ. Folowing [**LS**], Kottwitz and Shelstad conjectured that
the function $\widetilde{f}^G(\delta)$ belongs to the space $\mathcal{S}(G)$. The twisted fundamental lemma
asserts that if G is unramified, and $\widetilde{f}$ is the characteristic function of the open

subset $GL(N, \mathfrak{o}_F) \rtimes \widetilde{\theta}(N)$ of $\widetilde{G}(N, F)$, $\widetilde{f}^G$ is the image in $\mathcal{S}(G)$ of the characteristic function of a hyperspecial, maximal compact subgroup K_F of $G(F)$.

For archimedean F, Shelstad has recently completed a proof of the general twisted transfer conjecture [**She5**], using the explicit specialization to real groups of the twisted transfer factors of [**KS**]. This followed other recent papers [**She2**]–[**She4**], in which she reformulated much of her earlier work on ordinary (untwisted) endoscopy from the perspective of [**LS**].

For nonarchimedean F, the breakthrough was the geometric proof of the fundamental lemma by Ngô [**N**]. He combined the local geometric ideas of Goresky, Kottwitz and MacPherson [**GKM**] on affine Springer fibres with an analogue of the global Hitchin fibration to establish the fundamental lemma for a local field of positive characteristic. By earlier results on the independence of characteristic [**W2**], this gave the fundamental lemma also for the local field F of characteristic 0. The paper of Ngô treats the ordinary (untwisted) fundamental lemma, and a variant to which Waldspurger had reduced the general (twisted) case [**W3**]. It therefore resolves the fundamental lemma in complete generality. As for the Kottwitz-Langlands-Shelstad (KLS) transfer conjecture, Waldspurger had established some time ago that the ordinary transfer conjecture would follow from the fundamental lemma [**W1**]. His more recent papers [**W3**], [**W4**] extend this implication to the general case. The general results of Waldspurger therefore yield the nonarchimedean LSK-conjecture in all cases.

We will use the transfer mapping from $\widetilde{G}(N)$ to answer the question posed above on a spectral interpretation for the functions in $\mathcal{S}(G)$. We begin with a local parameter $\psi \in \Psi(N)$ for $GL(N)$. From the correspondence (4.11) derived from Theorem 4.4, we obtain a irreducible unitary representation $\pi_\psi \in \Pi_{\mathrm{unit}}(N)$ of $GL(N, F)$. Assume that ψ lies in the subset $\widetilde{\Psi}(N)$ of self-dual parameters. As we have noted, π_ψ then has an extension to the group $\widetilde{G}(N, F)^+$. However, the extension is determined a priori only up to the sign character on the semidirect factor $\widetilde{\theta}(N)^+$ of $\widetilde{G}(N, F)^+$. We need to define it uniquely.

It is the theory of Whittaker models that provides a canonical extension of π_ψ. This theory is well understood for general linear groups, and is expected to carry over to tempered L-packets for general quasisplit groups. In fact, for our group G, the conjectured properties (proposed by Shahidi in [**Sha**]) were established in §8.3 of [**ECR**]. In general, one must fix a Whittaker datum (B, χ), consisting of a rational Borel subgroup B of a given quasisplit group over F, and a nondegenerate character χ on the unipotent radical $N_B(F)$ of $B(F)$. (In the case of a twisted group, such as $\widetilde{G}(N)$, one must take (B, χ) to be stable under the relevant automorphism, $\widetilde{\theta}(N)$ in the case $\widetilde{G}(N)$.) A (B, χ)-Whittaker vector for an irreducible representation is then a nonzero vector in the underlying complex vector space, on which $N_B(F)$ is χ-equivariant. If it exists for the given representation, a (B, χ)-Whittaker vector is known to be unique up to a complex multiple. (See the brief introduction in [**ECR**, §2.5], for example.)

We will not review the theory of Whittaker models further, except to note its role in the choice of transfer factors. In general, a transfer factor is defined uniquely only up to a nonzero scalar multiple. But for our group G (and indeed, for any quasisplit group over F), Langlands and Shelstad attach a canonical transfer factor to any F-splitting [**LS**, §1.3] of G. The group $G_{\mathrm{ad}}(F)$ of F-points in the adjoint group of G acts simply transitively on the set of F-splittings of G. The associated transfer factors are equal (for all G') if and only if their splittings lie in the same orbit under the image $\big(G(F)\big)_{\mathrm{ad}}$ of $G(F)$ in $G_{\mathrm{ad}}(F)$. The finite quotient

$$(6.6) \qquad\qquad G_{\mathrm{ad}}(F)/\big(G(F)\big)_{\mathrm{ad}}$$

therefore acts simply transitively on the set of (families of) normalized transfer factors. These normalizations are really geometric, in that they lead to the simplest explicit formulas for the transfer of orbital integrals. Kottwitz and Shelstad, for their part, introduced a different normalization for the transfer factors of G (and for any quasisplit group over F). It was attached to any Whittaker datum for G. The group $G_{\mathrm{ad}}(F)$ also acts simply transitively on the set of Whittaker data, and the quotient (6.6) again acts simply transitively on the corresponding set of such normalizations. These are spectral normalizations, in that they are expected to lead to the simplest explicit formulas for the transfer of characters. We assume implicitly from now on that the transfer factors for G have been assigned the Whittaker normalization attached to a fixed Whittaker datum.

Similar remarks apply to twisted transfer factors, but there is no need to discuss them explicitly. We simply fix a $\widetilde{\theta}(N)$-stable Whittaker datum $\big(\widetilde{B}(N), \widetilde{\chi}(N)\big)$ for $GL(N)$, and work with the associated normalized twisted transfer factors. If $\psi = \phi$ lies in the subset $\widetilde{\Phi}_{\mathrm{bdd}}(N)$ of $\widetilde{\Psi}(N)$, the self-dual representation $\pi_\psi = \pi_\phi$ of $GL(N,F)$ is tempered, and therefore has a $\big(\widetilde{B}(N), \widetilde{\chi}(N)\big)$-Whittaker vector. We take $\widetilde{\pi}_\phi$ to be the unique extension of π_ϕ to the group $\widetilde{G}(N,F)^+$ such that the operator $\widetilde{\pi}_\phi(N) = \widetilde{\pi}_\phi\big(\widetilde{\theta}(N)\big)$ maps the Whittaker vector to itself. The definition carries over to our general parameter $\psi \in \widetilde{\Psi}(N)$, even though the nontempered Speh representation π_ψ need not have a Whittaker vector. For we can work with the induced representation ρ_ψ of which π_ψ is the Langlands quotient. This representation does have a Whittaker vector, which serves to define an extension $\widetilde{\rho}_\psi$ of ρ_ψ. Its quotient then gives an extension $\widetilde{\pi}_\psi$ of π_ψ. (See [**ECR**, §2.1].)

Given the extension $\widetilde{\pi}_\psi$ of π_ψ, we define a linear form

$$(6.7) \qquad\qquad \widetilde{f}_N(\psi) = \mathrm{tr}\big(\widetilde{\pi}_\psi(\widetilde{f})\big), \qquad \widetilde{f} \in \widetilde{\mathcal{H}}(N),$$

on $\widetilde{\mathcal{H}}(N)$. Does it transfer to G? More precisely, is it the image of a stable linear form on $\mathcal{H}(G)$ that is dual to the transfer $\widetilde{f} \to \widetilde{f}^G$ of functions? We would expect a necessary condition to be that as an L-homomorphism from $L_F \times SU(2)$ into $^L\big(GL(N)\big)$, ψ factors through the L-subgroup $^L G$. In other words, ψ should lie in the subset

$$\widetilde{\Psi}(G) = \Psi(G)/\mathrm{Out}_N(G)$$

of $\widetilde{\Psi}(N)$ attached to our group $G \in \widetilde{\mathcal{E}}_{\mathrm{sim}}(N)$. Now the dual transfer of functions takes $\widetilde{\mathcal{H}}(N)$ into the subspace $\widetilde{\mathcal{S}}(G)$ of functions in $\mathcal{S}(G)$ that are invariant under the finite group $\widetilde{\mathrm{Out}}_N(G)$. It is convenient also to write $\widetilde{\mathcal{H}}(G)$ for the subspace of functions in $\mathcal{H}(G)$ invariant under $\mathrm{Out}_N(G)$ (with the nontrivial element in $\mathrm{Out}_N(G)$, when it exists, identified as in §3 with an actual F-automorphism of G). The question above should then be whether (6.7) transfers to a stable linear form on the subspace $\widetilde{\mathcal{H}}(G)$ of $\mathcal{H}(G)$. The following theorem, which is a foundation for the local correspondence we will state in the next section, gives an affirmative answer.

THEOREM 6.1. *Suppose that F is local, that $G \in \widetilde{\mathcal{E}}_{\mathrm{ell}}(N)$, and that ψ lies in the set $\widetilde{\Psi}(G)$ attached to a fixed group $G \in \widetilde{\mathcal{E}}_{\mathrm{ell}}(N)$. Then there is a unique stable linear form*

$$(6.8) \qquad f \longrightarrow f^G(\psi), \qquad f \in \widetilde{\mathcal{H}}(G),$$

on $\widetilde{\mathcal{H}}(G)$ with the general property

$$(6.9) \qquad \widetilde{f}^G(\psi) = \widetilde{f}_N(\psi), \qquad \widetilde{f} \in \widetilde{\mathcal{H}}(N),$$

together with the secondary property

$$(6.10) \qquad f^G(\psi) = f^S(\psi_S) f^O(\psi_O), \qquad \widetilde{f} \in \widetilde{\mathcal{H}}(G),$$

in case

$$G = G_S \times G_O, \qquad G_\varepsilon \in \widetilde{\mathcal{E}}_{\mathrm{sim}}(N_\varepsilon),$$

$$\psi = \psi_S \times \psi_O, \qquad \psi_\varepsilon \in \widetilde{\Psi}(G_\varepsilon),$$

and

$$f^G = f^S \times f^O, \qquad f^\varepsilon \in \widetilde{\mathcal{S}}(G_\varepsilon), \ \varepsilon = O, S,$$

are composite.

REMARKS. 1. The primary case is for $G \in \widetilde{\mathcal{E}}_{\mathrm{sim}}(N)$ simple. It has dominated our past discussion, for the reason that many questions for composite twisted endoscopic data are amenable to induction. When G is simple, the mapping $\widetilde{f} \to \widetilde{f}^G$ takes $\widetilde{\mathcal{H}}(N)$ *onto* $\widetilde{\mathcal{S}}(G)$ [**ECR**, Corollary 2.2]. The uniqueness of the linear form (6.8) then follows from the formula (6.9) in this case.

2. If $G \in \widetilde{\mathcal{E}}_{\mathrm{ell}}(N)$ is composite, the uniqueness follows from the product formula (6.10). In this case, the symbol ψ on the right hand side of (6.9) is understood to be the image of the given composite parameter under the (not necessarily injective) mapping from $\widetilde{\Psi}(G)$ to $\widetilde{\Psi}(N)$. The problem here is to establish the compatibility condition represented by the two sides of (6.9).

3. Suppose that ψ lies in the set $\widetilde{\Psi}(G)$ attached to some G in the complement of $\widetilde{\mathcal{E}}_{\mathrm{ell}}(N)$ in the full set $\widetilde{\mathcal{E}}(N)$ of twisted endoscopic data. The assertions (6.8) and (6.9) of the theorem then follow easily by reduction to the Levi component of a proper, $\widetilde{\theta}(N)$-stable parabolic subgroup of $GL(N)$.

4. The notation $f^G(\psi)$ in (6.8) is deliberate, even if perhaps also slightly confusing. It reminds us that we are dealing with a linear form on $\widetilde{\mathcal{S}}(G)$. In particular, we can define $f^G(\phi)$ for any ϕ in the subset $\widetilde{\Phi}_{\mathrm{bdd}}(G)$ of $\widetilde{\Psi}(G)$, or by analytic continuation, any ϕ in the larger set $\widetilde{\Phi}(G)$. It can be shown that either of the two functions $\{f^G(\delta)\}$ or $\{f^G(\phi)\}$ attached to f determines the other. We can therefore regard any element f^G in $\widetilde{\mathcal{S}}(G)$ as a function of either δ or ϕ. This is in answer to the question above on a spectral interpretation of the function f^G (or rather, the slightly weaker question for the subspace $\widetilde{\mathcal{S}}(G)$ of $\mathcal{S}(G)$). The local classification theorem, which we state in the next section, expresses the stable character (6.8) of ϕ in terms of the characters of representations π in the L-packet of ϕ, in spectral analogy with the conjugacy classes γ in a stable conjugacy class δ.

7. Statement of theorems

We will now state our theorems of classification. We fix a (quasisplit, special) orthogonal or symplectic group $G \in \widetilde{\mathcal{E}}_{\mathrm{sim}}(N)$ over the field F. Our ultimate concern is the global classification of automorphic representations. However, this necessarily relies on an understanding of local irreducible representations. We therefore assume for the moment that F is local.

We would ideally like to attach a canonical L-packet Π_ϕ of irreducible representations $\pi \in \Pi(G)$ of $G(F)$ to any local Langlands parameter $\phi \in \Phi(G)$. As we shall explain presently, it would suffice to consider the case of bounded parameters $\phi \in \Phi_{\mathrm{bdd}}(G)$, which leads to L-packets Π_ϕ of tempered representations $\pi \in \Pi_{\mathrm{temp}}(G)$. It also represents a special case of our other family of parameters $\psi \in \Psi(G)$, which leads to packets Π_ψ of unitary representations $\pi \in \Pi_{\mathrm{unit}}(G)$, and will be the setting for the theorem we actually state. We do need to bear in mind that the packets Π_ψ are larger than the L-packets Π_{ϕ_ψ} attached to the (unbounded) images ϕ_ψ (4.10) of ψ in $\Phi(G)$. In particular, they are not in general a part of the local Langlands classification. Their role is rather to describe local constituents of automorphic representations.

We actually have to settle for something a little weaker. As we discussed at the end of the last section, the transfer mapping $\widetilde{f} \to \widetilde{f}^G$ from $\widetilde{\mathcal{H}}(N)$ to $\mathcal{S}(G)$ is not generally surjective. Its image is the subspace $\widetilde{\mathcal{S}}(G)$ of $\widetilde{\mathrm{Out}}_N(G)$-invariant functions in $\mathcal{S}(G)$. We have therefore to work with the set $\widetilde{\Psi}(G)$ of $\widetilde{\mathrm{Out}}_N(G)$-orbits in $\Psi(G)$. This fits into the sequence

$$\widetilde{\Phi}_{\mathrm{bdd}}(G) \subset \widetilde{\Psi}(G) \subset \widetilde{\Phi}(G)$$

of families of orbits of parameters. The group $\widetilde{\mathrm{Out}}_N(G)$ (of order 1 or 2) acts by outer automorphisms also on $G(F)$, and hence on equivalence classes of irreducible representations. We therefore have an associated sequence

$$\widetilde{\Pi}_{\mathrm{temp}}(G) \subset \widetilde{\Pi}_{\mathrm{unit}}(G) \subset \widetilde{\Pi}(G)$$

of families of orbits of representations. This qualification is only relevant to the case that G is type $\mathbf{D}_n$. If G is of type $\mathbf{B}_n$ or $\mathbf{C}_n$, $\widetilde{\mathrm{Out}}_N(G)$ is trivial, and the sets are unchanged from before.

We have thus to attach packets $\widetilde{\Pi}_\psi$ in $\widetilde{\Pi}_{\mathrm{unit}}(G)$ to parameters $\psi \in \widetilde{\Psi}(G)$. We expect them to be multiplicity free. If F is nonarchimedean, this is a deep theorem of Moeglin [**M2**]. If F is archimedean, however, it is unknown. We have therefore to formulate the local theorem as an assertion for packets with multiplicities.

If S is any set, a *set over* S, or an S-*set*, or even an S-*packet* will mean simply a set S_1 with a fibration

$$S_1 \longrightarrow S$$

over S. Any function on S, such as the character $f_G(\pi)$ on $\widetilde{\mathcal{H}}(G)$ in case S equals $\widetilde{\Pi}(G)$, will be identified with its pullback to a function on S_1. The order

$$m_1 : S \longrightarrow \mathbb{N} \cup \{0, \infty\}$$

of the fibres in S_1 represents a multiplicity function, which makes S into what can be called a multiset on S. If S_1 is multiplicity free, in that every element in S has multiplicity at most 1, it is of course just a subset of S.

THEOREM 7.1. *Assume that F is local and that $G \in \widetilde{\mathcal{E}}(N)$.*

(a) *For any local parameter $\psi \in \widetilde{\Psi}(G)$, there is a finite packet $\widetilde{\Pi}_\psi$ over $\widetilde{\Pi}_{\mathrm{unit}}(G)$, together with a mapping*

$$(7.1) \qquad \pi \longrightarrow \langle \cdot, \pi \rangle, \qquad \pi \in \widetilde{\Pi}_\psi,$$

from $\widetilde{\Pi}_\psi$ to the group $\widehat{\mathcal{S}}_\psi$ of irreducible characters on $\mathcal{S}_\psi$, with the following property: if s is a semisimple element in the centralizer $S_\psi = S_\psi(G)$ and (G', ψ') is the preimage of (ψ, s) under the local version of the correspondence (5.11) in §5, then

$$(7.2) \qquad f'(\psi') = \sum_{\pi \in \widetilde{\Pi}_\psi} \langle s_\psi x, \pi \rangle f_G(\pi), \qquad f \in \widetilde{\mathcal{H}}(G),$$

where x is the image of s in $\mathcal{S}_\psi$, and s_ψ is the image in $\mathcal{S}_\psi$ of the element (5.6).

(b) *If $\psi = \phi$ belongs to the subset $\widetilde{\Phi}_{\mathrm{bdd}}(G)$ of parameters in $\widetilde{\Psi}(G)$ that are trivial on the factor $SU(2)$, the elements in $\widetilde{\Pi}_\phi$ are tempered and multiplicity free, and the corresponding mapping from $\widetilde{\Pi}_\phi$ to $\widehat{\mathcal{S}}_\phi$ is injective. Moreover, every element in $\widetilde{\Pi}_{\mathrm{temp}}(G)$ belongs to exactly one packet $\widetilde{\Pi}_\phi$. Finally, if F is nonarchimedean, the mapping from $\widetilde{\Pi}_\phi$ to $\widehat{\mathcal{S}}_\phi$ is bijective.*

REMARKS. 1. The premise of this theorem depends on Theorem 6.1. To be precise, the left hand side of (7.2) is a product of linear forms (6.8) postulated by the earlier theorem, taken over the factors of the endoscopic group G'. Composed with the transfer mapping $f \to f'$, it represents a linear form on $\widetilde{\mathcal{H}}(G)$. As such, it determines the packet $\widetilde{\Pi}_\psi$ and the pairing $\langle x, \pi \rangle$ from the expression on the right hand side of (7.2).

2. The finite subsets $\widetilde{\Pi}_\phi$ of $\widetilde{\Pi}_{\mathrm{temp}}(G)$ in (b) represent the tempered L-packets. They are composed of $\widetilde{\mathrm{Out}}_N(G)$-orbits of tempered representations, which are parametrized by characters in $\widehat{\mathcal{S}}_\phi$. Since the theorem posits a disjoint union

$$(7.3) \qquad \widetilde{\Pi}_{\mathrm{temp}}(G) = \coprod_{\widetilde{\Phi}_{\mathrm{bdd}}(G)} \widetilde{\Pi}_\phi,$$

it can be regarded as an endoscopic classification of the irreducible tempered representations of $G(F)$. It amounts to the local Langlands correspondence for G if G is of type $\mathbf{B}_n$ or $\mathbf{C}_n$, and something slightly weaker if G is of type $\mathbf{D}_n$.

3. Suppose that F is archimedean. Shelstad has established a general classification

$$(7.4) \qquad \Pi_{\mathrm{temp}}(G) = \coprod_{\Phi_{\mathrm{bdd}}(G)} \Pi_\phi$$

of $\Pi_{\mathrm{temp}}(G)$ for any real group G, in terms of L-packets Π_ϕ that satisfy endoscopic character relations (7.2) (with $s_\psi = 1$). In the papers [**She1**]–[**She4**], she does not define the stable distributions on the left hand side of (7.2) (in case $G \in \widetilde{\mathcal{E}}_{\mathrm{sim}}(N)$ and $\psi = \phi$) in terms of twisted transfer from general linear groups. However, this property (and more) is established in the recent papers of Shelstad [**She5**] and Mezo [**Me**]. We depend on these results for our proof of the theorem for nonarchimedean F (and for the proof of the global theorems we will state presently). For general parameters ψ, Adams, Barbasch and Vogan [**ABV**] have constructed packets Π_ψ that satisfy relations (7.2), again for any real group G. However, it is not presently known whether the stable distributions they define for the left hand side of (7.2) match those we obtain by twisted transfer from general linear groups (in our case that $G \in \widetilde{\mathcal{E}}_{\mathrm{sim}}(N)$). The point is important for our purposes, because the global results at which the local packets $\widetilde{\Pi}_\psi$ are ultimately aimed are all proved by comparison with the twisted trace formula for $GL(N)$.

4. Suppose that F is nonarchimedean. The structure of the general packets $\widetilde{\Pi}_\psi$ is better known than in the archimedean case, thanks to the work of Moeglin. We have already mentioned her proof [**M2**] that the packets $\widetilde{\Pi}_\psi$ are multiplicity free. This is a consequence of a general algorithm [**M1**] for computing the Langlands parameters of elements in $\widetilde{\Pi}_\psi$, assuming the classification of the tempered representations $\widetilde{\Pi}_{\mathrm{temp}}(G)$ provided by part (b) of the theorem.

Suppose that ψ belongs to the larger set $\widetilde{\Psi}^+(G)$ of local parameters for G, defined without the condition that their restriction to L_F be bounded. Then ψ can be expressed rather simply as a composition

$$\psi = \xi_M \circ \psi_{M,\lambda},$$

where $\psi_{M,\lambda}$ is a twist of a parameter $\psi_M \in \widetilde{\Psi}(M)$ for a Levi subgroup M of G and ξ_M the embedding of $^L M$ into $^L G$ attached to a parabolic subgroup $P \in \mathcal{P}(M)$ of G with Levi component M. The twisting element λ lies in the open chamber defined by P in a certain real vector space

$$\mathfrak{a}_M^* = X(M)_F \otimes \mathbb{R}.$$

It can be identified with either a homomorphism from $L_F \times SL(2,\mathbb{C})$ into the connected component of 1 in $Z(\widehat{M})^{\Gamma_F}$, or a real quasicharacter on the group $M(F)$. With the former interpretation, we observe that S_ψ is contained in $\widehat{M}$, and hence that the mapping $x \to x_M$ of $\mathcal{S}_{\psi_M}$ into $\mathcal{S}_\psi$ is an isomorphism. With the latter interpretation, we define the packet of ψ as a corresponding set (of orbits of) induced representations, which we denote hesitantly as

$$(7.5) \qquad \widetilde{\Pi}_\psi = \left\{ \pi = \mathcal{I}_P(\pi_{M,\lambda}) : \ \pi_M \in \widetilde{\Pi}_{\psi_M} \right\}.$$

It is bijective with $\widetilde{\Pi}_{\psi_M}$, and comes with a pairing

$$\langle x, \pi \rangle = \langle x_M, \pi_M \rangle, \qquad x \in \mathcal{S}_\psi, \ \pi \in \widetilde{\Pi}_\psi,$$

with $\mathcal{S}_\psi$. The assertion (a) of Theorem 7.1 for the more general parameter ψ, with the understanding that the elements in the packet $\widetilde{\Pi}_\psi$ might now be reducible, then follows from its analogue for ψ_M, and the standard character formula for an induced representation.

Consider the case that the local parameter $\psi = \phi$ in $\widetilde{\Psi}^+(G)$ is trivial on the factor $SU(2)$. Then ϕ belongs to the set $\widetilde{\Phi}(G)$ of general (unbounded) Langlands parameters. It has a decomposition $\phi = \xi_M \circ \phi_{M,\lambda}$, as above. However, we shall denote the packet we introduced above differently, as

$$\widetilde{P}_\psi = \left\{ \rho = \mathcal{I}_P(\pi_{M,\lambda}) : \ \pi_M \in \widetilde{\Pi}_{\phi_M} \right\}.$$

(The P in $\widetilde{P}_\psi$ is to be understood as an upper case ρ.) We reserve the symbol $\widetilde{\Pi}_\psi$ for the packet

$$(7.6) \qquad \widetilde{\Pi}_\psi = \left\{ \pi = \pi_\rho : \ \rho \in \widetilde{P}_\psi \right\}$$

of irreducible Langlands quotients of representations in $\widetilde{P}_\psi$. It comes with the pairing

$$\langle x, \pi \rangle = \langle x, \pi_\rho \rangle = \langle x, \rho \rangle = \langle x_M, \pi_M \rangle, \qquad x \in \mathcal{S}_\psi, \ \pi \in \widetilde{\Pi}_\psi,$$

with $\mathcal{S}_\psi$ that it inherits from $\widetilde{P}_\psi$. However, it does not satisfy the endoscopic character relation (7.2), since the Langlands quotient π_ρ need not be induced. Nonetheless, the original Langlands classification [**L5**] for real groups (extended to p-adic groups in [**BW**]) tells us that

$$(7.7) \qquad\qquad \widetilde{\Pi}(G) = \coprod_{\phi \in \widetilde{\Phi}(G)} \widetilde{\Pi}_\phi.$$

We therefore obtain an explicit classification of general representations $\pi \in \widetilde{\Pi}(G)$ from the tempered case given by Theorem 7.1(b).

Incidentally, it would appear that the notation in (7.5) is in conflict with that of (7.6). We hope that it will not be! For we shall consider only parameters $\psi \in \widetilde{\Psi}^+(G)$ that arise from the local components of automorphic representations. That we have to take them in the larger set $\widetilde{\Psi}^+(G)$ is a necessary consequence of our not having at our disposal the generalized Ramanujan conjecture for $GL(N)$, as we have noted. But it is still a pretty stringent restriction, which we conjecture implies that the induced representations $\rho = \mathcal{I}_P(\pi_{M,\lambda})$ are irreducible (and unitary) [**ECR**, Conjecture 8.3.1]. If this is so, each ρ equals its Langlands quotient π_ρ, and there is no conflict with the notation (7.5).

Suppose now that the field F is global. There will be two global theorems for the group $G \in \widetilde{\mathcal{E}}_{\mathrm{sim}}(N)$ over F. The first is the central result. It gives a decomposition of the automorphic discrete spectrum of G in terms of global parameters in the subset

$$\widetilde{\Psi}_2(G) = \left\{ \psi \in \widetilde{\Psi}(G) : |S_\psi| < \infty \right\}$$

of $\widetilde{\Psi}(G)$ and the local objects of Theorem 7.1(a). It is best formulated in terms of the global Hecke algebra $\mathcal{H}(G)$ of functions on $G(\mathbb{A})$, with respect to a suitable maximal compact subgroup

$$K = \prod_v K_v$$

of $G(\mathbb{A})$.

By definition, $\mathcal{H}(G)$ is the space of finite linear combinations of products

$$\prod_v f_v, \qquad f_v \in \mathcal{H}(G_v),$$

such that f_v is the characteristic function of K_v for almost all v. We write $\widetilde{\mathcal{H}}(G)$ for the locally symmetric subalgebra, in which each f_v lies in the subalgebra $\widetilde{\mathcal{H}}(G_v)$ of $\mathcal{H}(G_v)$. For any function f in $\widetilde{\mathcal{H}}(G)$, and any admissible representation

$$\pi = \bigotimes_v \pi_v, \qquad \pi_v \in \Pi(G_v),$$

the character $f_G(\pi)$ depends only on the $\widetilde{\mathrm{Out}}_N(G_v)$-orbit of π_v in $\widetilde{\Pi}(G_v)$, for any v. We will have to describe the discrete spectrum of G as an $\widetilde{\mathcal{H}}(G)$-module, since the local packets consist of $\widetilde{\mathrm{Out}}_N(G_v)$-orbits π_v in $\widetilde{\Pi}_{\mathrm{unit}}(G_v)$. Of course if G is of type $\mathbf{B}_n$ or $\mathbf{C}_n$, the groups $\mathrm{Out}_N(G_v)$ are all trivial, and $\widetilde{\mathcal{H}}(G)$ equals $\mathcal{H}(G)$. We would then have a description as an $\mathcal{H}(G)$-module, or for that matter, a decomposition in terms of irreducible representations $\pi \in \Pi_{\mathrm{unit}}(G)$.

We are assuming the seed Theorems 5.1 and 5.2. The first of these was needed to define the global set $\widetilde{\Psi}(G)$ itself. The second led us to a localization mapping $\psi \to \psi_v$ from $\widetilde{\Psi}(G)$ to the local set $\widetilde{\Psi}_v^+(G)$. The local theorem we have just stated (together with the ensuing discussion) allows us to attach a local packet $\widetilde{\Pi}_{\psi_v}$ to ψ and v. We can thus attach a global packet

$$(7.8) \qquad \widetilde{\Pi}_\psi = \left\{ \pi = \bigotimes_v \pi_v \ : \ \pi_v \in \widetilde{\Pi}_{\psi_v}, \ \langle \cdot, \pi_v \rangle = 1 \text{ for almost all } v \right\}$$

of (orbits of) representations of $G(\mathbb{A})$ to any $\psi \in \widetilde{\Psi}(G)$. Any element π in the global packet $\widetilde{\Pi}_\psi$ determines a character

$$(7.9) \qquad \langle x, \pi \rangle = \prod_v \langle x_v, \pi_v \rangle, \qquad x \in \mathcal{S}_\psi,$$

on the global centralizer quotient $\mathcal{S}_\psi$. On the right hand side of (7.9), the product can be taken over a finite set, while $x \to x_v$ is the mapping from $\mathcal{S}_\psi$ to $\mathcal{S}_{\psi_v}$ we have discussed earlier.

THEOREM 7.2. *Assume that F is global and that $G \in \widetilde{\mathcal{E}}_{\mathrm{sim}}(N)$. Then there is an $\widetilde{\mathcal{H}}(G)$-module isomorphism*

$$(7.10) \qquad L^2_{\mathrm{disc}}\big(G(F)\backslash G(\mathbb{A})\big) \cong \bigoplus_{\psi \in \widetilde{\Psi}_2(G)} m_\psi \left(\bigoplus_{\pi \in \widetilde{\Pi}_\psi(\varepsilon_\psi)} \pi \right),$$

where m_ψ equals 1 or 2 and

$$\varepsilon_\psi : \ \mathcal{S}_\psi \ \longrightarrow \ \{\pm 1\}$$

is a linear character defined explicitly in terms of symplectic ε-factors, while $\widetilde{\Pi}_\psi(\varepsilon_\psi)$ is the subset of representations π in the global packet $\widetilde{\Pi}_\psi$ such that the character $\langle \cdot, \pi \rangle$ on $\mathcal{S}_\psi$ equals ε_ψ.

The statement will not be complete until we define the integer m_ψ and the character ε_ψ. The first of these is quite elementary. The global parameter ψ comes with the L-embedding

$$\widetilde{\psi}_G : \ \mathcal{L}_\psi \times SL(2, \mathbb{C}) \ \longrightarrow \ {}^L G,$$

determined as a $\widehat{G}$-orbit up to the action of the group $\widetilde{\mathrm{Out}}_N(G)$. We define m_ψ for any $\psi \in \widetilde{\Psi}(G)$ to be the number of $\widehat{G}$-orbits in the associated $\widetilde{\mathrm{Out}}_N(G)$-orbit. For an equivalent description, we write the parameter $\psi \in \widetilde{\Psi}_2(G)$ as

$$\psi = \psi_1 \boxplus \cdots \boxplus \psi_r,$$

for distinct, self dual factors $\psi_i \in \widetilde{\Psi}_{\mathrm{sim}}(N_i)$. One checks that m_ψ equals 1 unless G is of type $\mathbf{D}_n$ (or in other words, N is even and $\widehat{G} = SO(N, \mathbb{C})$), and the rank N_i of each of the components ψ_i is also even, in which case m_ψ equals 2. The integer m_ψ obviously bears on the question of the multiplicity with which a representation π occurs in the automorphic discrete spectrum, but one also needs information about the localizations ψ_v. For a full statement in the case that $\psi = \phi$ lies in the subset $\widetilde{\Phi}_{\mathrm{bdd}}(G)$ of $\widetilde{\Psi}(G)$, see [**A4**, §3(vii)].

The sign character ε_ψ is more interesting. We first make an observation on some general ε-factors. Suppose that $\psi \in \widetilde{\Psi}(N)$ is an arbitrary global parameter, and that r is an arbitrary finite dimensional representation of $\mathcal{L}_\psi$, subject only to the condition that its equivalence class is stable under the group $\mathrm{Aut}(\mathcal{L}_\psi)$. Then r pulls back to a well defined representation r_v of L_{F_v}, for any v. We can therefore define the global L-function

$$L(s,r) = \prod_v L(s, r_v)$$

by an Euler product that converges for the real part of s large. We do not know that it has analytic continuation and functional equation, since it really amounts to a rather general automorphic L-function, even though its local factors are arithmetic L-functions defined as in [**T**]. But we can still define the corresponding global ε-factor as a finite product

$$\varepsilon(s, r, \psi_F) = \prod_v \varepsilon(s, r_v, \psi_{F_v}).$$

Again, we cannot say that this function is independent of the nontrivial additive character ψ_F on $\mathbb{A}/F$. But if r is symplectic, by which we mean that it takes values in the symplectic subgroup of the underlying general linear group, the value of the local factor $\varepsilon(s, r_v, \psi_{F_v})$ at $s = \frac{1}{2}$ is known to equal $+1$ or -1, and to be independent of ψ_{F_v}. We therefore have a global sign

$$\varepsilon\big(\tfrac{1}{2}, r\big) = \varepsilon\big(\tfrac{1}{2}, r, \psi_F\big) = \pm 1$$

in this case, which is independent of ψ_F.

We can define ε_ψ if ψ is a general parameter in $\widetilde{\Psi}(G)$. We first define a representation

$$\tau_\psi : \ S_\psi \times \mathcal{L}_\psi \times SL(2, \mathbb{C}) \ \longrightarrow \ GL(\widehat{\mathfrak{g}})$$

on the Lie algebra of $\widehat{\mathfrak{g}}$ of $\widehat{G}$ by setting

$$\tau_\psi(s, g, h) = \mathrm{Ad}_G\big(s \cdot \widetilde{\psi}_G(g, h)\big), \qquad s \in S_\psi, \ (g, h) \in \mathcal{L}_\psi \times SL(2, \mathbb{C}),$$

where Ad_G is the adjoint representation of $^L G$. This representation is orthogonal, and hence self-dual, since it is invariant under the Killing form on $\widehat{\mathfrak{g}}$. Let

$$\tau_\psi = \bigoplus_\alpha \tau_\alpha = \bigoplus_\alpha (\lambda_\alpha \otimes \mu_\alpha \otimes \nu_\alpha)$$

be its decomposition into irreducible representations λ_α, μ_α and ν_α of the respective groups S_ψ, $\mathcal{L}_\psi$ and $SL(2, \mathbb{C})$. We then define

$$\varepsilon_\psi(x) = \prod_\alpha{}' \det\big(\lambda_\alpha(s)\big), \qquad s \in S_\psi,$$

where x is the image of s in $\mathcal{S}_\psi$, and $\prod'$ denotes the product over those indices α with μ_α symplectic and with

$$\varepsilon\big(\tfrac{1}{2}, \mu_\alpha\big) = -1.$$

The second global theorem is a supplement to the first. It gives a long conjectured L-function criterion for whether a self-dual cuspidal automorphic representation of $GL(N)$ is symplectic or orthogonal, in the sense that it is a functorial image from a group G whose L-group is symplectic or orthogonal. It also gives an automorphic analogue of a well known property [**FQ**], [**D**] of orthogonal (arithmetic) root numbers. The main point for us, however, is that this theorem has a critical role in the proofs of all of the theorems. It is an indispensable part of what governs the signs that arise in the comparison of trace formulas.

To place the second global theorem in context, we observe a property of certain Rankin-Selberg L-functions. The Rankin-Selberg representation of $GL(N, \mathbb{C}) \times GL(N, \mathbb{C})$ on the space of complex $(N \times N)$-matrices is irreducible, but its restriction to the diagonal image of $GL(N, \mathbb{C})$ is a direct sum $S^2 \oplus \Lambda^2$, where S^2 (resp. Λ^2) is the representation of $GL(N, \mathbb{C})$ on the space of symmetric (resp. skew-symmetric) $(N \times N)$-matrices. If π is a cuspidal automorphic representation of $GL(N)$, the diagonal Rankin-Selberg functions then break into formal products

$$(7.11) \qquad L(s, \pi \times \pi) = L(s, \pi, S^2) L(s, \pi, \Lambda^2)$$

and

$$(7.12) \qquad \varepsilon(s, \pi \times \pi) = \varepsilon(s, \pi, S^2, \psi_F)\, \varepsilon(s, \pi, \Lambda^2, \psi_F).$$

The two L-functions on the right hand side of (7.11) are among the cases of the Langlands-Shahidi method treated in [**Sha**]. In both cases, the local L-functions and ε-factors can be constructed, with the consequence that the formal products (7.11) and (7.12) become actual products, for which the resulting two global L-functions have analytic continuation with functional equation (4.15).

Suppose that $\phi \in \widetilde{\Phi}_{\mathrm{sim}}(N)$ is a simple generic parameter. For us, this amounts to a cuspidal automorphic representation $\pi = \pi_\phi$ of $GL(N)$, which is now self-dual. Theorem 5.1 asserts that ϕ belongs to the subset

$$\widetilde{\Phi}_{\mathrm{sim}}(G) = \widetilde{\Phi}_{\mathrm{sim}}(N) \cap \widetilde{\Psi}(G),$$

for a unique $G \in \mathcal{E}_{\mathrm{sim}}(N)$. We need to understand how G is related to L-functions and ε-factors. It is known that the Rankin-Selberg L-function

$$L(s, \phi \times \phi) = L(s, \pi \times \pi) = L(s, \pi \times \pi^\vee)$$

has a pole of order 1 at $s = 1$. It is also known that neither of the corresponding factors $L(s, \phi, S^2)$ and $L(s, \phi, \Lambda^2)$ on the right hand side of (7.12) has a zero at $s = 1$. It follows that exactly one of them has a pole at $s = 1$ (which will be of order 1). This motivates the assertion (a) of our second global theorem.

THEOREM 7.3. *Assume that F is global.*

(a) *Suppose that $G \in \widetilde{\mathcal{E}}_{\mathrm{sim}}(N)$ and that ϕ belongs to $\widetilde{\Phi}_{\mathrm{sim}}(G)$. Then $\widehat{G}$ is orthogonal if and only if the symmetric square L-function $L(s, \phi, S^2)$ has a pole at $s = 1$, while $\widehat{G}$ is symplectic if and only if the skew-symmetric square L-function $L(s, \phi, \Lambda^2)$ has a pole at $s = 1$.*

(b) *Suppose that for $i = 1, 2$, ϕ_i belongs to $\widetilde{\Phi}_{\mathrm{sim}}(G_i)$, for simple endoscopic data $G_i \in \widetilde{\mathcal{E}}_{\mathrm{sim}}(N_i)$. Then the associated Rankin-Selberg ε-factor satisfies*

$$\varepsilon\left(\tfrac{1}{2}, \phi_1 \times \phi_2\right) = 1,$$

if $\widehat{G}_1$ and $\widehat{G}_2$ are either both orthogonal or both symplectic.

We have completed the formal statements of the three main theorems of [**ECR**]. Representations of quasisplit special orthogonal and symplectic groups G have been studied from other points of view. There is a rather complete theory for the special case of representations with Whittaker models [**CKPS**], [**GRS**]. Since Theorems 7.1 and 7.2 give a classification of local and global representations, it is reasonable to ask which of these have Whittaker models. The answer is given in [**ECR**, §8.3], following a conjecture in [**Sha**]. The θ-correspondence has been a different source of results. (See [**Ku**], for example.) It would be very interesting to compare the relations it has provided with the classification of Theorems 7.1 and 7.2. Considerably more is known in the local case, thanks to the article [**M3**] and its predecessors. For global F, we refer the reader to [**Ji**] for examples and a description of some of the problems.

Some applications of the theorems were listed in [**A4**, §3]. There will no doubt be others. Some will follow from the extension of the theorems to orthogonal and symplectic groups that are not quasisplit. For a description of a proposed classification of representations for such groups, which remains conjectural, see [**ECR**, §9]. Other applications await an extension of the theorems to different groups, such as the split group $GSp(2n)$ of symplectic similitudes, for example.

8. Implications for functoriality

It remains to add some observations on functoriality (Principle 1.1). We consider the two cases

$$(8.1) \qquad (G, G', \rho) = \big(GL(N), G, \xi\big), \qquad G \in \widetilde{\mathcal{E}}_{\mathrm{sim}}(N),$$

and

$$(8.2) \qquad (G, G', \rho) = (G, G', \xi'), \qquad G \in \widetilde{\mathcal{E}}_{\mathrm{sim}}(N), \ G' \in \mathcal{E}_{\mathrm{ell}}(G),$$

discussed at the end of §1. What are the implications of the theorems we have stated?

We first observe that in both cases, we have a mapping

$$(8.3) \qquad \psi' \longrightarrow \psi = \rho \circ \psi' = \rho(\psi'), \qquad \psi' \in \widetilde{\Psi}(G'),$$

from $\widetilde{\Psi}(G')$ to $\widetilde{\Psi}(G)$. This is obvious if F is local, and would be so in the global case as well if our parameters were defined on the Langlands group L_F. As matters stand, we must appeal to the definitions of §5 for global F, which depend on the nontrivial Theorems 5.1 and 5.2. In any case, we do have the mapping (8.3) for any F. As usual, we have particular interest in its restriction to the subset $\widetilde{\Phi}_{\mathrm{bdd}}(G')$ of $\widetilde{\Psi}(G')$.

Suppose for the moment that F is local. Local functoriality in the first case (8.1) is given by Theorem 6.1 and the definition (6.7) of the linear form on the right hand side of (6.9). In the second case, it is given by the character identity (7.2) in Theorem 7.1(a) and the bijective correspondence (5.11). Because these two cases have their origins in the theory of endoscopy, we obtain an explicit description of the local functoriality correspondence. It is given by the relation

$$(8.4) \qquad \pi' \longrightarrow \pi, \qquad \pi' \in \widetilde{\Pi}(G'),$$

where for a given π', π ranges over the representations in the packet $\widetilde{\Pi}_\phi$ of the local parameter $\phi \in \widetilde{\Phi}(G)$ such that $\phi = \rho(\phi')$ is the image of the parameter $\phi' \in \widetilde{\Phi}(G')$ with $\pi' \in \widetilde{\Pi}_{\phi'}$. In other words, it is the correspondence

$$\widetilde{\Pi}_{\phi'} \longrightarrow \widetilde{\Pi}_\phi,$$

in which ϕ' ranges over the local parameters in $\widetilde{\Phi}(G')$ and $\phi' \to \phi$ is the mapping from $\widetilde{\Phi}(G')$ to $\widetilde{\Phi}(G)$ analogous to (8.3).

Functoriality is called a "principle" for a reason. It represents a phenomenon, which we can almost regard as a general law of nature, rather than a precise conjecture. It is capable of taking different forms, which depend on the context. This flexibility is built into its statement in §1, by postulating a correspondence rather than a mapping, whose precise nature is left unspecified. For example, in the local case above, we can consider packets $\widetilde{P}_\phi$ of induced representations ρ (sometimes known as *standard* representations) in place of packets $\widetilde{\Pi}_\phi$ of irreducible Langlands quotients π. Let $\widetilde{\Pi}_\phi^+$ be the set of irreducible constituents π of these representations, a packet that contains $\widetilde{\Pi}_\phi$, and equals $\widetilde{\Pi}_\phi$ if ϕ lies in the subset $\widetilde{\Phi}_{\mathrm{bdd}}(G)$ of $\widetilde{\Phi}(G)$. These larger packets are no longer disjoint, in contrast to the L-packets $\widetilde{\Pi}_\phi$. This leads to a more complicated version of local functoriality. We take it to be the coarser relation

$$(8.5) \qquad\qquad \pi' \longrightarrow \pi, \qquad \pi' \in \widetilde{\Pi}(G'),$$

where for a given π', π ranges over the representations in packets $\widetilde{\Pi}_\phi^+$ of local parameters $\phi \in \widetilde{\Phi}^+(G)$ such that $\phi = \rho(\phi')$ is the image as in (8.3) of a parameter $\phi' \in \widetilde{\Phi}(G')$ with $\pi' \in \widetilde{\Pi}_{\phi'}^+$. We should also not be distracted by the fact that $\widetilde{P}_\phi$ and $\widetilde{\Pi}_\phi$ consist of orbits of irreducible representations (unless $G = GL(N)$), rather than actual representations. This is a side issue, which is not present if G is of type $\mathbf{B}_n$ or $\mathbf{C}_n$, and is not particularly significant for what we are discussing here.

Assume now that F is global. Global functoriality is harder to describe. For among other things, the statement of Principle 1.1 is predicated on the broader, formal definition [**L3**] of automorphic representations, which we have not given. We can, however, formulate it without difficulty for the subsets

$$\widetilde{\Pi}_{\phi,\mathrm{aut}} \subset \widetilde{\Pi}_\phi, \qquad \phi \in \widetilde{\Phi}_{\mathrm{bdd}}(G),$$

of automorphic representations in the global packets attached to parameters in $\widetilde{\Phi}_{\mathrm{bdd}}(G)$. In this case, global functoriality is given by the correspondence

$$\widetilde{\Pi}_{\phi',\mathrm{aut}} \longrightarrow \widetilde{\Pi}_{\phi,\mathrm{aut}}$$

in which ϕ' ranges over global parameters in $\widetilde{\Phi}_{\mathrm{bdd}}(G')$, and $\phi' \to \phi$ is the mapping given by the restriction of (8.3) to the subset $\widetilde{\Phi}_{\mathrm{bdd}}(G')$ of $\widetilde{\Psi}(G')$. The correspondence follows from the definitions of §5, the multiplicity formula of Theorem 7.2, and the reduction by Eisenstein series of the full automorphic spectrum to relatively discrete automorphic spectra for Levi subgroups. This reduction also gives an explicit characterization of the subset $\widetilde{\Pi}_{\phi,\mathrm{aut}}$ of $\widetilde{\Pi}_\phi$.

More generally, consider a global parameter ϕ in the larger set $\widetilde{\Phi}(G)$. We can then form the global packet

$$\widetilde{\Pi}_\phi^+ = \left\{ \pi = \overset{\sim}{\bigotimes_v} \pi_v : \pi_v \in \widetilde{\Pi}_{\phi_v}^+,\ \pi_v \text{ unramified for almost all } v \right\},$$

in analogy with (2.2). The global parameter ϕ can be represented as the image of a "discrete" parameter $\phi_M \in \widetilde{\Phi}_2(M)$, for a Levi subgroup M of G, under the dual embedding $^L M \subset {}^L G$. Let $\widetilde{\Pi}_{\phi,\mathrm{aut}}^+$ be the subset of irreducible representations in $\widetilde{\Pi}_\phi^+$ obtained from parabolic induction from representations in the subset $\widetilde{\Pi}_{\phi_M,\mathrm{aut}}$ of $\widetilde{\Pi}_{\phi_M}$. It is then a consequence of the definitions [**L3**, p. 203] that $\widetilde{\Pi}_{\phi,\mathrm{aut}}^+$ is the subset of representations in $\widetilde{\Pi}_\phi^+$ that are automorphic in the extended sense of [**L3**]. In this more general setting, global functoriality is given by the correspondence

$$(8.6) \qquad\qquad \widetilde{\Pi}_{\phi',\mathrm{aut}}^+ \longrightarrow \widetilde{\Pi}_{\phi,\mathrm{aut}}^+,$$

where ϕ' ranges over global parameters in the larger set $\widetilde{\Phi}(G')$, and $\phi' \to \phi$ is the mapping from $\widetilde{\Phi}(G')$ to $\widetilde{\Phi}(G)$ analogous to (8.3). It is clearly compatible with the local functoriality correspondence (8.5).

The global functoriality correspondence (8.6) treats many automorphic representations (in the extended sense of [**L3**]), but it still represents a special case. For example, it does not include the subset $\widetilde{\Pi}_{\psi,\mathrm{cusp}}$ of *cuspidal* automorphic representations in a packet $\widetilde{\Pi}_\psi$, if ψ lies in the complement of $\widetilde{\Phi}_{\mathrm{bdd}}(G)$. This subset is a subtle object, which has been studied in depth by Moeglin. Leaving aside the question of how to characterize it explicity, we let $\widetilde{\Psi}^+(G)$ be the global analogue of the local set defined after the statement of Theorem 7.1. This will be the largest of our global sets of parameters, a family that contains both $\widetilde{\Psi}(G)$ and $\widetilde{\Phi}(G)$. As in the special case of the subset $\widetilde{\Phi}(G)$, a parameter ψ in $\widetilde{\Psi}^+(G)$ is the image of a discrete parameter $\psi_M \in \widetilde{\Psi}_2^+(M)$, for some M. We then write $\widetilde{\Pi}_{\psi,\mathrm{ic}}^+ = \widetilde{\Pi}_{\psi,\mathrm{ind\text{-}cusp}}^+$ for the packet of irreducible constituents of standard representations ρ obtained by parabolic induction from representations in the subset $\widetilde{\Pi}_{\psi_M,\mathrm{cusp}}$ of $\widetilde{\Pi}_{\psi_M}$. It lies in a chain

$$\widetilde{\Pi}_{\psi,\mathrm{ic}}^+ \subset \widetilde{\Pi}_{\psi,\mathrm{aut}}^+ \subset \widetilde{\Pi}_\psi^+,$$

for packets $\widetilde{\Pi}_{\psi,\mathrm{aut}}^+$ and $\widetilde{\Pi}_\psi^+$ defined in the same way, but with $\widetilde{\Pi}_{\psi_M,\mathrm{cusp}}$ replaced by the larger sets $\widetilde{\Pi}_{\psi_M,\mathrm{aut}}$ and $\widetilde{\Pi}_{\psi_M}$. The packet $\widetilde{\Pi}_\psi^+$, for example, is also equal to the set of restricted tensor products of representations in the local packets $\widetilde{\Pi}_{\psi_v}^+$, as in the special case of the global packet $\widetilde{\Pi}_\phi^+$ defined above. As ψ varies, the global packets $\widetilde{\Pi}_{\psi,\mathrm{ic}}^+$ are presumably disjoint, which would give a classification of the set $\widetilde{\Pi}_{\mathrm{aut}}(G)$ below as a disjoint union of these packets.

General global functoriality ought then to be given by the generalization

$$(8.7) \qquad\qquad \widetilde{\Pi}_{\psi',\mathrm{ic}}^+ \longrightarrow \widetilde{\Pi}_{\psi,\mathrm{ic}}^+, \qquad \psi' \in \widetilde{\Psi}^+(G'),$$

of (8.6). It would be compatible with a version of local functoriality that generalizes both the coarser relations (8.5) above and (8.11) below. I have not thought carefully about the implications of these constructions. Rather than pursue them further here, let me describe a simpler variant of global functoriality, which is easy to formulate.

We define

$$(8.8) \qquad\qquad \mathcal{C}_{\mathrm{aut}}(G) = \big\{ c(\pi) : \ \pi \in \Pi_{\mathrm{aut}}(G) \big\},$$

where $\Pi_{\mathrm{aut}}(G)$ is the set of representations

$$\pi = \widetilde{\bigotimes_v} \pi_v$$

of $G(\mathbb{A})$ that are automorphic in the sense of [**L3**]. Repeating what was implicit above, we write $\widetilde{\Pi}_{\mathrm{aut}}(G)$ for the set of orbits of representations of $G(\mathbb{A})$ under the group

$$\widetilde{\mathrm{Out}}_N(G_{\mathbb{A}}) = \prod_v \widetilde{\mathrm{Out}}_N(G_v)$$

that have a representative in $\Pi_{\mathrm{aut}}(G)$, for any group $G \in \widetilde{\mathcal{E}}_{\mathrm{sim}}(N)$. If $G = GL(N)$, we take $\widetilde{\Pi}_{\mathrm{aut}}(G)$ to be simply the set of self-dual representations in $\Pi_{\mathrm{aut}}(G)$. In either case, we can then define a set

$$\widetilde{\mathcal{C}}_{\mathrm{aut}}(G) = \big\{ c(\pi) : \ \pi \in \widetilde{\Pi}_{\mathrm{aut}}(G) \big\}.$$

It consists of $\widetilde{\mathrm{Out}}_N(G_{\mathbb{A}})$-orbits of (equivalence classes of) families of semisimple conjugacy classes with representatives in $\mathcal{C}_{\mathrm{aut}}(G)$ if $G \in \widetilde{\mathcal{E}}_{\mathrm{sim}}(N)$, and simply the subset of self-dual elements in $\mathcal{C}_{\mathrm{aut}}(N)$ if $G = GL(N)$. We then have a mapping

$$(8.9) \qquad c' \sim \{c'_v : \ v \notin S'\} \ \longrightarrow \ c = \rho(c') \sim \{\rho_v(c'_v) : \ v \notin S\}$$

from $\widetilde{\mathcal{C}}_{\mathrm{aut}}(G')$ to a larger set $\widetilde{\mathcal{C}}_{\mathbb{A}}(G)$, defined as above but without the condition of automorphy. We claim that this mapping takes $\widetilde{\mathcal{C}}_{\mathrm{aut}}(G')$ to the subset $\widetilde{\mathcal{C}}_{\mathrm{aut}}(G)$ of $\widetilde{\mathcal{C}}_{\mathbb{A}}(G)$. In other words, the image $\rho(c')$ of c' can be represented by an element c in $\mathcal{C}_{\mathrm{aut}}(G)$. This version of functoriality is less delicate than the others. We leave the reader to check that it follows from the various definitions and theorems above.

The mapping $c' \to c$ from $\widetilde{\mathcal{C}}_{\mathrm{aut}}(G')$ to $\widetilde{\mathcal{C}}_{\mathrm{aut}}(G)$ is obviously quite concrete, like the sets of Hecke eigenfamilies that comprise its domain and codomain. It also illustrates the term "functoriality". For we can build a category C from the groups we have been working with. Its objects are quasisplit groups over F, primarily the groups $G \in \widetilde{\mathcal{E}}_{\mathrm{sim}}(N)$ and the general linear groups $GL(N)$, but also direct products of such groups. The morphisms $\mathrm{hom}(G', G)$ between objects $G', G \in \mathrm{ob}(C)$ are L-homomorphisms

$$\rho : \ {}^L G' \ \longrightarrow \ {}^L G$$

given by (8.1) and (8.2), and whatever supplements are required for direct products. We then have a mapping

$$F : \ G \ \longrightarrow \ \widetilde{\mathcal{C}}_{\mathrm{aut}}(G), \qquad G \in \mathrm{ob}(C),$$

from objects $G \in \mathrm{ob}(C)$ to sets. The version of functoriality we have just formulated asserts that for any $\rho \in \mathrm{hom}(G', G)$, the mapping

$$F(\rho) : \ c' \ \longrightarrow \ \rho(c'), \qquad c' \in \widetilde{\mathcal{C}}_{\mathrm{aut}}(G'),$$

given by (8.9) takes the set $F(G') = \widetilde{\mathcal{C}}_{\mathrm{aut}}(G')$ to the set $F(G) = \widetilde{\mathcal{C}}_{\mathrm{aut}}(G)$. In other words, the mapping

$$F : \ \mathrm{ob}(C) \ \longrightarrow \ \mathrm{ob}(St)$$

comes also with a mapping

$$F : \ \mathrm{hom}(G', G) \ \longrightarrow \ \mathrm{hom}(F(G'), F(G)), \qquad G', G \in \mathrm{ob}(C),$$

and is therefore a functor from C to the category St of sets.

In this section, we have abandoned our informal characterization of automorphic representations, which we adopted for expository reasons in §1, for the broader definition in [**L3**]. We shall call the representations in this smaller class *globally tempered* automorphic representations, since we defined them in terms of global harmonic analysis. We have already agreed to denote them by $\Pi(G)$, for any one of our groups G over the global field F. We thus have an embedding

$$\Pi(G) \subset \Pi^+(G),$$

where $\Pi^+(G) = \Pi_{\mathrm{aut}}(G)$ is the set of general automorphic representations. This is parallel to the associated embedding

$$\Psi(G) \subset \Psi^+(G)$$

of global parameter sets. How would we formulate the principle of functoriality for globally tempered automorphic representations?

Given G, we have the associated sets $\widetilde{\Pi}(G) \subset \widetilde{\Pi}^+(G)$, which are parallel to the global parameter sets $\widetilde{\Psi}(G) \subset \widetilde{\Psi}^+(G)$. For any $\psi \in \Psi(G)$, we also have the subset

$$\widetilde{\Pi}_{\psi,\mathrm{aut}} = \widetilde{\Pi}_\psi \cap \widetilde{\Pi}(G) = \widetilde{\Pi}_\psi \cap \widetilde{\Pi}^+(G)$$

of automorphic representations in the packet $\widetilde{\Pi}_\psi$. Functoriality for globally tempered automorphic representations will then be the correspondence

$$(8.10) \qquad\qquad \widetilde{\Pi}_{\psi',\mathrm{aut}} \;\longrightarrow\; \widetilde{\Pi}_{\psi,\mathrm{aut}},$$

where ψ' ranges over the global parameters in $\widetilde{\Psi}(G')$, and $\psi' \to \psi$ is the mapping (8.3) from $\widetilde{\Psi}(G')$ to $\widetilde{\Psi}(G)$.

The local analogue of (8.10) will have to be a little different from (8.4) and (8.5). Taking F now to be local, we assume for simplicity that Conjecture 8.3.1 of [**ECR**] is valid (as we already have implicitly, in the notation of (8.10)). This was the assertion we mentioned in §7 that the local packets (7.5) and (7.6) are the same for the local parameters $\psi \in \widetilde{\Psi}^+(G)$ obtained from (globally tempered) automorphic representations. The conjecture was actually stated more generally for any ψ in the intermediate set

$$\widetilde{\Psi}(G) \subset \widetilde{\Psi}^+_{\mathrm{unit}}(G) \subset \widetilde{\Psi}^+(G),$$

defined in terms of unitary representations for $GL(N)$, following the statement of Theorem 1.5.1 of [**ECR**]. Now as ψ varies over even the smallest set $\widetilde{\Psi}(G)$, the associated packets $\widetilde{\Pi}_\psi$ are *not* disjoint, in contrast to the packets $\widetilde{\Pi}_\phi$ attached to parameters $\phi \in \widetilde{\Phi}(G)$. This complicates the local functoriality analogue of (8.10). We define it as the coarser relation

$$(8.11) \qquad\qquad \pi' \;\longrightarrow\; \pi, \qquad \pi' \in \widetilde{\Pi}(G'),$$

where for a given π', π ranges over the representations in packets $\widetilde{\Pi}_\psi$ of local parameters $\psi \in \widetilde{\Psi}^+_{\mathrm{unit}}(G)$ such that $\psi = \rho(\psi')$ is the image (8.3) of a parameter $\psi' \in \widetilde{\Psi}^+_{\mathrm{unit}}(G')$ with $\pi' \in \widetilde{\Pi}_{\psi'}$. The (globally tempered) functoriality correspondence (8.10) is then compatible with its local analogue (8.11).

There is one last point, which might be somewhat surprising. It is conceivable that for global F, there could be elements $\psi \in \widetilde{\Psi}_2(G)$ such that the set $\widetilde{\Pi}_{\psi,\mathrm{aut}}$ in (8.10) is empty. One sees easily that ψ cannot contribute to the continuous automorphic spectrum of G [**A5**, §3], so $\widetilde{\Pi}_{\psi,\mathrm{aut}}$ is equal to the set $\widetilde{\Pi}_\psi(\varepsilon_\psi)$ of Theorem 7.2. Whether it is empty or not therefore depends on the sign character ε_ψ. Examples of this phenomenon were found some years ago by Cogdell and Piatetski-Shapiro [**CP**], by different methods. The question for us here is whether there is a global parameter $\psi' \in \widetilde{\Psi}(G')$ with image $\psi = \rho(\psi')$ in $\widetilde{\Psi}(G)$ such that the set $\widetilde{\Pi}_{\psi,\mathrm{aut}}$ is empty, but $\widetilde{\Pi}_{\psi',\mathrm{aut}}$ is not. If the answer is affirmative, there will be a globally tempered automorphic representation π' of G' that is not in the domain of the global functoriality correspondence for ρ.

This last section has been written quite quickly. I hope that the discussion has not been too murky, and that it is essentially correct. It does seem to raise some interesting questions, which bear further reflection.

References

[ECR]　J. Arthur, *The Endoscopic Classification of Representations: Orthogonal and Symplectic Groups*, Colloquium Publications, 61, 2013, American Mathematical Society.

[ABV]　J. Adams, D. Barbasch, and D. Vogan, *The Langlands Classification and Irreducible Characters for Real Reductive Groups*, Progr. Math. **104**, Birkhauser, Boston, 1992.

[A1]　J. Arthur, *A stable trace formula III. Proof of the main theorems*, Annals of Math. **158** (2003), 769–873.

[A2]　———, *A note on the automorphic Langlands group*, Canad. Math. Bull. **45** (2002), 466–482.

[A3]　———, *An introduction to the trace formula*, in *Harmonic Analysis, the Trace Formula, and Shimura Varieties*, Clay Mathematics Proceedings, vol. 4, 2005, 1–263.

[A4]　———, *The Endoscopic Classification of Representations*, in *Automorphic Representations and L-functions*, Tata Institute of Fundamental Research, 2013, 1–22.

[A5]　———, *Eigenfamilies, characters and multiplicities*, preprint.

[BJ]　A. Borel and H. Jacquet, *Automorphic forms and automorphic representations*, in *Automorphic Forms, Representations and L-functions*, Proc. Symp. Pure Math. vol. 33, Part 1, Amer. Math. Soc., 1979, 189–202.

[BW]　A. Borel and N. Wallach, *Continuous Cohomology, Discrete Subgroups, and Representations of Reductive Groups*, Ann. of Math. Studies **94**, Princeton Univ. Press, Princeton, N.J., 1980.

[CKPS]　J. Cogdell, H. Kim, I. Piatetski-Shapiro, and F. Shahidi, *Functoriality for the classical groups*, Publ. Math. Inst. Hautes Études Sci. **99** (2004), 163–233.

[CP]　J. Cogdell and I. Piatetski-Shapiro, *On base change for odd orthogonal groups*, J. Amer. Math. **8** (1995), 975–996.

[D]　P. Deligne, *Les constantes locales de l'équation fonctionnelle de la fonction L d'Artin d'une représentation orthogonale*, Invent. Math. **35** (1976), 299–316.

[F]　D. Flath, *Decomposition of representations into tensor products*, in *Automorphic Forms, Representations and L-functions*, Proc. Sympos. Pure Math. vol. 33, Part 1, Amer. Math. Soc., 1979, 179–184.

[FLN]　E. Frenkel, R. Langlands and B.C. Ngo, *Formule des traces et functorialité: Le début d'un programme*, preprint.

[FQ]　A. Frohlich and J. Queyrot, *On the functional equations of the Artin L-function for characters of real representations*, Invent. Math. **20** (1973), 125–138.

[GRS]　D. Ginzburg, S. Rallis, and D. Soudry, *The descent map from automorphic representations of GL(n) to classical groups*, World Scientific Publishing Co. Pte. Ltd., Hackensack, NJ, 2011.

[GKM]　M. Goresky, R. Kottwitz, and R. MacPherson, *Homology of affine Springer fibres in the unramified case*, Duke Math. J. **121** (2004), 509–561.

[HT]　M. Harris and R. Taylor, *On the Geometry and Cohomology of Some Simple Shimura Varieties*, Ann. of Math. Studies **151**, Princeton Univ. Press, Princeton, N.J., 2001.

[H]　G. Henniart, *Une preuve simple des conjectures de Langlands de GL(n) sur un corps p-adique*, Invent. Math. **139** (2000), 439–455.

[JPS]　H. Jacquet, I. Piatetski-Shapiro, and J. Shalika, *Rankin-Selberg convolutions*, Amer. J. Math. **105** (1983), 367–464.

[JS]　H. Jacquet and J. Shalika, *On Euler products and the classification of automorphic representations* II, Amer. J. Math. **103** (1981), 777–815.

[Ji]　D. Jiang, *Automorphic integral transforms for classical groups* I: *endoscopy correspondences*, to appear in *Proceedings on the Conference on Automorphic Forms and Related Geometry: Assessing the Legacy of I.I. Piatetski-Shapiro*.

[K]   R. Kottwitz, *Stable trace formula: cuspidal tempered terms*, Duke Math. J. **51** (1984), 611–650.

[KS]   R. Kottwitz and D. Shelstad, *Foundations of Twisted Endoscopy*, Astérisque, vol. 255.

[Ku]   S. Kudla, *Notes on the local theta correspondence*, notes from the European School on Group Theory, September 1996.

[L1]   R. Langlands, *Problems in the theory of automorphic forms*, in *Lectures in Modern Analysis and Applications*, Lecture Notes in Math. 170, Springer, New York, 1970, 18–61.

[L2]   ———, *On the Functional Equations Satisfied by Eisenstein Series*, Lecture Notes in Math. 544, Springer, New York, 1976.

[L3]   ———, *On the notion of an automorphic representation. A supplement to the preceding paper*, in *Automorphic Forms, Representations and L-functions*, Proc. Sympos. Pure Math. vol. 33, Part 1, Amer. Math. Soc., 1979, 203–208.

[L4]   ———, *Automorphic representations, Shimura varieties, and motives. Ein Märchen*, in *Automorphic Forms, Representations and L-functions*, Proc. Sympos. Pure Math. vol. 33, Part 2, Amer. Math. Soc., 1979, 205–246.

[L5]   ———, *On the classification of irreducible representations of real algebraic groups*, in *Representation Theory and Harmonic Analysis on Semisimple Lie Groups*, AMS Mathematical Surveys and Monographs, vol. 31, 1989, 101–170.

[L6]   ———, *Un nouveau point de repère des formes automorphes*, Canad. Math. Bull. **50** (2007), 243–267.

[LS]   R. Langlands and D. Shelstad, *On the definition of transfer factors*, Math. Ann. **278** (1987), 219–271.

[Me]   P. Mezo, *Tempered spectral transfer in the twisted endoscopy of real groups*, preprint.

[M1]   C. Moeglin, *Paquets d'Arthur discrets pour un groupe classique p-adique*, Contemp. Math. **489** (2009), 179–257.

[M2]   ———, *Multiplicité 1 dans les paquets d'Arthur aux places p-adiques*, in *On Certain L-functions*, Clay Mathematics Proceedings, vol. 13, 2011, 233–274.

[M3]   ———, *Conjecture d'Adams pour la correspondence de Howe et filtration de Kudla*, in *Arithmetic Geometry and Automorphic Forms*, 445–503, Adv. Lect. Math. (ALM), 19, Int. Press, Somerville, MA, 2011.

[MW1]   C. Moeglin and J.-L. Waldspurger, *Le spectre résiduel de GL(n)*, Ann. Scient. Éc. Norm. Sup. 4$^\mathrm{e}$ série **22** (1989), 605–674.

[MW2]   ———, *La partie géométrique de la formule des traces tardue*, preprint.

[Mok]   C.P. Mok, *Endoscopic classification of representations of quasisplit unitary groups*, to appear in Memoires of the American Mathematical Society.

[N]   B.C. Ngô, *Le lemme fondamental pour les algèbres de Lie*, Publ. Math. Inst. Hautes Études Sci. 111, 1–269.

[Sch]   P. Scholze, *The local Langlands correspondence for GL_n over p-adic fields*, to appear in Invent. Math.

[Sha]   F. Shahidi, *A proof of Langlands' conjecture on Plancherel measures; Complementary series for p-adic groups*, Annals of Math. **132** (1990), 273–330.

[She1]   D. Shelstad, *L-indistinguishability for real groups*, Math. Ann. **259** (1982), 385–430.

[She2]   ———, *Tempered endoscopy for real groups I: geometric transfer with canonical factors*, Contemp. Math., **472** (2008), 215–246.

[She3]   ———, *Tempered endoscopy for real groups II: spectral transfer factors*, to appear in *Autmorphic Forms and the Langlands Program*, Higher Education Press/International Press, 243–283.

[She4]   ———, *Tempered endoscopy for real groups III: inversion of transfer and L-packet structure*, Representation Theory **12** (2008), 369–402.

[She5]   ———, *On geometric transfer in real twisted endoscopy*, preprint.

[T]   J. Tate, *Fourier Analysis in Number Fields and Hecke's Zeta Functions*, in *Algebraic Number Theory*, Thompson, Washington, D.C., 1967, 305–347.

[W1]   J.-L. Waldspurger, *Le lemme fondamental implique le transfer*, Compositio Math. **105** (1997), 153–236.

[W2]   ———, *Endoscopie et changement de caractéristique*, J. Inst. Math. Jussieu **5** (2006), 423–525.

[W3]   ———, *L'endoscopie tordue n'est pas si tordue*, Memoirs of AMS **908** (2008).

[W4]   ———, *A propos du lemme fondamental pondéré tordu*, Math. Ann. **343** (2009), 103–174.

[W5]   ———, *Préparation à la stabilisation de la formule des traces tordue I: endoscopie tordue sur un corps local*, preprint.

[W6]   ———, *Préparation à la stabilisation de la formule des traces tordue II: intégrales orbitales et endoscopie sur un corps non-archimédien*, preprint

[W7]   ———, *Préparation à la stabilisation de la formule des traces tordue III: intégrales orbitales et endoscopie sur un corps archimédien*, preprint.

DEPARTMENT OF MATHEMATICS, UNIVERSITY OF TORONTO, TORONTO, CANADA M5S 2E4

E-mail address: `arthur@math.toronto.edu`

Universality for random matrices and log-gases

László Erdős

ABSTRACT. Eugene Wigner's revolutionary vision predicted that the energy levels of large complex quantum systems exhibit a universal behavior: the statistics of energy gaps depend only on the basic symmetry type of the model. These universal statistics show strong correlations in the form of level repulsion and they seem to represent a new paradigm of point processes that are characteristically different from the Poisson statistics of independent points.

Simplified models of Wigner's thesis have recently become mathematically accessible. For mean field models represented by large random matrices with independent entries, the celebrated Wigner-Dyson-Gaudin-Mehta (WDGM) conjecture asserts that the local eigenvalue statistics are universal. For invariant matrix models, the eigenvalue distributions are given by a log-gas with potential V and inverse temperature $\beta = 1, 2, 4$. corresponding to the orthogonal, unitary and symplectic ensembles. For $\beta \notin \{1, 2, 4\}$, there is no natural random matrix ensemble behind this model, but the analogue of the WDGM conjecture asserts that the local statistics are independent of V.

In these lecture notes we review the recent solution to these conjectures for both invariant and non-invariant ensembles. We will discuss two different notions of universality in the sense of (i) local correlation functions and (ii) gap distributions. We will demonstrate that the local ergodicity of the Dyson Brownian motion is the intrinsic mechanism behind the universality. In particular, we review the solution of Dyson's conjecture on the local relaxation time of the Dyson Brownian motion. Additionally, the gap distribution requires a De Giorgi-Nash-Moser type Hölder regularity analysis for a discrete parabolic equation with random coefficients. Related questions such as the local version of Wigner's semicircle law and delocalization of eigenvectors will also be discussed. We will also explain how these results can be extended beyond the mean field models, especially to random band matrices.

CONTENTS

2010 *Mathematics Subject Classification.* 15B52, 82B44.

Key words and phrases. β-ensemble, local semicircle law, Dyson Brownian motion. De Giorgi-Nash-Moser theory.

1. Introduction

1.1. The pioneering vision of Wigner.

> Perhaps I am now too courageous when I try to guess the distribution of the distances between successive levels (of energies of heavy nuclei). Theoretically, the situation is quite simple if one attacks the problem in a simpleminded fashion. The question is simply what are the distances of the characteristic values of a symmetric matrix with random coefficients.

Eugene Wigner on the Wigner surmise, 1956

Large complex systems often exhibit remarkably simple universal patterns as the number of degrees of freedom increases. The simplest example is the central limit theorem: the fluctuation of the sums of independent random scalars, irrespective of their distributions, follows the Gaussian distribution. The other cornerstone of probability theory identifies the Poisson point process as the universal limit of many independent point-like events in space or time. These mathematical descriptions assume that the original system has independent (or at least weakly dependent) constituents. What if independence is not a realistic approximation and strong correlations need to be modelled? Is there a universality for strongly correlated models?

At first sight this seems an impossible task. While independence is a unique concept, correlations come in many different forms; a-priori there is no reason to believe that they all behave similarly. Nevertheless they do, according to the pioneering vision of Wigner [87] at least if they originate from certain physical systems and if the "right" question is asked. The actual correlated system he studied was the energy levels of heavy nuclei. Looking at spectral measurement data, it is obvious that the eigenvalue density (or density of states, as it is called in physics) heavily depends on the system. But Wigner asked a different question: what about the distribution of the rescaled energy gaps? He discovered that the difference of consecutive energy levels, after rescaling with the local density, shows a surprisingly universal behavior. He even predicted a universal law, given by the simple formula (called the *Wigner surmise*),

$$(1.1) \qquad \mathbb{P}\Big(\widetilde{E}_j - \widetilde{E}_{j-1} = s + \mathrm{d}s\Big) \approx \frac{\pi s}{2} \exp\Big(-\frac{\pi}{4}s^2\Big)\mathrm{d}s,$$

where $\widetilde{E}_j = \varrho E_j$ denote the rescaling of the actual energy levels E_j by the density of states ϱ near the energy E_j. This law is characteristically different from the gap distribution of the Poisson process which is the exponential distribution, $e^{-s}\mathrm{d}s$. The prefactor s in (1.1) indicates a *level repulsion* for the point process $\widetilde{E}_j$, i.e. the eigenvalues are strongly correlated.

Comparing measurement data from various experiments, Wigner's pioneering vision was that the energy gap distribution (1.1) of complicated quantum systems is essentially universal; it depends only on the basic symmetries of model (such as time-reversal invariance). This thesis has never been rigorously proved for any realistic physical system but experimental data and extensive numerics leave no doubt on its correctness (see [65] for an overview).

Wigner not only predicted universality in complicated systems, but he also discovered a remarkably simple mathematical model for this new phenomenon: the eigenvalues of large random matrices. For practical purposes, Hamilton operators of quantum models are often approximated by large matrices that are obtained from some type of discretization of the original continuous model. These matrices have specific forms dictated by physical rules. Wigner's bold step was to neglect all details and consider the simplest random matrix whose entries are independent and identically distributed. The only physical property he retained was the basic symmetry class of the system; time reversal physical models were modelled by real symmetric matrices, while systems without time reversal symmetry (e.g. with magnetic fields) were modelled by complex Hermitian matrices. As far as the gap statistics are concerned, this simple-minded model reproduced the behavior of the complex quantum systems! The universal behavior extends to the joint statistics of several consecutive gaps which are essentially equivalent to the local correlation functions of the point process $\widetilde{E}_j$. From mathematical point of view, a universal strongly correlated point process was found. The natural representatives of these universality classes are the random matrices with independent identically distributed Gaussian entries. These are called the *Gaussian orthogonal ensemble (GOE)* and the *Gaussian unitary ensemble (GUE)* in case of real symmetric and complex Hermitian matrices, respectively.

Since Wigner's discovery random matrix statistics are found everywhere in physics and beyond, wherever nontrivial correlations prevail. Among many other applications, random matrix theory (RMT) is present in chaotic quantum systems in physics, in principal component analysis in statistics, in communication theory and even in number theory. In particular, the zeros of the Riemann zeta function on the critical line are expected to follow RMT statistics due to a spectacular result of Montgomery [69].

In retrospect, Wigner's idea should have received even more attention. For centuries, the primary territory of probability theory was to model uncorrelated or weakly correlated systems. The surprising ubiquity of random matrix statistics is a strong evidence that it plays a similar fundamental role for correlated systems as Gaussian distribution and Poisson point process play for uncorrelated systems. RMT seems to provide essentially the only universal and generally computable pattern for complicated correlated systems.

In fact, a few years after Wigner's seminal paper [87], Gaudin [53] has discovered another remarkable property of this new point process: the correlation functions have a determinantal structure, at least if the distributions of the matrix elements are Gaussian. The algebraic identities within the determinantal form opened up the route to calculations and to obtain explicit formulas for local correlation functions. In particular, the gap distribution for the complex Hermitian case is given by a Fredholm determinant involving Hermite polynomials. In fact, Hermite polynomials were first introduced in the context of random matrices by Mehta and Gaudin [67] earlier. Dyson and Mehta [24, 26, 66] have later extended this exact calculation to correlation functions and to other symmetry classes. When compared with the exact formula, the Wigner surmise (1.1), based upon a simple 2×2 matrix model, turned out to be quite accurate. While the determinantal structure is present only in Gaussian Wigner matrices, the paradigm of local universality predicts that the formulas for the local eigenvalue statistics obtained in the Gaussian case hold for general distributions as well.

1.2. Physical models. The ultimate mathematical goal is to prove Wigner's vision for a very large class of realistic quantum mechanical models. This is extremely hard, since the local statistics involve tracking individual eigenvalues in the bulk spectrum. Wigner's original model, the energy levels of heavy nuclei, is a strongly interacting many-body quantum system. The rigorous analysis of such model with the required precision is beyond the reach of current mathematics.

A much simpler question is to neglect all interactions and to study the natural one-body quantum model, the Schrödinger operator $-\Delta + V$ with a potential V on $\mathbb{R}^d$. The complexity comes from assuming that V is generic in some sense, in particular to exclude models with additional symmetries that may lead to non-universal eigenvalue correlations. Two well-studied examples are (i) the random Schrödinger operators where $V = V(x)$ is a random field with a short range correlation, and (ii) quantum chaos models, where V is generic but fixed and the statistical ensemble is generated by sampling the spectrum in small spectral windows at high energies (an alternative formulation uses the semiclassical limit).

Unfortunately, there are essentially no rigorous results on local spectral universality even in these one-body models. Random Schrödinger operators are conjectured to exhibit a metal-insulator transition that was discovered by Anderson [4]. The high disorder regime is relatively well understood since the seminal work of Fröhlich and Spencer [51] (an alternative proof is given by Aizenman and Molchanov [1]). However, in this regime the eigenfunctions are localized and thus eigenfunctions belonging to neighboring eigenvalues are typically spatially separated, hence uncorrelated. Therefore, due to localization, the system does not have sufficient correlation to fall into the RMT universality class; in fact the local eigenvalue statistics follow the Poisson process [68]. In contrast, in the low disorder regime, starting from three spatial dimension and away from the spectral edges, the eigenfunctions are conjectured to be delocalized (*extended states conjecture*). Spatially overlapping eigenfunctions introduce correlations among eigenvalues and it is expected that the local statistics are given by RMT. In the theoretical physics literature, the existence of the delocalized regime and its RMT statistics are considered as facts, supported both by non-rigorous arguments and numerics. One of the most intriguing approach is via supersymmetric (SUSY) functional integrals that remarkably reproduce all formulas obtained by the determinantal calculations in much more general setup but in a non-rigorous way due to neglecting highly oscillatory terms. The rigorous mathematics seriously lags behind these developments; even the existence of the delocalized regime is not proven, let alone detailed spectral statistics.

Judged from the horizons of theoretical physics, rigorous mathematics does not fare much better in the quantum chaos models either. The grand vision is that the quantization of an integrable classical Hamiltonian system exhibits Poisson eigenvalue statistics and a chaotic classical system gives rise to RMT statistics [8, 10]. While Poisson statistics have been shown to emerge some specific integrable models [64, 77, 80], there is no rigorous result on the RMT statistics. Recently there has been a remarkable mathematical progress in *quantum unique ergodicity (QUE)* that predicts that *all* eigenfunctions of chaotic systems are uniformly distributed all over the space, at least in some macroscopic sense. For arithmetic domains QUE has been proved in [**62**]. For general manifolds much less is known, but a lower bound on the topological entropy of the support of the limiting densities of eigenfunctions excludes that eigenfunctions are supported only on a periodic orbit [**2**]. Very roughly, QUE can be considered as the analogue of the extended states for random Schrödinger operators. Theoretically, the overlap of eigenfunctions should again lead to correlations between neighboring eigenvalues, but their direct quantitative analysis would require a much more precise understanding of the eigenfunctions.

1.3. Random matrix ensembles. In these lectures we consider even simpler models to test Wigner's universality hypothesis, namely the random matrix ensemble itself. The main goal is to show that their eigenvalues follow the local statistics of the Gaussian Wigner matrices which have earlier been computed explicitly by Dyson, Gaudin and Mehta. The statement that the local eigenvalue statistics is independent of the law of the matrix elements is generally referred to as the *universality conjecture of random matrices* and we will call it the *Wigner-Dyson-Gaudin-Mehta conjecture*. It was first formulated in Mehta's treatise on random matrices [**65**] in 1967 and has remained a key question in the subject ever since. The goal of these lecture notes is to review the recent progress that has led to the proof of this conjecture and we sketch some important ideas. We will, however, not be able to present all aspects of random matrices and we refer the reader to recent comprehensive books [**3, 18, 20**].

1.3.1. *Wigner ensembles.* To make the problem simpler, we restrict ourselves to either real symmetric or complex Hermitian matrices so that the eigenvalues are real. The standard model consists of $N \times N$ square matrices $H = (h_{ij})$ with matrix elements having mean zero and variance $1/N$, i.e.,

$$(1.2) \qquad \mathbb{E}\, h_{ij} = 0, \quad \mathbb{E}|h_{ij}|^2 = \frac{1}{N} \qquad i,j = 1,2,\ldots,N.$$

The matrix elements h_{ij}, $i,j = 1,\ldots,N$, are real or complex independent random variables subject to the symmetry constraint $h_{ij} = \overline{h}_{ji}$. These ensembles of random matrices are called *(standard) Wigner matrices*. We will always consider the limit as the matrix size goes to infinity, i.e., $N \to \infty$. Every quantity related to H depends on N, so we should have used the notation $H^{(N)}$ and $h_{ij}^{(N)}$, etc., but for simplicity we will omit N in the notation.

In Section 2 we will also consider generalizations of these ensembles, where we allow the matrix elements h_{ij} to have different distributions (but retaining independence). The main motivation is to depart from the mean-field character of the standard Wigner matrices, where the quantum transition amplitudes h_{ij} between any two sites i, j have the same statistics. The most prominent example is the random band matrix ensemble (see Example 2.1) that naturally interpolates between standard Wigner matrices and random Schrödinger operators with a short range hopping mechanism (see [**81**] for an overview).

The first rigorous result about the spectrum of a random matrix of this type is the famous *Wigner semicircle law* [**87**] which states that the empirical density of the eigenvalues, $\lambda_1, \lambda_2, \ldots, \lambda_N$, under the normalization (1.2), is given by

$$(1.3) \qquad \varrho_N(x) := \frac{1}{N} \sum_{j=1}^{N} \delta(x - \lambda_j) \rightharpoonup \varrho_{sc}(x) := \frac{1}{2\pi} \sqrt{(4 - x^2)_+}$$

in the weak limit as $N \to \infty$. The limit density is independent of the details of the distribution of h_{ij}.

The Wigner surmise (1.1) is a much finer problem since it concerns individual eigenvalues and not only their behavior on macroscopic scale. To understand it, we introduce correlation functions. If $p_N(\lambda_1, \lambda_2, \ldots, \lambda_N)$ denotes the joint probability density of the (unordered) eigenvalues, then the n-point correlation functions (marginals) are defined by

$$(1.4) \quad p_N^{(n)}(\lambda_1, \lambda_2, \ldots, \lambda_n) := \int_{\mathbb{R}^{N-n}} p_N(\lambda_1, \ldots, \lambda_n, \lambda_{n+1}, \ldots \lambda_N) d\lambda_{n+1} \ldots d\lambda_N.$$

To keep this introduction simple, we state the corresponding results in terms of the eigenvalue correlation functions for Hermitian $N \times N$ matrices. In the Gaussian case (GUE) the joint probability density of the eigenvalues can be expressed explicitly as

$$(1.5) \qquad p_N(\lambda_1, \lambda_2, \ldots, \lambda_N) = \text{const.} \prod_{i<j} (\lambda_i - \lambda_j)^2 \prod_{j=1}^{N} e^{-\frac{1}{2}N\lambda_j^2},$$

where the normalization constant can be computed explicitly. The Vandermonde determinant structure allows one to compute the k-point correlation functions in the large N limit via Hermite polynomials that are the orthogonal polynomials with respect to the Gaussian weight function.

The result of Dyson, Gaudin and Mehta asserts that for any fixed energy E in the bulk of the spectrum, i.e., $|E| < 2$, the small scale behavior of $p_N^{(n)}$ is given explicitly by

$$(1.6)$$
$$\frac{1}{[\varrho_{sc}(E)]^n} p_N^{(n)} \left(E + \frac{\alpha_1}{N\varrho_{sc}(E)}, E + \frac{\alpha_2}{N\varrho_{sc}(E)}, \ldots, E + \frac{\alpha_n}{N\varrho_{sc}(E)} \right) \to \det \left(K(\alpha_i - \alpha_j) \right)_{i,j=1}^{n}$$

where K is the celebrated sine kernel

$$(1.7) \qquad K(x, y) = \frac{\sin \pi(x - y)}{\pi(x - y)}.$$

Note that the limit in (1.6) is independent of the energy E as long as it lies in the bulk of the spectrum. The rescaling by a factor N^{-1} of the correlation functions in (1.6) corresponds to the typical distance between consecutive eigenvalues and we will refer to the law under such scaling as *local statistics*. Note that the correlation functions do not factorize, i.e. the eigenvalues are strongly correlated despite that the matrix elements are independent. Similar but more complicated formulas were obtained for symmetric matrices and also for the self-dual quaternion random matrices which is the third symmetry class of random matrix ensembles.

The convergence in (1.6) holds for each fixed $|E| < 2$ and uniformly in $(\alpha_1, \ldots, \alpha_n)$ in any compact subset of $\mathbb{R}^n$. Fix now k compact subsets $A_1, \ldots A_k$ in $\mathbb{R}$. From (1.6) one can compute the distribution of the number n_j of the rescaled eigenvalues $\widetilde{\lambda}_\alpha := N(\lambda_\alpha - E)\varrho_{sc}(E)$ in A_j around a fixed energy $|E| < 2$. The limit of the joint probabilities

$$(1.8) \qquad \mathbb{P}\Big(\#\{\widetilde{\lambda}_\alpha \in A_j\} = n_j, \; j = 1, 2, \ldots, k\Big)$$

is given as derivatives of a Fredholm determinant involving the sine kernel. Clearly (1.8) gives a complete local description of the rescaled eigenvalues as a point process around a fixed energy E. In particular it describes the distribution of the eigenvalue gap that contains a *fixed energy E*. However, (1.8) does not determine the distribution of the gap with a *fixed label*, e.g. the gap $\lambda_{N/2+1} - \lambda_{N/2}$. Only the cumulative statistics of many consecutive gaps can be deduced, see [18] for a precise formulation. The slight discrepancy between the statements at fixed energy and with fixed label leads to involved technical complications.

1.3.2. *Invariant ensembles.* The explicit formula (1.5) is special for Gaussian Wigner matrices; if h_{ij} are independent but non-Gaussian, then no analogous explicit formula is known for the joint probability density. Gaussian Wigner matrices have this special property because their distribution is *invariant* under base transformation. The derivation of (1.5) relies on the fact that in the diagonalization $H = U\Lambda U^*$ of H, where Λ is diagonal and U is unitary, the distributions of U and Λ decouple. The Gaussian measure of h_{ij} with the normalization (1.2) can also be expressed as

$$(1.9) \qquad \exp\Big(-\frac{1}{2}N\operatorname{Tr}H^2\Big)\mathrm{d}H = \exp\Big(-\frac{1}{2}N\operatorname{Tr}\Lambda^2\Big)\mathrm{d}(U\Lambda U^*),$$

where $\mathrm{d}H$ is the Lebesgue measure on hermitian matrices. The Vandermonde determinant in (1.5) originates from the integrating the Jacobian $\mathrm{d}(U\Lambda U^*)/\mathrm{d}\Lambda$ over the unitary group. Similar argument holds for real symmetric matrices with orthogonal conjugations, the only difference is the exponent 2 of the Vandermonde determinant becomes 1. The exponent is 4 for the third symmetry class of Wigner matrices, the self-dual quaternion matrices with symmetry group being the symplectic matrices (*Gaussian symplectic ensemble, GSE*).

Starting from (1.5), there are two natural generalizations of Gaussian Wigner matrices. One direction is the Wigner matrices with non-Gaussian but independent entries that we have already introduced in Section 1.3.1. Another direction is to consider a more general real function $V(H)$ of H instead of the quadratic H^2 in (1.9). Since invariance still holds, $\operatorname{Tr}V(H) = \operatorname{Tr}V(U\Lambda U^*) = \operatorname{Tr}V(\Lambda)$, the same argument gives (1.5), with $V(\lambda_i)$ instead of $\lambda_j^2/2$, for the correlation functions of $\exp(-N\operatorname{Tr}V(H))$. These are called *invariant ensembles* with potential V. Their matrix elements are in general correlated except in the Gaussian case.

Invariant ensembles in all three symmetry classes can be given simultaneously by the probability measure

$$Z^{-1}e^{-\frac{1}{2}N\beta\operatorname{Tr}V(H)}\mathrm{d}H,$$

where N is the size of the matrix H, V is a real valued potential and $Z = Z_N$ is the normalization constant. The positive parameter β is determined by the symmetry class, its value is 1, 2 or 4, for real symmetric, complex hermitian and self-dual quaternion matrices, respectively. The Lebesgue measure $\mathrm{d}H$ is understood over the matrices in the same class. The probability distribution of the eigenvalues $\boldsymbol{\lambda} =$

$(\lambda_1, \ldots, \lambda_N)$ is given by the explicit formula (c.f. (1.5))

$$(1.10) \qquad \mu_{\beta,V}^{(N)}(\boldsymbol{\lambda})\mathrm{d}\boldsymbol{\lambda} \sim e^{-\beta N \mathcal{H}(\boldsymbol{\lambda})}\mathrm{d}\boldsymbol{\lambda}$$

$$\text{with Hamiltonian } \mathcal{H}(\boldsymbol{\lambda}) := \sum_{k=1}^{N} \frac{1}{2}V(\lambda_k) - \frac{1}{N}\sum_{1 \leqslant i < j \leqslant N} \log(\lambda_j - \lambda_i).$$

The key structural ingredient of this formula, the logarithmic interaction that gives rise to the the Vandermonde determinant, is the same as in the Gaussian case, (1.5). Thus all previous computations, developed for the Gaussian case, can be carried out for $\beta = 1, 2, 4$, provided that the Gaussian weight function for the orthogonal polynomials is replaced with the function $e^{-\beta V(x)/2}$. The analysis of the correlation functions depends critically on the the asymptotic properties of the corresponding orthogonal polynomials.

While the asymptotics of the Hermite polynomial for the Gaussian case are well-known, the extension of the necessary analysis to a general potential is a demanding task; important progress was made since the late 1990's by Fokas-Its-Kitaev [50], Bleher-Its [9], Deift *et. al.* [18, 21, 22], Pastur-Shcherbina [72, 73] and more recently by Lubinsky [63]. These results concern the simpler $\beta = 2$ case. For $\beta = 1, 4$, the universality was established only quite recently for analytic V with additional assumptions [19, 20, 60, 79] using earlier ideas of Widom [86]. The final outcome of these sophisticated analyses is that universality holds for the measure (1.10) in the sense that the short scale behavior of the correlation functions is independent of the potential V (with appropriate assumptions) provided that β is one of the classical values, i.e., $\beta \in \{1, 2, 4\}$, that corresponds to an underlying matrix ensemble.

Notwithstanding matrix ensembles or orthogonal polynomials, the measure (1.10) on N points $\lambda_1, \ldots, \lambda_N$ is perfectly well defined for any $\beta > 0$. It can be interpreted as the Gibbs measure for a system of particles with external potential $\frac{1}{2}V$ and with a logarithmic interaction (log-gas) at inverse temperature β. From this point of view β is a continuous parameter and the classical values $\beta = 1, 2, 4$ play apparently no distinguished role. It is therefore natural to extend the universality problem to all non-classical β but the orthogonal polynomial methods are difficult to apply for this case. For any $\beta > 0$ the local statistics for the Gaussian case $V(x) = x^2/2$ is given by a point process, denoted by Sine_β. It can be obtained from a rescaling of the Airy_β process as $\lim_{a \to \infty} \sqrt{a}(\mathrm{Airy}_\beta + a) = \mathrm{Sine}_\beta$. The Airy process itself is the low lying eigenvalues of the one dimensional Schrödinger operator $-\frac{\mathrm{d}^2}{\mathrm{d}x^2} + x + \frac{2}{\sqrt{\beta}}b'_x$ on the positive half line, where b'_x is the white noise. The relation between Gaussian random matrices and random Schrödinger operators is derived from a tridiagonal matrix representation [23]. Another convenient representation of the Sine_β process is given by the "Brownian carousel" [76, 85].

Beyond random matrices, the log-gas can also be viewed as the only interacting particle model with a scale-invariant interaction and with a single relevant parameter, the inverse temperature β. It is believed to be the canonical model for strongly correlated systems and thus to play a similarly fundamental role in probability theory as the Poisson process or the Brownian motion. Nevertheless, we still have very little information about its properties. Unlike the universality problem that is inherently analytical, many properties of the log-gas are destined, at the first sight, to be revelead by smart algebraic identities. Despite many trials by physicists and mathematicians, the log-gas with a general β seems to defy all algebraic attempts.

We do not really understand why the algebraic approach is suitable for $\beta = 2$, and to a lesser extent for $\beta = 1, 4$, but it fails for any other β, while from an analytical point of view there is no difference between various values of β. To understand this fascinating ensemble, a main goal is to develop general analytical methods that work for any β.

1.4. Universality of the local statistics: the main results. All universality results reviewed in the previous sections rely on some version of the explicit formula (1.10) that is not available for Wigner matrices with non-Gaussian matrix elements. The only result prior 2009 towards universality for Wigner matrices was the proof of Johansson [**57**] (extended by Ben Arous-Péché [**7**]) for complex Hermitian Wigner matrices with a substantial Gaussian component. The hermiticity is necessary, since the proof still relies on an algebraic formula, a modification of the Harish-Chandra/Itzykson/Zuber integral observed first by Brézin and Hikami in this context [**14**].

To indicate the restrictions imposed by the usage of explicit formulas, we note that previous methods were not suitable to deal even with very small perturbations of the Gaussian Wigner case. For example, universality was already not known if only a few matrix elements of H had a distribution different from Gaussian.

Given this background, the main challenge a few years ago was to develop a new approach to universality that does not rely on any algebraic identity. We believe that the genuine reason behind Wigner's universality is of analytic nature. Algebraic computations may be used to obtain explicit formulas for the most convenient representative of a universality class (typically the Gaussian case), but only analytical methods have the power to deal with the general case. In light of the two main classes of random matrix ensembles, we set the following two main problems.

Problem 1: Prove the Wigner-Dyson-Gaudin-Mehta conjecture, i.e. the universality for Wigner matrices with a general distribution for the matrix elements.

Problem 2: Prove the universality of the local statistics for the log-gas (1.10) for all $\beta > 0$.

We were able to solve Problem 1 for a very general class of distributions. As for Problem 2, we solved it for the case of real analytic potentials V assuming that the equilibrium measure is supported on a single interval, which, in particular, holds for any convex potential. We will give a historical overview of related results in Section 1.5.3.

The original universality conjectures, as formulated in Mehta's book [**65**], do not specify the type of convergence in (1.6). We focus on two types of results for both problems. First we show that universality holds in the sense that local correlation functions around an energy E converge weakly if E is averaged on a small interval of size $N^{-1+\varepsilon}$. Second, we prove the universality of the joint distribution of consecutive gaps with fixed labels.

We note that universality of the *cumulative* statistics of N^ε gaps directly follows from the weak convergence of the correlation functions but our result on a single gap requires a quite different approach. From the point of view of Wigner's original vision on the ubiquity of the random matrix statistics in seemingly disparate ensembles and physical systems, the issue of cumulative gap statistics versus single gap statistics is minuscule. Our main reason of pursuing the single gap universality is less for the result itself; more importantly, we develop new methods to analyze the structure of the log-gases, which seem to represent the universal statistics of strongly correlated systems. In the next two sections we state the results precisely.

1.4.1. *Generalized Wigner matrices.* Our main results hold for a larger class of ensembles than the standard Wigner matrices, which we will call *generalized Wigner matrices.*

DEFINITION 1.1. ([**46**]) The real symmetric or complex Hermitian matrix ensemble H with centred and independent matrix elements $h_{ij} = \overline{h}_{ji}$, $i \leqslant j$, is called generalized Wigner matrix if the following assumptions hold on the variances of the matrix elements $s_{ij} = \mathbb{E}|h_{ij}|^2$:

(A) For any j fixed

$$(1.11) \qquad \sum_{i=1}^{N} s_{ij} = 1.$$

(B) There exist two positive constants, C_1 and C_2, independent of N such that

$$(1.12) \qquad \frac{C_1}{N} \leqslant s_{ij} \leqslant \frac{C_2}{N}.$$

The result on the correlation functions is the following theorem:

THEOREM 1.2 (Wigner-Dyson-Gaudin-Mehta conjecture for averaged correlation functions). *[**33**, Theorem 7.2] Suppose that $H = (h_{ij})$ is a complex Hermitian (respectively, real symmetric) generalized Wigner matrix. Suppose that for some constants $\varepsilon > 0$, $C > 0$,*

$$(1.13) \qquad \mathbb{E}\left|\sqrt{N}h_{ij}\right|^{4+\varepsilon} \leqslant C.$$

Let $n \in \mathbb{N}$ and $O : \mathbb{R}^n \to \mathbb{R}$ be compactly supported and continuous. Let E satisfy $|E| < 2$ and let $\xi > 0$. Then for any sequence b_N satisfying $N^{-1+\xi} \leqslant b_N \leqslant ||E| - 2|/2$ we have

$$(1.14) \quad \lim_{N \to \infty} \int_{E-b_N}^{E+b_N} \frac{\mathrm{d}x}{2b_N} \int_{\mathbb{R}^n} \mathrm{d}\alpha_1 \cdots \mathrm{d}\alpha_n\, O(\alpha_1, \ldots, \alpha_n)$$

$$\times \frac{1}{\varrho_{sc}(E)^n} \left(p_N^{(n)} - p_{\mathrm{G},N}^{(n)} \right) \left(x + \frac{\alpha_1}{N\varrho_{sc}(E)}, \ldots, x + \frac{\alpha_n}{N\varrho_{sc}(E)} \right) = 0.$$

Here ϱ_{sc} is the semicircle law defined in (1.3), $p_N^{(n)}$ is the n-point correlation function of the eigenvalue distribution of H (1.4), and $p_{\mathrm{G},N}^{(n)}$ is the n-point correlation function of an $N \times N$ GUE (respectively, GOE) matrix.

The condition (1.12) can be relaxed, see Corollary 8.3 [**31**]. For example, the lower bound can be changed to $N^{-9/8+\varepsilon}$. Alternatively, the upper bound $C_2 N^{-1}$ can replaced with $N^{-1+\varepsilon_n}$ for some $\varepsilon_n > 0$. For band matrices, the upper and lower bounds can be simultaneously relaxed.

We remark that for the complex Hermitian case the convergence of the correlation functions can be strengthened to a convergence at each fixed energy, i.e. for any fixed $|E| < 2$ we have that

$$(1.15) \quad \int_{\mathbb{R}^n} \mathrm{d}\alpha_1 \cdots \mathrm{d}\alpha_n \, O(\alpha_1, \ldots, \alpha_n)$$

$$\times \frac{1}{\varrho_{sc}(E)^n} \left(p_N^{(n)} - p_{\mathrm{G},N}^{(n)} \right) \left(E + \frac{\alpha_1}{N\varrho_{sc}(E)}, \ldots, E + \frac{\alpha_n}{N\varrho_{sc}(E)} \right) = 0.$$

The main ideas leading to the results (1.14) and (1.15) have been developed in a series of papers. We will give a short overview of the key methods in Section 1.5.1 and of the related results in Section 1.5.3.

The second result on generalized Wigner matrices asserts that the local gap statistics in the bulk of the spectrum are universal for any general Wigner matrix, in particular they coincide with those of the Gaussian case. To formulate the statement, we need to introduce the notation γ_j for the j-th quantile of the semicircle density, i.e. $\gamma_j = \gamma_j^{(N)}$ is defined by

$$(1.16) \qquad \frac{j}{N} = \int_{-2}^{\gamma_j} \varrho_{sc}(x)\mathrm{d}x.$$

We also introduce the notation $[\![A, B]\!] := \{A, A+1, \ldots, B\}$ for any integers $A < B$.

THEOREM 1.3 (Gap universality for Wigner matrices). *[45, Theorem 2.2] Let H be a generalized real symmetric or complex Hermitian Wigner matrix with subexponentially decaying matrix elements, i.e. we assume that*

$$(1.17) \qquad \mathbb{P}\big(\sqrt{N}|h_{ij}| \geqslant x\big) \leqslant C_0 \exp(-x^\vartheta)$$

holds for any $x > 0$ with some C_0, ϑ positive constants. Fix a positive number $\alpha > 0$, an integer $n \in \mathbb{N}$ and a smooth, compactly supported function $O : \mathbb{R}^n \to \mathbb{R}$. There exists an $\varepsilon > 0$ and $C > 0$, depending only on C_0, ϑ, α and O such that

$$(1.18) \quad \left| [\mathbb{E} - \mathbb{E}^\mu] O\big(N(x_j - x_{j+1}), N(x_j - x_{j+2}), \ldots, N(x_j - x_{j+n})\big) \right| \leqslant CN^{-\varepsilon},$$

for any $j \in [\![\alpha N, (1-\alpha)N]\!]$ and for any sufficiently large $N \geqslant N_0$, where N_0 depends on all parameters of the model, as well as on n and α. Here $\mathbb{E}$ and $\mathbb{E}^\mu$ denotes the expectation with respect to the Wigner ensemble H and the Gaussian equilibrium measure (see (1.5) for the Hermitian case), respectively.

More generally, for any $k, m \in [\![\alpha N, (1 - \alpha)N]\!]$ we have

$$(1.19)$$

$$\left| \mathbb{E} O\big((N\varrho_k)(x_k - x_{k+1}), (N\varrho_k)(x_k - x_{k+2}), \ldots, (N\varrho_k)(x_k - x_{k+n})\big) \right.$$

$$\left. - \mathbb{E}^\mu O\big((N\varrho_m)(x_m - x_{m+1}), (N\varrho_m)(x_m - x_{m+2}), \ldots, (N\varrho_m)(x_m - x_{m+n})\big) \right|$$

$$\leqslant CN^{-\varepsilon},$$

where the local density ϱ_k is defined by $\varrho_k := \varrho_{sc}(\gamma_k)$.

As it was already mentioned, the gap universality with a certain local averaging, i.e. for the cumulative statistics of N^ε consecutive gaps, follows directly from the universality of the correlation functions, Theorem 1.2. The gap distribution for Gaussian random matrices, with a local averaging, can then be explicitly expressed via a Fredholm determinant, see [18–20]. The first result for a single gap, i.e. without local averaging, was only achieved recently in the special case of the Gaussian unitary ensemble (GUE) in [84], which statement then easily implies the same results for complex Hermitian Wigner matrices satisfying the four moment matching condition.

1.4.2. *Log-gases.* In the case of invariant ensembles, it is well-known that for V satisfying certain mild conditions the sequence of one-point correlation functions, or densities, associated with $\mu = \mu^{(N)}$ from (1.10) has a limit as $N \to \infty$ and the limiting equilibrium density $\varrho_V(s)$ can be obtained as the unique minimizer of the functional

$$I(\nu) = \int_{\mathbb{R}} V(t)\nu(t)\mathrm{d}t - \int_{\mathbb{R}}\int_{\mathbb{R}} \log|t - s|\nu(s)\nu(t)\mathrm{d}t\mathrm{d}s.$$

We assume that $\varrho = \varrho_V$ is supported on a single compact interval, $[A, B]$ and $\varrho \in C^2(A, B)$. Moreover, we assume that V is *regular* in the sense that ϱ is strictly positive on (A, B) and vanishes as a square root at the endpoints, see (1.4) of [12]. It is known that these condition are satisfied if, for example, V is strictly convex. In this case ϱ_V satisfies the equation

$$(1.20) \qquad \frac{1}{2}V'(t) = \int_{\mathbb{R}} \frac{\varrho_V(s)\mathrm{d}s}{t - s}$$

for any $t \in (A, B)$. For the Gaussian case, $V(x) = x^2/2$, the equilibrium density is given by the semicircle law, $\varrho_V = \varrho_{sc}$, see (1.3).

The following result was proven in Corollary 2.2 of [11] for convex potential V and it was generalized in Theorem 1.2 of [12] for the non-convex case.

THEOREM 1.4 (Bulk universality of β-ensemble). *Assume V is real analytic with* $\inf_{x \in \mathbb{R}} V''(x) > -\infty$. *Let* $\beta > 0$. *Consider the β-ensemble* $\mu = \mu^{(N)}_{\beta,V}$ *given in* (1.10) *and let* $p^{(n)}_N$ *denote the n-point correlation functions of μ, defined analogously to* (1.4). *For the Gaussian case, $V(x) = x^2/2$, the correlation functions are denoted by* $p^{(n)}_{G,N}$. *Let* $E \in (A, B)$ *lie in the interior of the support of ϱ and similarly let* $E' \in (-2, 2)$ *be inside the support of ϱ_{sc}. Let $O : \mathbb{R}^n \to \mathbb{R}$ be a smooth, compactly supported function. Then for* $b_N = N^{-1+\xi}$ *with any* $0 < \xi \leqslant 1/2$ *we have*

$$(1.21)$$

$$\lim_{N \to \infty} \int \mathrm{d}\alpha_1 \cdots \mathrm{d}\alpha_n\, O(\alpha_1, \ldots, \alpha_n)$$

$$\left[\int_{E-b_N}^{E+b_N} \frac{\mathrm{d}x}{2b_N} \frac{1}{\varrho(E)^n} p^{(n)}_N\left(x + \frac{\alpha_1}{N\varrho(E)}, \ldots, x + \frac{\alpha_n}{N\varrho(E)}\right) \right.$$

$$\left. - \int_{E'-b_N}^{E'+b_N} \frac{\mathrm{d}x}{2b_N} \frac{1}{\varrho_{sc}(E')^n} p^{(n)}_{G,N}\left(x + \frac{\alpha_1}{N\varrho_{sc}(E')}, \ldots, x + \frac{\alpha_n}{N\varrho_{sc}(E')}\right) \right] = 0,$$

i.e. the appropriately normalized correlation functions of the measure $\mu^{(N)}_{\beta,V}$ at the level E in the bulk of the limiting density asymptotically coincide with those of the Gaussian case. In particular, they are independent of the value of E.

For the corresponding theorem on the single gap we need to define the classical location of the j-th particle $\gamma_{j,V}$ by

$$(1.22) \qquad \frac{j}{N} = \int_A^{\gamma_{j,V}} \varrho_V(x)\mathrm{d}x,$$

similarly to the quantiles γ_j of the semicircle law, see (1.16). We set

$$(1.23) \qquad \varrho_j^V := \varrho_V(\gamma_{j,V}), \quad \text{and} \quad \varrho_j := \varrho_{sc}(\gamma_j)$$

to be the limiting densities at the classical location of the j-th particle. Our main theorem on the β-ensembles is the following.

THEOREM 1.5 (Gap universality for β-ensembles). *[45, Theorem 2.3] Let $\beta \geqslant 1$ and V be a real analytic potential with $\inf V'' > -\infty$, such that ϱ_V is supported on a single compact interval, $[A, B]$, $\varrho_V \in C^2(A, B)$, and that V is regular. Fix a positive number $\alpha > 0$, an integer $n \in \mathbb{N}$ and a smooth, compactly supported function $O : \mathbb{R}^n \to \mathbb{R}$. Let $\mu = \mu_V = \mu_{\beta,V}^{(N)}$ be given by (1.10) and let μ_G denote the same measure for the Gaussian case, $V(x) = \frac{1}{2}x^2$. Then there exist an $\varepsilon > 0$, depending only on α, β and the potential V, and a constant C depending on O such that*

$$(1.24)$$

$$\left| \mathbb{E}^{\mu_V} O\Big((N\varrho_k^V)(x_k - x_{k+1}), (N\varrho_k^V)(x_k - x_{k+2}), \ldots, (N\varrho_k^V)(x_k - x_{k+n}) \Big) \right.$$

$$\left. - \mathbb{E}^{\mu_G} O\Big((N\varrho_m)(x_m - x_{m+1}), (N\varrho_m)(x_m - x_{m+2}), \ldots, (N\varrho_m)(x_m - x_{m+n}) \Big) \right|$$

$$\leqslant CN^{-\varepsilon}$$

for any $k, m \in \llbracket \alpha N, (1-\alpha)N \rrbracket$ and for any sufficiently large $N \geqslant N_0$, where N_0 depends on V, β, as well as on n and α. In particular, the distribution of the rescaled gaps w.r.t. μ_V does not depend on the index k in the bulk.

We point out that Theorem 1.4 holds for any $\beta > 0$, but Theorem 1.5 requires $\beta \geqslant 1$. Most likely this is only a technical restriction related to a certain condition in the De Giorgi-Nash-Moser regularity theory that is the backbone of our proof.

1.5. Some remarks on the general strategy and on related results.

1.5.1. *Strategy for the universality of correlation functions.* The proof of Theorem 1.2 consists of the following three steps, discussed in Sections 2, 3.1 and 3.2, respectively. This three-step strategy was first introduced in [**35**].

Step 1. *Local semicircle law and delocalization of eigenvectors:* It states that the density of eigenvalues is given by the semicircle law not only as a weak limit on macroscopic scales (1.3), but also in a strong sense and down to short scales containing only N^ε eigenvalues for all $\varepsilon > 0$. This will imply the *rigidity of eigenvalues*, i.e., that the eigenvalues are near their classical location in the sense to be made clear in Section 3.1. We also obtain precise estimates on the matrix elements of the Green function which in particular imply complete delocalization of eigenvectors.

Step 2. *Universality for Gaussian divisible ensembles:* The Gaussian divisible ensembles are complex or real Hermitian matrices of the form

$$H_t = e^{-t/2} H_0 + \sqrt{1 - e^{-t}} U,$$

where H_0 is a Wigner matrix and U is an independent GUE/GOE matrix. The parametrization of H_t reflects that H_t is most conveniently obtained by an Ornstein-Uhlenbeck process. There are two methods and both methods imply the bulk universality of H_t for $t = N^{-\tau}$ for the entire range of $0 < \tau < 1$ with different estimates.

2a. *Proposition 3.1 of [35] which uses an extension of Johansson's formula [57].*
2b. *Local ergodicity of the Dyson Brownian motion (DBM).*

The approach in 2a yields a slightly stronger estimate (no local averaging in the energy) than the approach in 2b, but it works only in the complex Hermitian case. In these notes, we will focus on 2b. As time evolves, the eigenvalues of H_t evolve according to a system of stochastic differential equations, the Dyson Brownian motion. The distribution of the eigenvalues of H_t will be written as $f_t\mu$, where μ is the equilibrium measure (1.5). We will study the evolution equation $\partial_t f_t = \mathscr{L} f_t$, where $\mathscr{L}$ is the generator to the Dirichlet form $\int |\nabla f|^2 \mathrm{d}\mu$. As time goes to infinity, f_t converges to constant, i.e. to equilibrium. The key technical question is the speed to local equilibrium.

Step 3. *Approximation by Gaussian divisible ensembles:* It is a simple density argument in the space of matrix ensembles which shows that for any probability distribution of the matrix elements there exists a Gaussian divisible distribution with a small Gaussian component, as in Step 2, such that the two associated Wigner ensembles have asymptotically identical local eigenvalue statistics. The first implementation of this approximation scheme was via a reverse heat flow argument [35]; it was later replaced by the *Green function comparison theorem* [46] that was motivated by the four moment matching condition of [82].

The proof of Theorem 1.4 consists of the following two steps that will be presented in Sections 4.1 and 4.2.

Step 1. Rigidity of eigenvalues. This establishes that the location of the eigenvalues are not too far from their classical locations $\gamma_{j,V}$ determined by the equilibrium density ϱ_V, see (1.22). At this stage the analyticity of V is necessary since we make use of the loop equation from Johansson [58] and Shcherbina [79].

Step 2. Uniqueness of local Gibbs measures with logarithmic interactions. With the precision of eigenvalue location estimates from the Step 1 as an input, the eigenvalue spacing distributions are shown to be given by the corresponding Gaussian ones. (We will take the uniqueness of the spacing distributions as our definition of the uniqueness of Gibbs state.)

There are several similarities and differences between the proofs of Theorem 1.2 and 1.4. Both start with rigidity estimates on eigenvalues and then establish that the local spacing distributions are the same as in the Gaussian cases. The Gaussian divisible ensembles, which play a key role in our theory for noninvariant ensembles, are completely absent for invariant ensembles. The key connection between the two methods, however, is the usage of DBM (or its analogue) in the Steps 2. In Section 3.1, we will first present this idea.

1.5.2. *Strategy for gap universality.* The proofs of Theorems 1.3 and 1.5 require several new ideas. The focus is to analyze the local conditional measures $\mu_{\mathbf{y}}$ and $f_{t,\mathbf{y}}\mu_{\mathbf{y}}$ instead of the equilibrium measure μ and the DBM evolved measure $f_t\mu$. They are obtained by fixing all but $\mathcal{K}$ consecutive points, denoted by $\mathbf{y}$. The local measures are Gibbs measures on $\mathcal{K}$ points, denoted by $\mathbf{x}$, that are confined to an interval $J = J_{\mathbf{y}}$ determined by the boundary points of $\mathbf{y}$. The external potential, $V_{\mathbf{y}}$, of the local measure contains not only the external potential V from μ, but also the interactions between $\mathbf{x}$ and $\mathbf{y}$.

The first step is again to establish rigidity, but this time with respect to the conditional measures $\mu_{\mathbf{y}}$ and $(f_t \mu)_{\mathbf{y}} =: f_{t,\mathbf{y}}\mu_{\mathbf{y}}$, at least for most boundary conditions $\mathbf{y}$. Due to the logarithmic interactions, $V_{\mathbf{y}}$ is not analytic any more and the loop equation is not available, but the rigidity information can still be extracted from the rigidity with respect to the global measure with some additional arguments.

In the second step, which is the key part of the argument, we establish the universality of the gap distribution w.r.t $\mu_{\mathbf{y}}$ by interpolating between $\mu_{\mathbf{y}}$ and $\mu_{\widetilde{\mathbf{y}}}$ with two different boundary conditions $\mathbf{y}$ and $\widetilde{\mathbf{y}}$. This amounts to estimating the correlation between a gap observable, say $O(x_i - x_{i+1})$, and $V_{\mathbf{y}} - V_{\widetilde{\mathbf{y}}}$. The correlation between particles in log-gases decay only logarithmically, i.e. extremely slowly:

$$(1.25) \qquad \frac{\langle x_i; x_j \rangle}{\sqrt{\langle x_i; x_i \rangle \langle x_j; x_j \rangle}} \sim \frac{1}{\log |i-j|}$$

at least if i, j are far from the boundaries. Here $\langle \cdot \, ; \, \cdot \rangle$ denotes the covariance with respect to $\mu_{\mathbf{y}}$. The key observation is that correlation between a *gap* $x_i - x_{i+1}$ and a particle x_j decays much faster

$$(1.26) \qquad \frac{\langle x_i - x_{i+1}; x_j \rangle}{\sqrt{\langle x_i - x_{i+1}; x_i - x_{i+1} \rangle \langle x_j; x_j \rangle}} \sim \frac{1}{|i-j|},$$

because it is essentially the derivative in i of (1.25). The decay of the gap-gap correlation is even faster.

While the formulas (1.25)–(1.26) are plausible, their rigorous proof is extremely difficult due to the very strong correlations in $\mu_{\mathbf{y}}$. We are able to prove a much weaker version of (1.26), practically a decay of order $|i-j|^{-\varepsilon}$ for some small $\varepsilon > 0$, which is sufficient for our purposes. Even the proof of this weaker decay requires quite heavy tools.

We start with a classical observation by Helffer and Sjöstrand [56] that the covariance of any two observables f, g with respect to a Gibbs measure $\mu = \exp(-\mathcal{H}(\mathbf{x}))\mathrm{d}\mathbf{x}$ can be expressed as

(1.27)

$$\langle f(\mathbf{x}); g(\mathbf{x}) \rangle_\mu = \int_0^\infty \langle \mathbf{h}_t(\mathbf{x}), \nabla g(\mathbf{x}) \rangle_\mu \mathrm{d}t, \qquad \partial_t \mathbf{h}_t = -(\mathcal{L} + \mathcal{H}'')\mathbf{h}_t, \qquad \mathbf{h}_0 = \nabla f,$$

where $\mathcal{L} \geqslant 0$ is the generator to the Dirichlet form $\int |\nabla f|^2 \mathrm{d}\mu$ and $\mathcal{H}''$ is the Hessian of the Hamiltonian. The generator $\mathcal{L}$ in the heat equation in (1.27) creates a time dependent random environment $\mathbf{x}(t)$ that makes the matrix entries $(x_i - x_j)^{-2}$ of $\mathcal{H}''$ time dependent. The solution h_t to the equation in (1.27) can be thus represented as a random walk in a time dependent random environment, where the jump rate from site i to j is given by $(x_i(t) - x_j(t))^{-2}$ at time t. On large scales and for typical realizations of $\mathbf{x}(t)$, this jump rate is close to a discretization of the $\sqrt{-\Delta}$ operator. A discrete version of Di Giorgi-Nash-Moser partial regularity theory [15] then guarantees that the neighboring components of $\mathbf{h}_t$ are close, which renders the covariance $\langle \mathbf{h}_t(\mathbf{x}), \nabla g(\mathbf{x}) \rangle_\mu$ small, assuming that g is a function of $x_i - x_{i+1}$. In more general terms, the correlation decay (1.26) with $|i-j|^{-\varepsilon}$ is equivalent to the Hölder regularity a discrete parabolic PDE with random coefficients. This approach has a considerable potential to study log-gases since it connects the problem with one of the deepest phenomena in PDE.

Finally, in the third step, we pass the information on the universality of the gap w.r.t. local measures to the global ones. For the invariant ensemble this step is fairly straighforward, while for the Wigner ensemble we need to use an approximation step similar to Step 3 in Section 1.5.1.

1.5.3. *Historical remarks.* The method of the proof of Theorem 1.2 is extremely general and the result holds for a much larger class of matrix ensembles with independent entries. Adjacency matrices of the Erdős-Rényi graphs are also covered as long as the matrix is not too sparse, namely more than $N^{2/3}$ entries of each row are non-zero on average [32, 33]. Although Theorem 1.2 in its current form was proved in [33], the key ideas have been developed through several important steps in [35, 41, 46–48]. In particular, the Wigner-Dyson-Gaudin-Mehta (WDGM) conjecture for complex Hermitian matrices in the form of (1.15) was first proved in Theorem 1.1 of [35]. This result holds whenever the distributions of the matrix elements are smooth. The smoothness requirement for (1.15) was partially removed in [82] and completely removed in [36] but only in the averaged convergence sense (1.14). For a general distribution (1.15) was proved in Theorem 5 in [83]. Although the proof in [83] took a slightly different path, this generalization is an immediate corollary of previous results [44]. These arguments are restricted to the complex Hermitian case since they still use some explicit formula.

The WDGM conjecture for real symmetric matrices in the averaged form of (1.14) was resolved in [41] (a special case, under a restrictive third moment matching condition, was treated in [82]). In [41], a novel idea based on Dyson Brownian motion was introduced. The most difficult case, the real symmetric Bernoulli matrices, was solved in [47], where a "Fluctuation Averaging Lemma" (Theorem 2.16 of the current paper) exploiting cancellation of matrix elements of the Green function was first introduced. A more detailed historical review on Theorem 1.2 was given in Section 11 of [43].

For $\beta = 2$, Theorem 1.4 was proved for very general potentials, the best results for $\beta = 1, 4$ [19, 60, 79] are still restricted to analytic V with additional conditions. Prior to Theorem 1.4 there was no result for general β, except for the Gaussian case [85].

Given the historical importance of the Wigner surmise, it is somewhat surprising that single gap universality did not receive much attention until very recently. This is probably because our understanding of the Wigner-Dyson-Gaudin-Mehta universality became sufficiently sophisticated only in the last few years to realize the subtle difference between *fixed energy* and *fixed label* universality. In fact, even the GUE case was not known until the very recent paper by Tao [84]. In this work, the complex Hermitian Wigner case was also covered under the condition that the distribution matches that of the GUE to fourth order. Theorem 1.3 is considerable more general, as it applies to any symmetry classes and does not require moment matching. Finally, the single gap universality of the invariant ensembles has not been considered before Theorem 1.5.

1.5.4. *What will not be discussed.* In these lecture notes we focus on the four universality results, Theorem 1.2–1.5, and the necessary background material. There are many related questions on random matrix universality and several of them can be studied with the methods we present here. Here we just list them and give a few relevant references.

- Edge universality for Wigner matrices. See Section 9 of [43] for a summary and also the recent paper [61] that gives the the optimal moment condition. The analogues of Theorems 1.3 and 1.5 are presented in [13].

- Universality of eigenvectors. See [**59**].
- Universality for sample covariance matrices. See [**42, 74, 75**].
- Sparse matrices and adjancency matrices of Erdős-Rényi graphs. See Section 10 of [**43**].

1.5.5. *Structure of the lecture notes.* A large part of presentation in these lecture notes is borrowed from other papers and reviews written on the subject [**27, 31, 43**] and sometimes whole paragraphs of the original articles are verbatim taken over. The overlap is especially large with the review paper [**43**]; Sections 3.1–4.2 on the Dyson Brownian motion, the Green function comparison theorem and on the analysis of the β-ensemble are repeated without much changes. The local semicircle law (Section 2) is presented here more generally than in [**43**], following the recent paper [**31**]. For pedagogical reasons, we will give the proof in a simplified form in Section 2.4 and we only comment on the general proof in Section 2.5. Sections 2.3.3–2.3.4 cover new results on random band matrices based upon the recent work [**34**]. Section 5 presents an extensive outline of the proofs of Theorems 1.3 and 1.5 on the single gap universality following the very recent paper [**45**].

We will use the convention that C and c denote generic positive constants whose actual values are irrelevant and may change from line to line. For two N-dependent quantities A_N and B_N we use the notation $A_N \asymp B_N$ to express that $c \leqslant A_N/B_N \leqslant C$.

Acknowledgments. The results in these lecture notes were obtained in collaboration with Horng-Tzer Yau, Benjamin Schlein, Jun Yin, Antti Knowles and Paul Bourgade and in some work, also with Jose Ramirez and Sandrine Péche. This article reports the joint progress with these authors.

This work was partially supported by SFB-TR 12 Grant of the German Research Council.

2. Local semicircle law for general Wigner-type matrices

2.1. Setup and the main results. Let $(h_{ij} : i \leqslant j)$ be a family of independent, complex-valued random variables satisfying $\mathbb{E}h_{ij} = 0$ and $h_{ii} \in \mathbb{R}$ for all i. For $i > j$ we define $h_{ij} := \bar{h}_{ji}$, and denote by $H = (h_{ij})_{i,j=1}^N$ the $N \times N$ matrix with entries h_{ij}. By definition, H is Hermitian: $H = H^*$. (Note that this setup also includes the case of a real symmetric matrix H.) Such ensembles will be called *general Wigner-type matrices*. Note that we allow for the matrix elements having different distributions. This class of matrices is a natural generalization of the *standard real symmetric Wigner matrices* for which $h_{ij} \in R$ are identical distributed, and the *standard complex Hermitian Wigner matrices* for which the off-diagonal elements $h_{ij} \in \mathbb{C}$ are identically distributed and the diagonal elements $h_{ii} \in \mathbb{R}$ have their own, but still identical distribution.

The fundamental data of the model is the $N \times N$ matrix of variances $S = (s_{ij})$, where

$$s_{ij} := \mathbb{E} |h_{ij}|^2.$$

We introduce the parameter $M := \left[\max_{i,j} s_{ij} \right]^{-1}$ that expresses the maximal size of s_{ij}:

$$(2.1) \qquad s_{ij} \leqslant M^{-1}$$

for all i and j. We regard N as the fundamental parameter and $M = M_N$ as a function of N.

$$(2.2) \qquad N^\delta \leqslant M \leqslant N$$

for some fixed $\delta > 0$. We assume that S is (doubly) stochastic:

$$(2.3) \qquad \sum_j s_{ij} = 1$$

for all i. For standard Wigner matrices, h_{ij} are identically distributied, hence $s_{ij} = \frac{1}{N}$ and $M = N$. In this presentation, we allow for the matrix elements having different distributions but independence (up to the Hermitian symmetry) is always assumed.

EXAMPLE 2.1. Random band matrices are characterized by translation invariant variances of the form

$$(2.4) \qquad s_{ij} = \frac{1}{W} f\left(\frac{|i-j|_N}{W}\right)$$

where f is a smooth, symmetric probability density on $\mathbb{R}$, W is a large parameter, called the *band width*, and $|i-j|_N$ denotes the periodic distance on the discrete torus $\mathbb{T}$ of length N. The generalization in higher spatial dimensions is straighforward, in this case the rows and columns of H are labelled by a discrete d dimensional torus $\mathbb{T}_L^d$ of length L with $N = L^d$.

For convenience we assume that the normalized entries

$$(2.5) \qquad \zeta_{ij} := s_{ij}^{-1/2} h_{ij}$$

have a polynomial decay of arbitrary high degree, i.e. for all $p \in \mathbb{N}$ there is a constant μ_p such that

$$(2.6) \qquad \mathbb{E}|\zeta_{ij}|^p \leqslant \mu_p$$

for all N, i, and j. We make this assumption to streamline notation, but in fact, our results hold, with the same proof, provided (2.6) is valid for some large but fixed p. If we strengthen it to uniform subexponential decay, (1.17), then certain estimates will become stronger. In this paper we work with (2.6) for simplicity, but we remark that most of our previous work used (1.17).

Throughout the following we use a spectral parameter $z \in \mathbb{C}$ satisfying $\operatorname{Im} z > 0$. We shall use the notation

$$z = E + i\eta$$

without further comment. The eigenvalues of H in the $N \to \infty$ limit are distributed by the celebrated Wigner semicircle law,

$$(2.7) \qquad \varrho(x) = \varrho_{sc}(x) := \frac{1}{2\pi}\sqrt{(4 - x^2)_+}\,.$$

and its Stieltjes transform at spectral parameter z is defined by

$$(2.8) \qquad m(z) := \int_{\mathbb{R}} \frac{\varrho(x)}{x - z}\, \mathrm{d}x\,.$$

To avoid confusion, we remark that m was denoted by m_{sc} and ϱ by ϱ_{sc} in most of our previous papers. In this section we drop the subscript referring to "semicircle".

It is well known that the Stieltjes transform m is the unique solution of

$$(2.9) \qquad m(z) + \frac{1}{m(z)} + z = 0$$

with $\operatorname{Im} m(z) > 0$ for $\operatorname{Im} z > 0$. Thus we have

$$(2.10) \qquad m(z) = \frac{-z + \sqrt{z^2 - 4}}{2}.$$

We define the *resolvent* of H through

$$G(z) := (H - z)^{-1},$$

and denote its entries by $G_{ij}(z)$. The Stieltjes transform of the empirical spectral measure

$$\varrho_N(\mathrm{d}x) = \frac{1}{N} \sum_\alpha \delta(\lambda_\alpha - x)\mathrm{d}x$$

for the eigenvalues $\lambda_1 \leqslant \lambda_2 \leqslant \ldots \leqslant \lambda_N$ of H is

$$(2.11) \qquad m_N(z) := \int_{\mathbb{R}} \frac{\varrho_N(\mathrm{d}x)}{x - z} = \frac{1}{N} \operatorname{Tr} G(z).$$

An important parameter of the model is

$$(2.12) \qquad \Gamma(z) := \left\| \frac{1}{1 - m^2(z)S} \right\|_{\infty \to \infty}.$$

Note that S, being a stochastic matrix, satisfies $-1 \leqslant S \leqslant 1$, and 1 is an eigenvalue with eigenvector $\mathbf{e} = N^{-1/2}(1, 1, \ldots 1)$, $S\mathbf{e} = \mathbf{e}$. We assume that 1 is simple for convenience. Another important parameter is

$$(2.13) \qquad \widetilde{\Gamma}(z) := \left\| \frac{1}{1 - m^2(z)S}\bigg|_{\mathbf{e}^\perp} \right\|_{\infty \to \infty},$$

i.e. the norm of $1 - m^2 S$ restricted to the subspace orthogonal to the constants. Clearly $\widetilde{\Gamma} \leqslant \Gamma$.

For standard Wigner matrices we easily obtain that

$$(2.14) \qquad \Gamma(z) = \frac{1}{|1 - m^2(z)|} \asymp \frac{1}{\sqrt{\kappa_E + \eta}}, \qquad \widetilde{\Gamma}(z) = 1,$$

where $\kappa_E := \big||E| - 2\big|$ denotes the distance of E to the spectral edges. For generalized Wigner matrices (Definition 1.1) essentially the same relations hold:

$$(2.15) \qquad \Gamma(z) = \frac{1}{|1 - m^2(z)|} \asymp \frac{1}{\sqrt{\kappa_E + \eta}}, \qquad \widetilde{\Gamma}(z) \asymp 1.$$

The following definition introduces a notion of a high-probability bound that is suited for our purposes.

DEFINITION 2.2 (Stochastic domination). Let

$$X = \big(X^{(N)}(u) : N \in \mathbb{N}, u \in U^{(N)}\big), \qquad Y = \big(Y^{(N)}(u) : N \in \mathbb{N}, u \in U^{(N)}\big)$$

be two families of nonnegative random variables, where $U^{(N)}$ is a possibly N-dependent parameter set. We say that X is *stochastically dominated by Y, uniformly in u,* if for all (small) $\varepsilon > 0$ and (large) $D > 0$ we have

$$\sup_{u \in U^{(N)}} \mathbb{P}\Big[X^{(N)}(u) > N^\varepsilon Y^{(N)}(u)\Big] \leqslant N^{-D}$$

for large enough $N \geqslant N_0(\varepsilon, D)$. Unless stated otherwise, throughout this paper the stochastic domination will always be uniform in all parameters apart from the parameter δ in (2.2) and the sequence of constants μ_p in (2.6); thus, $N_0(\varepsilon, D)$ also depends on δ and μ_p. If X is stochastically dominated by Y, uniformly in u, we use the notation $X \prec Y$. Moreover, if for some complex family X we have $|X| \prec Y$ we also write $X = O_\prec(Y)$.

For example, using Chebyshev's inequality and (2.6) one easily finds that

$$(2.16) \qquad h_{ij} \prec (s_{ij})^{1/2} \prec M^{-1/2},$$

uniformly in i and j, so that we may also write $h_{ij} = O_\prec((s_{ij})^{1/2})$. An easy exercise shows that the relation $\prec$ satisfies the familiar algebraic rules of order relations, e.g. such relations can be added and multiplied. The definition of $\prec$ with the polynomial factors $N^{-\varepsilon}$ and N^{-D} are taylored for the assumption (2.6). We remark that if (1.17) is assumed, a stronger form of stochastic domination can be introduced but we will not pursue this direction here.

Since
$$\lim_{\eta \to 0^+} \operatorname{Im} m(E + i\eta) = \pi \varrho(E),$$

the convergence of the Stieltjes transform $m_N(z)$ to $m(z)$ as $N \to \infty$ will show that the empirical local density of the eigenvalues around the energy E in a window of size η converges to the semicircle law $\varrho(E)$. Therefore the key task is to control $m_N(z)$ for small $\eta = \operatorname{Im} z$.

We now define the lower threshold for η that depends on the energy $E \in [-10, 10]$

$$(2.17)$$
$$\widetilde{\eta}_E := \min\left\{ \eta \; : \; \frac{1}{M\eta} \leqslant \min\left\{ \frac{M^{-\gamma}}{\widetilde{\Gamma}(z)^3}, \; \frac{M^{-2\gamma}}{\widetilde{\Gamma}(z)^4 \operatorname{Im} m(z)} \right\} \text{ for all } z \in [E + i\eta, E + 10i] \right\}.$$

Here $\gamma > 0$ is a parameter that can be chosen arbitrarily small; for all practical purposes the reader can neglect it. For generalized Wigner matrices, $M \asymp N$, from (2.15) we have

$$\widetilde{\eta}_E \leqslant C N^{-1+2\gamma},$$

i.e. we will get the local semicircle law on the smallest possible scale $\eta \gg N^{-1}$, modulo a polynomial correction with an arbitrary small exponent. We remark that if we assume subexponential decay (1.17) instead of the polynomial decay (2.6), then the small polynomial correction can be replaced with a logarithmic correction factor.

Finally we define our fundamental control parameter

$$(2.18) \qquad \Pi(z) := \sqrt{\frac{\operatorname{Im} m(z)}{M\eta}} + \frac{1}{M\eta}.$$

We can now state the main result of this section, which in this form appeared in [31].

THEOREM 2.3 (Local semicircle law). *Uniformly in the energy $|E| \leqslant 10$ and $\eta \in [\widetilde{\eta}_E, 10]$ we have the bounds*

$$(2.19) \qquad \left| G_{ij}(z) - \delta_{ij} m(z) \right| \prec \Pi(z) = \sqrt{\frac{\operatorname{Im} m(z)}{M\eta}} + \frac{1}{M\eta}, \qquad z = E + i\eta$$

uniformly in i, j, as well as

$$(2.20) \qquad \left| m_N(z) - m(z) \right| \prec \frac{1}{M\eta}.$$

We point out two remarkable features of these bounds. The error term for the resolvent entries behaves essentially as $(M\eta)^{-1/2}$, with an improvement near the edges where $\operatorname{Im} m$ vanishes. The error bound for the Stieltjes transform, i.e. for the average of the diagonal resolvent entries, is one order better, $(M\eta)^{-1}$, but without improvement near the edge.

Various local semicircle laws have a long history. For standard Wigner matrices (i.e. $M = N$ and $\widetilde{\Gamma} = 1$), the optimal threshold for the smallest possible $\eta \gg 1/N$ has first been achieved in [**39**] in the bulk after an intermediate result on scale $\eta \gg N^{-2/3}$ in [**38**]. The first effective result near the edge was given in [**40**]. The optimal power $(N\eta)^{-1}$ for $m_N - m$ in the bulk has first been obtained in [**47**] where the first version of the *fluctuation averaging mechanism* has appeared. The optimal behavior near the edge was first derived in [**48**]. The case of $M \ll N$ has first been studied in [**46**] where the threshold $\eta \gg 1/M$ in the bulk spectrum, $|E| < 2$, has been achieved. The optimal power $(M\eta)^{-1}$ is proved in [**31**], where the technique of [**46**] was combined with the fluctuation averaging mechanism. The edge behavior, i.e. the deterioration of the threshold $\widetilde{\eta}_E$ near the edge has also been extensively studied in [**31**] and it led to the power -3 of $\widetilde{\Gamma}$ in the definition of $\widetilde{\eta}_E$, but it is yet unclear whether this power is optimal.

In Section 2.3 we demonstrate that all proofs of the local semicircle law rely on some version of a self-consistent equation. At the beginning this was a scalar equation for m_N. The self-consistent vector equation for $v_i = G_{ii} - m$ (see Section 2.3.2) first appeared in [**46**]. This allowed us to deviate from the identical distributions for h_{ij} and opened up the route to estimates on individual resolvent matrix elements. Finally, the self-consistent matrix equation for $\mathbb{E}|G_{xy}|^2$ first appeared in [**34**] and it yielded diffusion profile for the resolvent (Section 2.3.4).

We now list a few consequences of the local semicircle law. It is an elementary property of the Stieltjes transform that once $m_N \approx m$ is established for all spectral parameter z with $\operatorname{Im} z \geqslant \widetilde{\eta}_E$, then ϱ_N and ϱ coincide on scales larger than $\widetilde{\eta}_E$ near the energy E. This means that $\int f \varrho_N \to \int f \varrho$ for test functions on scale $\widetilde{\eta}_E$ i.e. $|f'| \ll 1/\widetilde{\eta}_E$. In particular, if $m_N(z) \to m(z)$ uniformly on the half planes $\{z : \operatorname{Im} z \geqslant \varepsilon\}$ for any fixed ε, then ϱ_N converges to ϱ weakly. The bound (2.20) asserts much more: it identifies ϱ_N with ϱ on scales of order $1/M$. In the standard Wigner case, $M = N$, it is basically the optimal scale since below scale $1/N$ the empirical measure ϱ_N strongly fluctuates due to individual eigenvalues.

Once the local density is identified, we can deduce results on the location of individual eigenvalues and on the counting function. Here we formulate the corresponding statements only for the simpler case when s_{ij} is comparable with $1/N$ as they were stated in [**48**] (apart from the fact that in [**48**] a subexponential decay was assumed). The precise results in the general case are somewhat more complicated and they can be found in [**31**].

Let $\gamma_\alpha = \gamma_{\alpha,N}$ denote the location of the α-th point under the semicircle law, i.e., γ_α is defined by

$$(2.21) \qquad N \int_{-\infty}^{\gamma_\alpha} \varrho(x)\mathrm{d}x = \alpha, \qquad 1 \leqslant \alpha \leqslant N.$$

We will call γ_α the *classical location* of the α-th point. Furthermore, for any real energy E, let

$$\mathfrak{n}_N(E) := \frac{1}{N}\#\{\lambda_\alpha \leqslant E\}, \qquad n(E) := \int_{-\infty}^{E} \varrho(x)\mathrm{d}x$$

be the *empirical counting function* of the eigenvalues and its classical counterpart.

COROLLARY 2.4 (Rigidity of eigenvalues and limit of the counting function). [**48**, *Theorem 2.2*] *For generalized Wigner matrices (Definition 1.1) we have*

$$(2.22) \qquad |\lambda_\alpha - \gamma_\alpha| \prec \frac{1}{N^{2/3}\widehat{\alpha}^{1/3}}, \qquad \widehat{\alpha} := \min\{\alpha, N+1-\alpha\},$$

uniformly in $\alpha \in \{1, 2, \ldots, N\}$. *Furthermore,*

$$(2.23) \qquad |\mathfrak{n}_N(E) - n(E)| \prec \frac{1}{N}.$$

uniformly in $E \in \mathbb{R}$.

We remark that under the stronger decay condition (1.17) instead of (2.6), the N^ε factors implicitly present in the notation $\prec$ can be improved to logarithmic factors, see [**48**].

Corollary 2.4 is a simple consequence of the Helffer-Sjöstrand formula which translates information on the Stieltjes transform of the empirical measure first to the counting function and then to the locations of eigenvalues. The formula yields the representation

$$(2.24) \qquad f(\lambda) = \frac{1}{2\pi} \int_{\mathbb{R}^2} \frac{\partial_{\bar{z}}\widetilde{f}(x+iy)}{\lambda - x - iy}\mathrm{d}x\mathrm{d}y$$

$$= \frac{1}{2\pi} \int_{\mathbb{R}^2} \frac{iyf''(x)\chi(y) + i(f(x) + iyf'(x))\chi'(y)}{\lambda - x - iy}\mathrm{d}x\mathrm{d}y$$

for any real valued C^2 function f on $\mathbb{R}$, where $\chi(y)$ is any smooth cutoff function with bounded derivatives and supported in $[-1, 1]$ with $\chi(y) = 1$ for $|y| \leqslant 1/2$. In the applications, f will be a smoothed version of the characteristic functions of spectral intervals so that $\sum_j f(\lambda_j)$ counts eigenvalues in that interval. From (2.24) we have

$$\frac{1}{N}\sum_j f(\lambda_j) = \frac{1}{2\pi} \int_{\mathbb{R}^2} \Big(iyf''(x)\chi(y) + i(f(x) + iyf'(x))\chi'(y)\Big)m_N(x+iy)\mathrm{d}x\mathrm{d}y,$$

and then m_N can be approximated by $m(x + iy)$. The details of the argument can be found in [**37**]. Once (2.23) is established, it is an elementary argument to translate it into the rigidity of the eigenvalues (2.22). The powers $2/3$ and $1/3$ in (2.22) stem from the fact that $\varrho(x)$ has a square root singularity near the spectral edges $x \approx \pm 2$, therefore $n(x) \sim (x+2)_+^{3/2}$ for x near -2.

Although Wigner's semicircle law and its local version only concerns $m_N = \frac{1}{N}\operatorname{Tr} G$, we remark that the resolvent matrix elements, G_{ii} and G_{ij} also carry important information. For example a good bound on G_{ii} implies delocalization of the eigenvectors. Indeed, by the spectral decomposition, we have

$$\operatorname{Im} G_{ii}(z) = \eta \sum_\alpha \frac{|u_\alpha(i)|^2}{(\lambda_\alpha - E)^2 + \eta^2},$$

where $\mathbf{u}_\alpha = (u_\alpha(1), \ldots u_\alpha(N))$ is the (normalized) eigenvector belonging to the eigenvalue λ_α. Choosing the energy E in the η-vicinity of λ_α, we obtain $|u_\alpha(i)|^2 \leqslant \eta \operatorname{Im} G_{ii}(z)$. Therefore, if $|G_{ii}(z)| \leqslant C$ can be shown uniformly for any z with $\operatorname{Im} z \geqslant \eta(N)$ for some N-dependent threshold $\eta(N)$, then we conclude

$$\max_\alpha \|\mathbf{u}_\alpha\|_\infty^2 \leqslant C\eta(N).$$

In the Wigner case, the threshold $\eta(N)$ is almost $1/N$, thus we obtain the complete delocalization of the eigenvectors:

COROLLARY 2.5. *For the ℓ^2-normalized eigenvectors $\mathbf{u}_\alpha$, $\alpha = 1, 2 \ldots, N$, of the standard Wigner matrix, we have*

$$(2.25) \qquad \|\mathbf{u}_\alpha\|_\infty \prec N^{-1/2}.$$

This result was first proven in [**39**] without bounding G_{ii}. In contrast to the argument in [**39**], the proof via G_{ii} can also be easily extended to a general class of Wigner-type matrices. For example, an elementary argument shows that if there are two positive constants c and C such that $c \leqslant N s_{ij} \leqslant C$ for all i, j, then $\widetilde{\Gamma}(z) \leqslant C$ uniformly in z, thus $\widetilde{\eta}_E \leqslant C N^{-1+\gamma}$ and (2.25) holds.

2.2. Tools. In this subsection we collect some basic definitions and facts.

DEFINITION 2.6 (Minors). For $\mathbb{T} \subset \{1, \ldots, N\}$ we define $H^{(\mathbb{T})}$ by

$$(H^{(\mathbb{T})})_{ij} := \mathbf{1}(i \notin \mathbb{T})\mathbf{1}(j \notin \mathbb{T})h_{ij}.$$

Moreover, we define the resolvent of $H^{(\mathbb{T})}$ and its normalized trace through

$$G_{ij}^{(\mathbb{T})}(z) := (H^{(\mathbb{T})} - z)_{ij}^{-1}, \qquad m^{(\mathbb{T})}(z) := \frac{1}{N} \operatorname{Tr} G^{(\mathbb{T})}(z).$$

We also set

$$\sum_i^{(\mathbb{T})} := \sum_{i\,:\,i\notin\mathbb{T}}.$$

DEFINITION 2.7 (Partial expectation and independence). Let $X \equiv X(H)$ be a random variable. For $i \in \{1, \ldots, N\}$ define the operations P_i and Q_i through

$$P_i X := \mathbb{E}(X|H^{(i)}), \qquad Q_i X := X - P_i X.$$

We call P_i *partial expectation* in the index i. Moreover, we say that X is *independent of* $\mathbb{T} \subset \{1, \ldots, N\}$ if $X = P_i X$ for all $i \in \mathbb{T}$.

We shall frequently make use of Schur's well-known complement formula, which we write as

$$(2.26) \qquad \frac{1}{G_{ii}^{(\mathbb{T})}} = h_{ii} - z - \sum_{k,l}^{(\mathbb{T}i)} h_{ik} G_{kl}^{(\mathbb{T}i)} h_{li},$$

where $i \notin \mathbb{T} \subset \{1, \ldots, N\}$.

The following resolvent identities form the backbone of all of our calculations. The idea behind them is that a resolvent matrix element G_{ij} depends strongly on the i-th and j-th columns of H, but weakly on all other columns. The first identity determines how to make a resolvent matrix element G_{ij} independent of an additional index $k \neq i, j$. The second identity expresses the dependence of a resolvent matrix element G_{ij} on the matrix elements in the i-th or in the j-th column of H. We added a third identity that relates sums of off-diagonal resolvent entries with a diagonal one. The proofs are elementary.

LEMMA 2.8 (Resolvent identities). *For any real or complex Hermitian matrix* H *and* $\mathbb{T} \subset \{1, \ldots, N\}$ *the following identities hold. If* $i, j, k \notin \mathbb{T}$ *and* $i, j \neq k$ *then*

$$(2.27) \qquad G_{ij}^{(\mathbb{T})} = G_{ij}^{(\mathbb{T}k)} + \frac{G_{ik}^{(\mathbb{T})} G_{kj}^{(\mathbb{T})}}{G_{kk}^{(\mathbb{T})}}, \qquad \frac{1}{G_{ii}^{(\mathbb{T})}} = \frac{1}{G_{ii}^{(\mathbb{T}k)}} - \frac{G_{ik}^{(\mathbb{T})} G_{ki}^{(\mathbb{T})}}{G_{ii}^{(\mathbb{T})} G_{ii}^{(\mathbb{T}k)} G_{kk}^{(\mathbb{T})}}.$$

If $i, j \notin \mathbb{T}$ *satisfy* $i \neq j$ *then*

$$(2.28) \qquad G_{ij}^{(\mathbb{T})} = -G_{ii}^{(\mathbb{T})} \sum_{k}^{(\mathbb{T}i)} h_{ik} G_{kj}^{(\mathbb{T}i)} = -G_{jj}^{(\mathbb{T})} \sum_{k}^{(\mathbb{T}j)} G_{ik}^{(\mathbb{T}j)} h_{kj}.$$

Moreover, we have

$$(2.29) \qquad \sum_{j} |G_{ij}^{(\mathbb{T})}|^2 = \frac{1}{\eta} \operatorname{Im} G_{ii}^{(\mathbb{T})},$$

which is sometimes called the Ward identity.

Finally, in order to estimate large sums of independent random variables as in (2.26) and (2.28), we will need a large deviation estimate for linear and quadratic functionals of independent random variables:

THEOREM 2.9 (Large deviation bounds). *Let* $\left(X_i^{(N)}\right)$ *and* $\left(Y_i^{(N)}\right)$ *be independent families of random variables and* $\left(a_{ij}^{(N)}\right)$ *and* $\left(b_i^{(N)}\right)$ *be deterministic; here* $N \in \mathbb{N}$ *and* $i, j = 1, \ldots, N$. *Suppose that all entries* $X_i^{(N)}$ *and* $Y_i^{(N)}$ *are independent and satisfy*

$$(2.30) \qquad \mathbb{E}X = 0, \qquad \mathbb{E}|X|^2 = 1, \qquad (\mathbb{E}|X|^p)^{1/p} \leqslant \mu_p$$

for all $p \in \mathbb{N}$ *and some constants* μ_p. *Then we have the bounds*

$$(2.31) \qquad \sum_{i} b_i X_i \prec \left(\sum_{i} |b_i|^2\right)^{1/2},$$

$$(2.32) \qquad \sum_{i,j} a_{ij} X_i Y_j \prec \left(\sum_{i,j} |a_{ij}|^2\right)^{1/2},$$

$$(2.33) \qquad \sum_{i \neq j} a_{ij} X_i X_j \prec \left(\sum_{i \neq j} |a_{ij}|^2\right)^{1/2}.$$

Sketch of the proof. The estimates (2.31), (2.32), and (2.33) follow from estimating high moments of the left hand sides combined with Chebyshev's inequality. The high moments of (2.31) directly follow from the Marcinkiewicz-Zygmund martingale inequality. High moments of (2.32), and (2.33) are computed by reducing them to (2.31) with a decoupling argument. The details are found in Lemmas B.2, B.3, and B.4 of [**30**].   □

2.3. Self-consistent equations on three levels. By the Schur complement formula (2.26), we have

$$(2.34) \qquad G_{ii} = \frac{1}{h_{ii} - z - \sum_{k,l}^{(i)} h_{ik} G_{kl}^{(i)} h_{li}}.$$

The partial expectation with respect to the index i gives

$$P_i \sum_{k,l}^{(i)} h_{ik} G_{kl}^{(i)} h_{li} = \sum_{k}^{(i)} s_{ik} G_{kk}^{(i)} = \sum_{k}^{(i)} s_{ik} G_{kk} + \sum_{k}^{(i)} s_{ik} \frac{G_{ik} G_{ki}}{G_{ii}},$$

where in the second step we used (2.27). Introducing the notation

$$v_i := G_{ii} - m$$

and recalling (2.3), we get the following self-consistent equation for v_i:

$$(2.35) \qquad v_i = \frac{1}{-z - m - \left(\sum_k s_{ik} v_k - \Upsilon_i\right)} - m,$$

where
(2.36)

$$\Upsilon_i := A_i + h_{ii} - Z_i, \qquad A_i := \sum_k s_{ik} \frac{G_{ik} G_{ki}}{G_{ii}}, \qquad Z_i := Q_i \sum_{k,l}^{(i)} h_{ik} G_{kl}^{(i)} h_{li}.$$

All these quantities depend on z, which fact is suppressed in the notation. We will show that Υ is a lower order error term. This is clear about h_{ii} by (2.16). The term A_i will be small since off-diagonal resolvent entries are small. Finally, Z_i will be small by a large deviation estimate Theorem 2.9. Before we present more details, we heuristically show the power of the self-consistent equations.

2.3.1. *A scalar self-consistent equation.* Introduce the notation

$$(2.37) \qquad [a] = \frac{1}{N} \sum_i a_i$$

for the *average* of a vector $(a_i)_{i=1}^N$. Consider the standard Wigner case, $s_{ij} = 1/N$. Then

$$\sum_k s_{ik} v_k = \frac{1}{N} \sum_k v_k = [v] \qquad \left(= m_N - m \right).$$

Neglecting Υ_i in (2.35) and taking the average of this relation for each i, we get

$$(2.38) \qquad [v] \approx \frac{1}{-z - m - [v]} - m.$$

Since

$$m = \frac{1}{-z - m}$$

by the defining equation of m, see (2.9), and this equation is stable under small perturbations, at least away from the spectral edges $z = \pm 2$, we obtain from (2.38) that $[v] \approx 0$. This means that $m_N \approx m$ and hence $\varrho_N \approx \varrho$, i.e. we obtained the Wigner's original semicircle law.

Historically the semicircle law was first found via the moment method [87] by computing $\frac{1}{N} \mathbb{E} \operatorname{Tr} H^k$, $k = 1, 2, \ldots$ in the $N \to \infty$ limit, and identifying them with the moments of the semicircle measure $\varrho(x)\mathrm{d}x$. In this approach the semicircle law emerges as a result of a somewhat tedious, albeit elementary calculation. A more direct approach is to take the average in i of the Schur's formula (2.34) which immediately gives

$$m_N \approx \frac{1}{-z - m_N}$$

after neglecting the error terms Υ_i. This identifies the limit of m_N immediately with m, the (unique) solution to (2.9). Taking the inverse Stieltjes transform then yields the semicircle law in a very direct way.

2.3.2. *A vector self-consistent equation.* If the variances s_{ij} are not constant or we are interested in individual resolvent matrix elements G_{ii} instead of their average, $\frac{1}{N}\operatorname{Tr} G$, then the scalar equation (2.38) discussed in the previous section is not sufficient. We have to consider (2.35) as a system of equations for the components of the vector $\mathbf{v} = (v_1, \ldots, v_N)$.

From the explicit formula for m (2.10), we know that $|m + z| \geqslant 1$. Assuming temporarily that

$$(2.39) \qquad \left| \sum_k s_{ik} v_k - \Upsilon_i \right| \leqslant \frac{1}{2},$$

we can expand the right-hand side of (2.35) around $-z - m$ up to second order and using the identity (2.9) we obtain

$$(2.40) \qquad v_i = m^2 \left(\sum_k s_{ik} v_k - \Upsilon_i \right) + O\left[\left(\sum_k s_{ik} v_k - \Upsilon_i \right)^2 \right].$$

This is the key equation to study v_i. Considering all higher order terms and Υ_i as errors we get a self-consistent equation

$$v_i = m^2 \sum_k s_{ik} v_k + \text{Error}$$

or, with matrix notation for the vector $\mathbf{v}$:

$$(2.41) \qquad \mathbf{v} = m^2 S \mathbf{v} + \mathcal{E},$$

where $\mathcal{E}$ represent the vector of error terms. Thus

$$\mathbf{v} = \frac{1}{1 - m^2 S}\mathcal{E}, \qquad \text{hence} \quad \|\mathbf{v}\|_\infty \leqslant \left\| \frac{1}{1 - m^2 S} \right\|_{\infty \to \infty} \|\mathcal{E}\|_\infty = \Gamma \|\mathcal{E}\|_\infty,$$

and this relation shows how the quantity Γ emerges. If the error term is indeed small and Γ is bounded, then we obtain that $\|\mathbf{v}\|_\infty = \max |G_{ii} - m|$ is small.

2.3.3. *A matrix self-consistent equation.* The off-diagonal resolvent matrix elements, G_{ij}, are strongly oscillating quantities and they are not expected to have a deterministic limit. However the local averages of their squares,

$$T_{xy} := \sum_i s_{xi} |G_{iy}|^2,$$

are expected to behave regularly. Note that in the Wigner case, using the identity (2.29), the quantity

$$T_{xy} = \sum_i s_{xi} |G_{iy}|^2 = \frac{1}{N} \sum_i G_{yi} G_{iy}^* = \frac{1}{N} \left(|G|^2 \right)_{yy} = \frac{1}{N\eta} \operatorname{Im} G_{yy}$$

is independent of x, but in the general case T_{xy} carries information on the localization length of the eigenfunctions. In particular, if T_{xy} decays only beyond a scale

ℓ, i.e. T_{xy} remains comparable with T_{xx} for $|x - y| \ll \ell$, then most eigenfunctions have a localization length at least ℓ.

The self-consistent equation for T_{xy} can be derived from (2.28):

$$G_{iy} = -G_{ii} \sum_{k}^{(i)} h_{ik} G_{ky}^{(i)}.$$

Replacing G_{ii} with m and taking the square, we have

$$|G_{iy}|^2 \approx |m|^2 \sum_{m,k}^{(i)} h_{ik} G_{ky}^{(i)} \overline{h}_{im} \overline{G_{my}^{(i)}}.$$

Taking partial expectation yields

$$P_i |G_{iy}|^2 \approx |m|^2 \sum_{k}^{(i)} s_{ik} |G_{ky}^{(i)}|^2 \approx |m|^2 \sum_{k} s_{ik} |G_{ky}|^2 = |m|^2 T_{iy},$$

where in the second step we used (2.27) to remove the upper index i and we used that the off-diagonal elements are of smaller order. This formula holds for $i \neq y$, for the special case $i = y$ we have a diagonal element, i.e.

$$P_i |G_{iy}|^2 \approx |m|^2 T_{iy} + |m|^2 \delta_{iy}.$$

Thus we have

$$(2.42) \quad T_{xy} = \sum_{i} s_{xi} |G_{iy}|^2 = \sum_{i} s_{xi} P_i |G_{iy}|^2 + \widetilde{T}_{xy}, \qquad \widetilde{T}_{xy} := \sum_{i} s_{xi} Q_i |G_{iy}|^2.$$

The term $\widetilde{T}_{xy}$ is lower order by a *fluctuation averaging mechanism*, see the explanation after Theorem 2.16. Hence, neglecting this term we have

$$T_{xy} \approx \sum_{i} s_{xi} P_i |G_{iy}|^2 = |m|^2 \sum_{i} s_{xi} T_{iy} + |m|^2 s_{xy},$$

i.e. the matrix T satisfies the self-consistent matrix equation

$$(2.43) \qquad T = |m|^2 ST + |m|^2 S + \mathcal{E},$$

where $\mathcal{E}$ is an error matrix. The solution is

$$(2.44) \qquad T = \frac{|m|^2 S}{1 - |m|^2 S} + \frac{1}{1 - |m|^2 S} \mathcal{E},$$

where the first term, given explicitly in terms of the variance matrix S, gives the leading order behavior for T:

$$T_{xy} \approx \Theta_{xy}, \qquad \Theta := \frac{|m|^2 S}{1 - |m|^2 S}.$$

2.3.4. *Application: diffusion profile for random band matrices.* Depending on the structure of S, in some cases Θ can be computed. Consider, for example, case of the random band matrices, (2.4). Here s_{xy} and hence Θ_{xy} are translation invariant, $\Theta_{xy} = \theta_{x-y}$, and the Fourier transform of θ is approximately given by

$$(2.45) \qquad \theta(p) \approx \frac{1}{\alpha \eta + W^2 D p^2}, \qquad D := \frac{1}{2} \int x^2 f(x) \mathrm{d}x, \qquad \alpha := \frac{2}{\sqrt{4 - E^2}}.$$

This means that the profile of Θ is approximately given by a diffusion profile on scale W with diffusion constant D.

The analysis of the error term in (2.44) requires estimating the norm of $(1 - |m|^2 S)^{-1}$. Note that unlike in the analysis of (2.41), here $1 - |m|^2 S$ and not $1 - m^2 S$ has to be inverted. Since $|m|^2 \sim 1 - C\eta$ for small η and S has eigenvalue 1, the inverse of $1 - |m|^2 S$ is very unstable. Fortunately, one can subtract the constant mode in (2.43) before solving the equation, thus eventually only the norm of $(1 - |m|^2 S)^{-1}$ on the subspace orthogonal to the constants is relevant. Thus the spectral gap of S plays an important role and we use that for band matrices the gap is of order $(W/N)^2$. The details are found in [34], where, among other results, the following theorem was shown:

THEOREM 2.10 (Diffusion profile). *[34, Theorem 2.4] Let H be a random band matrix with band width W, i.e. the variances are given by (2.4). Suppose that $N \ll W^{5/4}$ and $(W/N)^2 \leqslant \eta \leqslant 1$. Then*

$$(2.46) \quad |T_{xy} - \Theta_{xy}| \prec \frac{1}{N\eta}, \qquad \left| P_x |G_{xy}|^2 - \delta_{xy}|m|^2 - |m|^2\Theta_{xy} \right| \prec \frac{1}{N\eta} + \frac{\delta_{xy}}{\sqrt{W}}.$$

All estimates are uniform in $x, y \in \mathbb{T}$ and in the spectral parameter $z = E + i\eta$ for $|E| \leqslant 2 - \kappa$ and for any fixed $\kappa > 0$.

This theorem identifies $|G_{xy}|^2$ in two different senses of averaging; T_{xy} averages in one of the indices, while P_x takes partial expectation. In both cases the result is essentially Θ_{xy}.

The behavior of $\theta_{x-y} = \Theta_{xy}$ can be analyzed by inverse Fourier transform from (2.45). If $\eta \ll (W/N)^2$ then θ is essentially a constant, i.e. the profile is flat. Conversely, if $\eta \gg (W/N)^2$ then we get an exponentially decaying profile on the scale $|x| \sim W\eta^{-1/2}$. The shape of the profile is therefore nontrivial if and only if $\eta \gg (W/N)^2$. The total mass of the profile

$$(2.47) \qquad \qquad \sum_{x \in \mathbb{T}} \theta_x = \frac{\mathrm{Im}\, m}{\eta}(1 + O(\eta)) = O(\eta^{-1}),$$

and the average height of the profile is of order $(N\eta)^{-1}$. The peak of the exponential profile has height of order $(W\sqrt{\eta})^{-1}$, which dominates over the average height if and only if $\eta \gg (W/N)^2$. The regime $\eta \gg (W/N)^2$ corresponds to the regime where η is sufficiently large that the complete delocalization has not taken place, and the profile is mostly concentrated in the region $|x - y| \leqslant W\eta^{-1/2} \ll N$.

These scenarios are best understood in a dynamical picture in which η is decreased down from 1. The ensuing dynamics of θ corresponds to the diffusion approximation, where the quantum problem is replaced with a random walk of stepsize of order W. On a configuration space consisting of N sites, such a random walk will reach an equilibrium beyond time scales $(N/W)^2$. Here η^{-1} plays essentially the role of time t, so that in this dynamical picture equilibrium is reached for $t \sim \eta^{-1} \gg (N/W)^2$. Figure 2.1 illustrates this diffusive spreading of the profile for different values of η.

One important consequence of Theorem 2.10 is that it proves delocalization for band matrices with band width $W \gg N^{4/5}$ (see Corollary 2.3 of [34] for the precise statement). This improves the earlier result from [28, 29] where delocalization for $W \gg N^{6/7}$ was proved with very different methods. We remark that from the other side it is known that narrow band matrices with $W \ll N^{1/8}$ are in the localized regime [78]. The conjectured threshold for the phase transition is $W \sim \sqrt{N}$, see [52].

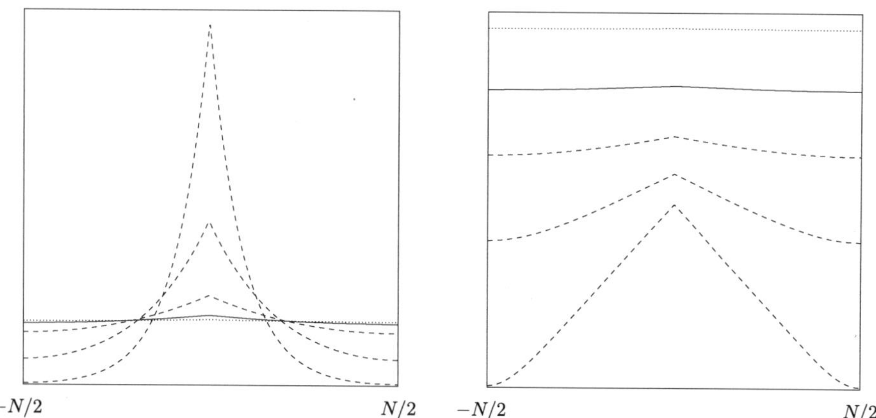

FIGURE 2.1. A plot of the diffusion profile function at five different values of η, where the argument x ranges over the torus $\mathbb{T}$. Left: the graph $x \mapsto \eta\theta_x$ (see (2.47) for the choice of normalization). Right: the graph $x \mapsto \log\theta_x$. Here we chose $N = 25W$ and $\eta = 5^{-k}$ for $k = 1, 2, 3, 4, 5$. The cases $k = 1, 2, 3$ (where $\eta > (W/N)^2$) are drawn using dashed lines, the case $k = 4$ (where $\eta = (W/N)^2$) using solid lines, and the case $k = 5$ (where $\eta < (W/N)^2$) using dotted lines.

2.4. Proof of the local semicircle law without using the spectral gap. In this section we sketch the proof of a weaker version of Theorem 2.3, namely we replace threshold $\widetilde{\eta}_E$ with a larger threshold η_E defined as

(2.48)
$$\eta_E := \min\left\{ \eta : \frac{1}{M\eta} \leqslant \min\left\{ \frac{M^{-\gamma}}{\Gamma(z)^3}, \frac{M^{-2\gamma}}{\Gamma(z)^4 \operatorname{Im} m(z)} \right\} \text{ for all } z \in [E + i\eta, E + 10i] \right\}.$$

This definition is exactly the same as (2.17), but $\widetilde{\Gamma}$ is replaced with the larger quantity Γ, in other words we do not make use of the spectral gap in S. This will pedagogically simplify the presentation and in Section 2.5 we will comment on the modifications for the stronger result.

We recall that here is no difference between Γ and $\widetilde{\Gamma}$ away from the edges (both are of order 1), so readers interested in the local semicircle law only in the bulk should be content with the simpler proof. Near the spectral edges, however, there is a substantial difference. Note that even in the Wigner case (see (2.14)), η_E is much larger near the spectral edges than the optimal threshold $\widetilde{\eta}_E \sim 1/N$.

DEFINITION 2.11. We call a deterministic nonnegative function $\Psi \equiv \Psi^{(N)}(z)$ an *admissible control parameter* if we have

$$cM^{-1/2} \leqslant \Psi \leqslant M^{-c}$$

for some constant $c > 0$ and large enough N. Moreover, we call any (possibly N-dependent) subset

$$\mathbf{D} = \mathbf{D}^{(N)} \subset \left\{ z \ : \ |E| \leqslant 10, \eta \geqslant M^{-1+\gamma} \right\}$$

a spectral domain.

In this section we will mostly use the spectral domain

$$\mathbf{S} := \Big\{ z \: : \: |E| \leqslant 10, \ \eta \in [\eta_E, 10] \Big\}.$$

Define the random control parameters
(2.49)
$$\Lambda_o := \max_{i \neq j}|G_{ij}|, \quad \Lambda_d := \max_i |G_{ii} - m|, \quad \Lambda := \max(\Lambda_o, \Lambda_d), \quad \Theta := |m_N - m|.$$

In the typical regime we will work, all these quantities are small. The key quantity is Λ and we will develop an iterative argument to control it. The first step is an apriori bound:

PROPOSITION 2.12. *We have* $\Lambda \prec M^{-\gamma/3}\Gamma^{-1}$ *uniformly in* $\mathbf{S}$.

The main estimate behind the proof of Theorem 2.3 for $\eta \geqslant \eta_E$ is the following iteration statement:

PROPOSITION 2.13. *Let* Ψ *be a control parameter satisfying*
(2.50)
$$cM^{-1/2} \leqslant \Psi \leqslant M^{-\gamma/3}\Gamma^{-1}.$$

and fix $\varepsilon \in (0, \gamma/3)$. *Then on the domain* $\mathbf{S}$ *we have the implication*
(2.51)
$$\Lambda \prec \Psi \quad \implies \quad \Lambda \prec F(\Psi),$$

where we defined

$$F(\Psi) := M^{-\varepsilon}\Psi + \sqrt{\frac{\operatorname{Im} m}{M\eta}} + \frac{M^\varepsilon}{M\eta}.$$

The proofs of these two propositions are postponed, we first complete the proof of the local semicircle law.

It is easy to check that, on the domain $\mathbf{S}$, if Ψ satisfies (2.50) then so does $F(\Psi)$. We may therefore iterate (2.51). This yields a bound on Λ that is essentially the fixed point of the map $\Psi \mapsto F(\Psi)$, which is given by Π, defined in (2.18), (up to the factor M^ε). More precisely, the iteration is started with $\Psi_0 := M^{-\gamma/3}\Gamma^{-1}$; the initial hypothesis $\Lambda \prec \Psi_0$ is provided by Proposition 2.12. For $k \geqslant 1$ we set $\Psi_{k+1} := F(\Psi_k)$. Hence from (2.51) we conclude that $\Lambda \prec \Psi_k$ for all k. Choosing $k := \lceil \varepsilon^{-1} \rceil$ yields

$$\Lambda \prec \sqrt{\frac{\operatorname{Im} m}{M\eta}} + \frac{M^\varepsilon}{M\eta}.$$

Since ε was arbitrary, we have proved that
(2.52)
$$\Lambda \prec \Pi,$$

which is (2.19).

To prove (2.20), i.e. to estimate Θ, we rewrite (2.35) as
(2.53)
$$-\sum_k s_{ik} v_k + \Upsilon_i = \frac{1}{m + v_i} - \frac{1}{m},$$

and expand the right hand side. Since $|m| \geqslant c$ and $|v_i| \leqslant \Lambda$, the expansion is possible on the event where $\Lambda \ll 1$, which occurs with very high probability by Proposition 2.12. On this event we get
(2.54)
$$m^2 \Big(-\sum_k s_{ik} v_k + \Upsilon_i \Big) = -v_i + O(\Lambda^2).$$

Averaging in (2.54) yields

$$(2.55) \qquad m^2\big(-[v]+[\Upsilon]\big) \;=\; -[v]+O_\prec(\Lambda^2)\,.$$

We will show in Lemma 2.15 in the next section that $|\Upsilon_i| \prec \Pi$, but in fact the average $[\Upsilon]$ is one order better. This is due to the fluctuation averaging phenomenon, and we have

PROPOSITION 2.14. *Suppose that* $\Lambda_o \prec \Psi_o$ *for some deterministic control parameter* Ψ_o *satisfying* $M^{-1/2} \leqslant \Psi \leqslant M^{-c}$. *Then* $[\Upsilon] = O_\prec(\Psi_o^2)$.

We will explain the proof in Section 2.4.4. Using this proposition and (2.52), we get

$$[v] \;=\; m^2[v] + O_\prec(\Pi^2)\,.$$

Therefore

$$|[v]| \;\prec\; \frac{\Pi^2}{|1-m^2|} \;\leqslant\; \left(\frac{\operatorname{Im} m}{|1-m^2|} + \frac{1}{|1-m^2|M\eta}\right)\frac{2}{M\eta} \;\leqslant\; \left(C + \frac{\Gamma}{M\eta}\right)\frac{2}{M\eta} \;\leqslant\; \frac{C}{M\eta}\,.$$

Here in the third step we used the elementary explicit bound $\operatorname{Im} m \leqslant C|1-m^2|$, and the bound $\Gamma \geqslant |1-m^2|^{-1}$ which follows from the definition of Γ by applying the matrix $(1-m^2 S)^{-1}$ to the constant vector. The last step follows from the definition of $\mathbf{S}$. Since $\Theta = |[v]|$, this concludes the proof of (2.20), and hence of Theorem 2.3 in the regime $\mathbf{S}$, i.e. for $\eta \geqslant \eta_E$. $\qquad\square$

In the next sections we explain the proofs of the three propositions used in this argument. We first control the off-diagonal elements, i.e. Λ_o, then we turn to the proof of Propositions 2.12, 2.13 and 2.14.

2.4.1. *Basic estimates for* Λ_o *and* Υ_i.

LEMMA 2.15. *The following statements hold for any spectral domain* $\mathbf{D}$ *and admissible control parameter* Ψ. *If* $\Lambda \prec \Psi$ *then*

$$(2.56) \qquad \Lambda_o + |\Upsilon_i| \;\prec\; \sqrt{\frac{\operatorname{Im} m + \Psi}{M\eta}}\,.$$

Moreover, for any fixed (N-independent) $\eta > 0$ *we have*

$$(2.57) \qquad \Lambda_o + |\Upsilon_i| \;\prec\; M^{-1/2}\,,$$

uniformly in $z \in \{w \in \mathbf{D} \,:\, \operatorname{Im} w = \eta\}$.

We remark that we could have written (2.56) as

$$(2.58) \qquad \Lambda_o + |\Upsilon_i| \;\prec\; \sqrt{\frac{\operatorname{Im} m + \Lambda}{M\eta}}\,,$$

but this formulation, while it carries the essence, is literally incorrect since it holds only if $\Lambda \prec M^{-c}$ has been apriori established.

Proof of Lemma 2.15. We first observe that $\Lambda \prec \Psi \ll 1$ and the positive lower bound $|m(z)| \geqslant c$ implies that

$$(2.59) \qquad \frac{1}{|G_{ii}|} \;\prec\; 1\,.$$

A simple iteration of the expansion formulas (2.27) concludes that

$$(2.60) \qquad |G_{ij}^{(\mathbb{T})}| \prec \Psi, \quad \text{for } i \neq j, \qquad |G_{ii}^{(\mathbb{T})}| \prec 1, \qquad \frac{1}{|G_{ii}^{(\mathbb{T})}|} \prec 1$$

for any subset $\mathbb{T}$ of fixed cardinality.

We begin with the first statement in Lemma 2.15. First we estimate Z_i, which we split as

$$(2.61) \qquad |Z_i| \leqslant \left| \sum_k^{(i)} (|h_{ik}|^2 - s_{ik}) G_{kk}^{(i)} \right| + \left| \sum_{k \neq l}^{(i)} h_{ik} G_{kl}^{(i)} h_{li} \right|.$$

We estimate each term using Theorem 2.9 by conditioning on $G^{(i)}$ and using the fact that the family $(h_{ik})_{k=1}^N$ is independent of $G^{(i)}$. By (2.31) the first term of (2.61) is stochastically dominated by

$$\left[\sum_k^{(i)} s_{ik}^2 |G_{kk}^{(i)}|^2 \right]^{1/2} \prec M^{-1/2},$$

where (2.60), (2.1) and (2.3) were used. For the second term of (2.61) we apply Theorem 2.9 (ii) with $a_{kl} = s_{ik}^{1/2} G_{kl}^{(i)} s_{li}^{1/2}$ and $X_k = \zeta_{ik}$ (see (2.5)). We find

$$(2.62)$$
$$\sum_{k,l}^{(i)} s_{ik} |G_{kl}^{(i)}|^2 s_{li} \leqslant \frac{1}{M} \sum_{k,l} s_{ik} |G_{kl}^{(i)}|^2 = \frac{1}{M\eta} \sum_k^{(i)} s_{ik} \operatorname{Im} G_{kk}^{(i)} \prec \frac{\operatorname{Im} m + \Psi}{M\eta},$$

where in the last step we used (2.27) and the estimate $1/G_{ii} \prec 1$. Thus we get

$$(2.63) \qquad |Z_i| \prec \sqrt{\frac{\operatorname{Im} m + \Psi}{M\eta}},$$

where we absorbed the bound $M^{-1/2}$ on the first term of (2.61) into the right-hand side of (2.63), using $\operatorname{Im} m \geqslant c\eta$ as follows from an explicit estimate.

Next, we estimate Λ_o. We can iterate (2.28) once to get, for $i \neq j$,

$$(2.64) \qquad G_{ij} = -G_{ii} \sum_k^{(i)} h_{ik} G_{kj}^{(i)} = -G_{ii} G_{jj}^{(i)} \left(h_{ij} - \sum_{k,l}^{(ij)} h_{ik} G_{kl}^{(ij)} h_{lj} \right).$$

The term h_{ij} is trivially $O_\prec(M^{-1/2})$. In order to estimate the other term, we invoke Theorem 2.9 (iii) with $a_{kl} = s_{ik}^{1/2} G_{kl}^{(ij)} s_{lj}^{1/2}$, $X_k = \zeta_{ik}$, and $Y_l = \zeta_{lj}$. As in (2.62), we find

$$\sum_{k,l}^{(i)} s_{ik} |G_{kl}^{(ij)}|^2 s_{lj} \prec \frac{\operatorname{Im} m + \Psi}{M\eta},$$

and thus

$$(2.65) \qquad \Lambda_o \prec \sqrt{\frac{\operatorname{Im} m + \Psi}{M\eta}},$$

where we again absorbed the term $h_{ij} \prec M^{-1/2}$ into the right-hand side.

In order to estimate A_i and h_{ii} in the definition of Υ_i, we use (2.60) to estimate

$$|A_i| + |h_{ii}| \prec \Lambda_o^2 + M^{-1/2} \leqslant \Lambda_o + C\sqrt{\frac{\operatorname{Im} m}{M\eta}} \prec \sqrt{\frac{\operatorname{Im} m + \Lambda}{M\eta}},$$

where the second step follows from $\operatorname{Im} m \geqslant c\eta$. Collecting (2.63), (2.65), this completes the proof of (2.56).

The proof of (2.57) is almost identical to that of (2.56). The quantities $\left|G_{kk}^{(i)}\right|$ and $\left|G_{kk}^{(ij)}\right|$ are estimated by the trivial deterministic bound $\eta^{-1} = O(1)$. We omit the details. $\qquad\square$

2.4.2. *Sketch of the proof of Proposition 2.12.* The core of the proof is a *continuity argument*. Its basic idea is to establish a *gap* in the range of Λ by proving

Claim 1. On the event $\Lambda \leqslant M^{-\gamma/4}\Gamma^{-1}$ we actually have the stronger bound $\Lambda \prec M^{-\gamma/2}\Gamma^{-1}$.

In other words, for all $z \in \mathbf{S}$, with high probability either $\Lambda \leqslant M^{-\gamma/2}\Gamma^{-1}$ or $\Lambda \geqslant M^{-\gamma/4}\Gamma^{-1}$. The second step is to show that $\Lambda \leqslant M^{-\gamma/2}\Gamma^{-1}$ holds for z with a large imaginary part η:

Claim 2. We have $\Lambda \prec M^{-1/2}$ uniformly in $z \in [-10, 10] + 2\mathrm{i}$.

Thus, for large η the parameter Λ is below the gap. Using the fact that Λ is continuous in $\eta = \operatorname{Im} z$ and hence cannot jump from one side of the gap to the other, we then conclude that Λ is below the gap for all $z \in \mathbf{S}$ and this is Proposition 2.12. See Figure 2.2 for an illustration of this argument.

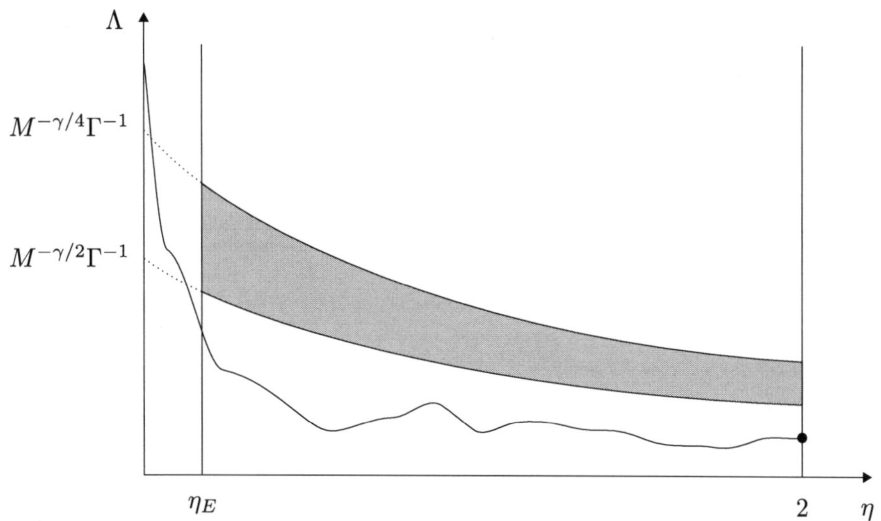

FIGURE 2.2. The (η, Λ)-plane for a fixed E with the graph of $\eta \to \Lambda(E + \mathrm{i}\eta)$. The shaded region is forbidden with high probability by Claim 1. The initial estimate, Claim 2, is marked with a black dot. The graph of $\Lambda = \Lambda(E + \mathrm{i}\eta)$ is continuous and lies beneath the shaded region. Note that this method does not control $\Lambda(E + \mathrm{i}\eta)$ in the regime $\eta \leqslant \eta_E$.

Now we explain Claim 1. We will work on the event $\Lambda \leqslant M^{-\gamma/4}\Gamma^{-1} \leqslant M^{-c}$, where we may invoke (2.56) to estimate Λ_o and Υ_i. In order to estimate Λ_d, we expand the right-hand side of (2.53) in v_i to get (2.54). Using (2.56) to estimate Υ_i, we therefore have

$$(2.66) \qquad v_i - m^2 \sum_k s_{ik} v_k = O_{\prec}\left(\Lambda^2 + \sqrt{\frac{\operatorname{Im} m + \Lambda}{M\eta}}\right).$$

We write the left-hand side as $[(1 - m^2 S)\mathbf{v}]_i$ with the vector $\mathbf{v} = (v_i)_{i=1}^N$. Inverting the operator $1 - m^2 S$, we therefore conclude that

$$(2.67) \qquad \Lambda_d = \max_i |v_i| \prec \Gamma\left(\Lambda^2 + \sqrt{\frac{\operatorname{Im} m + \Lambda}{M\eta}}\right).$$

Together with (2.56) and $\Gamma \geqslant c$, we therefore get

$$(2.68) \qquad \Lambda \prec \Gamma\left(\Lambda^2 + \sqrt{\frac{\operatorname{Im} m + \Lambda}{M\eta}}\right).$$

On the event $\Lambda \leqslant M^{-\gamma/4}\Gamma^{-1}$ we may estimate

$$\Gamma\Lambda^2 \leqslant M^{-\gamma/2}\Gamma^{-1}.$$

Moreover, by definition of $\mathbf{S}$, we have $\Gamma^3 \Pi \leqslant M^{-\gamma}$. Using the definition of Π (2.18), we therefore get

$$\Gamma\sqrt{\frac{\operatorname{Im} m + \Lambda}{M\eta}} \leqslant \Gamma\Pi + \Gamma\sqrt{\Gamma^{-1}\Pi} \leqslant CM^{-\gamma/2}\Gamma^{-1}.$$

Plugging this into (2.68) yields $\Lambda \prec M^{-\gamma/2}\Gamma^{-1}$, which is Claim 1.

Finally, we explain Claim 2. We write (2.53) as

$$(2.69) \qquad v_i = \frac{m\left(\sum_k s_{ik} v_k - \Upsilon_i\right)}{\left(m^{-1} - \sum_k s_{ik} v_k + \Upsilon_i\right)}.$$

In the regime $\eta = 2$, all resolvent entries are bounded by $1/2$, thus $|v_i| \leqslant 1$. Since $|\Upsilon_i| \prec M^{-1/2}$ by (2.57) and $|m^{-1}| \geqslant 2$ we find

$$\left| m^{-1} + \sum_k s_{ik} v_k - \Upsilon_i \right| \geqslant 1 + O_{\prec}(M^{-1/2}).$$

Using $|m| \leqslant 1/2$ we therefore conclude from (2.69) that

$$\Lambda_d \leqslant \frac{\Lambda_d + O_{\prec}(M^{-1/2})}{2 + O_{\prec}(M^{-1/2})} = \frac{\Lambda_d}{2} + O_{\prec}(M^{-1/2}),$$

i.e. $\Lambda_d = O_{\prec}(M^{-1/2})$. Together with the estimate on Λ_o from (2.57), we obtained Claim 2. $\qquad \square$

2.4.3. *Sketch of the proof of Proposition 2.13.* This argument is very similar to the proof of Claim 1 in Section 2.4.2. We can work on the event $\Lambda \leqslant M^{-\gamma/4}$, then the bound (2.56) is available to estimate Λ_o and Υ_i . Next, we estimate Λ_d. We expand the right-hand side of (2.53), we get

$$\psi|v_i| \;\leqslant\; C\psi\left|\sum_k s_{ik}v_k - \Upsilon_i\right| + C\psi\Lambda^2 \,.$$

Using the fluctuation averaging estimate (2.75) explained in the next section, as well as (2.56), we find

$$(2.70) \qquad \Lambda_d \;\prec\; \Gamma\Psi^2 + \sqrt{\frac{\operatorname{Im} m + \Psi}{M\eta}}\,,$$

which, combined with (2.56), yields

$$(2.71) \qquad \Lambda \;\prec\; \Gamma\Psi^2 + \sqrt{\frac{\operatorname{Im} m + \Psi}{M\eta}} \leqslant M^{-\varepsilon}\Psi + \sqrt{\frac{\operatorname{Im} m}{M\eta}} + \frac{M^\varepsilon}{M\eta} = F(\Psi),$$

where in the last step we used the assumption $\Psi \leqslant M^{-\gamma/3}\Gamma^{-1}$ and $\varepsilon \leqslant \gamma/3$. $\qquad\square$

2.4.4. *Fluctuation averaging: proof of Proposition 2.14.* The leading error in the self-consistent equation (2.54) for v_i is Υ_i. Among the three summands of Υ_i, see (2.36), typically Z_i is the largest, thus we typically have

$$(2.72) \qquad\qquad v_i \prec Z_i$$

in the regime where Γ is bounded. The large deviation bounds in Theorem 2.9 show that $Z_i \prec \Lambda_o$ and a simple second moment calculation shows that this bound is essentially optimal. On the other hand, (2.29) shows that the typical size of the off-diagonal resolvent matrix elements is at least of order $(N\eta)^{-1/2}$, thus the estimate

$$Z_i \prec \Lambda_o \prec \frac{1}{\sqrt{N\eta}}$$

is essentially optimal in the standard Wigner case ($M = N$). Together with (2.72) this shows that the natural bound for Λ is $(N\eta)^{-1/2}$, which is also reflected in the bound (2.19).

However, the bound (2.20) for the average, $m_N - m = [v]$, is of order $\Lambda^2 \sim (N\eta)^{-1}$, i.e. it is one order better than the bound for v_i. For the purpose of $[v]$, it is the average $[\Upsilon]$ of the leading errors Υ_i that matters, see (2.55). Since Z_i, the leading term in Υ_i, is a fluctuating quantity with zero expectation, the improvement comes from the fact that fluctuations cancel out in the average. The basic example of this phenomenon is the central limit theorem. In our case, however, Z_i are not independent. In fact, their correlations do not decay, at least in the Wigner case where all indices i play symmetric role. Thus standard results on central limit theorems for weakly correlated random variables do not apply.

Here we formulate a version of the fluctuation averaging mechanism, taken from [31], that is the most useful for this discussion and we comment on the history afterwards.

We shall perform the averaging with respect to a family of weights $T = (t_{ik})$ satisfying

$$(2.73) \qquad\qquad 0 \leqslant t_{ik} \leqslant M^{-1}, \qquad \sum_k t_{ik} = 1.$$

Typical example weights are $t_{ik} = s_{ik}$ and $t_{ik} = N^{-1}$. Note that in both of these cases T commutes with S.

THEOREM 2.16 (Fluctuation averaging). *Fix a spectral domain* $\mathbf{D}$ *and deterministic admissible control parameters* $\Psi, \Psi_o \leqslant M^{-c}$. *Suppose that* $\Lambda \prec \Psi$, $\Lambda_o \prec \Psi_o$ *and the weight* $T = (t_{ik})$ *satisfies* (2.73). *Then we have*

$$(2.74) \qquad \sum_k t_{ik} Q_k \frac{1}{G_{kk}} = O_{\prec}(\Psi_o^2), \qquad \sum_k t_{ik} Q_k G_{kk} = O_{\prec}(\Psi^2).$$

If in addition T *commutes with* S *then*

$$(2.75) \qquad \sum_k t_{ik} v_k = O_{\prec}(\Gamma \Psi^2), \qquad \sum_k t_{ik}(v_k - [v]) = O_{\prec}(\widetilde{\Gamma} \Psi^2).$$

The estimates (2.74) *and* (2.75) *are uniform in the index* i.

The first version of the fluctuation averaging mechanism appeared in [47] for the Wigner case, where $[Z] = N^{-1} \sum_k Z_k$ was bounded by Λ_o^2. Since $Q_k[G_{kk}]^{-1}$ is essentially Z_k, see (2.26), this corresponds to the first bound in (2.74). A different proof (with a better bound on the constants) was given in [48]. A conceptually streamlined version of the original proof was extended to sparse matrices [32] and to sample covariance matrices [74]. Finally, an extensive analysis in [30] treated the fluctuation averaging of general polynomials of resolvent entries and identified the order of cancellations depending on the algebraic structure of the polynomial. Moreover, in [30] an additional cancellation effect was found for the quantity $Q_i|G_{ij}|^2$. This improvement plays a key role in the proof of Theorem 2.10, see (2.42).

All proofs of the fluctuation averaging theorems rely on computing expectations of high moments of the averages and carefully estimating various terms of different combinatorial structure. In [30] we have developed a Feynman diagrammatic representation for bookkeeping the terms, but this is necessary only for the case of general polynomials. For the special cases stated in Theorem 2.16, the proof is relatively simple and it is presented in Appendix B of [31]. Here we will not repeat the proof, we only indicate the main mechanism by estimating the second moment of the first term in (2.74). The actual proof requires estimating all moments and then use Chebysev inequality to translate the moment estimates to probabilistic estimates standing behind the notation $O_{\prec}(\Psi_o^2)$ in (2.74).

Second moment calculation. First we claim that

$$(2.76) \qquad \left| Q_k \frac{1}{G_{kk}} \right| \prec \Psi_o.$$

Indeed, from Schur's complement formula (2.26) we get $|Q_k(G_{kk})^{-1}| \leqslant |h_{kk}| + |Z_k|$. The first term is estimated by $|h_{kk}| \prec M^{-1/2} \leqslant \Psi_o$. The second term is estimated exactly as in (2.61) and (2.62), giving $|Z_k| \prec \Psi_o$. In fact, the same bound (2.76) holds if G_{kk} is replaced with $G_{kk}^{(\mathbb{T})}$ as long as $|\mathbb{T}|$ is bounded.

Abbreviate $X_k := Q_k(G_{kk})^{-1}$ and compute the variance

$$(2.77) \quad \mathbb{E}\left| \sum_k t_{ik} X_k \right|^2 = \sum_{k,l} t_{ik} t_{il} \mathbb{E} X_k \overline{X_l} = \sum_k t_{ik}^2 \mathbb{E} X_k \overline{X_k} + \sum_{k \neq l} t_{ik} t_{il} \mathbb{E} X_k \overline{X_l}.$$

Using the bounds (2.73) on t_{ik} and (2.76), we find that the first term on the right-hand side of (2.77) is $O_{\prec}(M^{-1}\Psi_o^2) = O_{\prec}(\Psi_o^4)$, where we used that Ψ_o is admissible.

Let us therefore focus on the second term of (2.77). Using the fact that $k \neq l$, we apply (2.27) to X_k and X_l to get

$$(2.78) \qquad \mathbb{E}X_k\overline{X_l} = \mathbb{E}Q_k\left(\frac{1}{G_{kk}}\right)Q_l\overline{\left(\frac{1}{G_{ll}}\right)}$$

$$= \mathbb{E}Q_k\left(\frac{1}{G_{kk}^{(l)}} - \frac{G_{kl}G_{lk}}{G_{kk}G_{kk}^{(l)}G_{ll}}\right)Q_l\overline{\left(\frac{1}{G_{ll}^{(k)}} - \frac{G_{lk}G_{kl}}{G_{ll}G_{ll}^{(k)}G_{kk}}\right)}.$$

We multiply out the parentheses on the right-hand side. The crucial observation is that if the random variable Y is independent of i (see Definition 2.7) then $\mathbb{E}Q_i(X)Y = \mathbb{E}Q_i(XY) = 0$. Hence out of the four terms obtained from the right-hand side of (2.78), the only nonvanishing one is

$$\mathbb{E}Q_k\left(\frac{G_{kl}G_{lk}}{G_{kk}G_{kk}^{(l)}G_{ll}}\right)Q_l\overline{\left(\frac{G_{lk}G_{kl}}{G_{ll}G_{ll}^{(k)}G_{kk}}\right)} \prec \Psi_o^4,$$

where we used that the denominators are harmless, see (2.59). Together with (2.73), this concludes the proof of $\mathbb{E}\left|\sum_k t_{ik}X_k\right|^2 \prec \Psi_o^4$, which means that $\sum_k t_{ik}X_k$ is bounded by Ψ_o^2 in second moment sense. $\qquad\square$

Finally, Proposition 2.14 directly follows from the first estimate in (2.74) with $t_{ik} = 1//N$, since from (2.26) we have

$$Q_k\frac{1}{G_{kk}} = h_{kk} - Z_k = \Upsilon_k - A_k = \Upsilon_k + O_\prec(\Psi_o^2).$$

Taking the average over k, we get $[\Upsilon] = O_\prec(\Psi_o^2)$. $\qquad\square$

2.5. Remark on the proof of the local semicircle law with using the spectral gap. In Section 2.4 we proved the local semicircle law, Theorem 2.3, uniformly for $\eta \geqslant \eta_E$ instead of the larger regime $\eta \geqslant \tilde{\eta}_E$. The difference between these two thresholds stems from the difference between Γ and $\tilde{\Gamma}$, see (2.17) and (2.48).

The bound Γ on the norm of $(1 - m^2 S)^{-1}$ entered the proof when the self-consistent equation (2.66) was solved. The key idea is to solve the self-consistent equation (2.66) separately on the subspace of constants (the span of the vector $\mathbf{e}$) and on its orthogonal complement $\mathbf{e}^\perp$. Roughly speaking, we obtain

$$(2.79) \qquad |v_i - [v]| \prec \tilde{\Gamma}\left(\Lambda^2 + r(\Lambda)\right), \qquad r(\Lambda) := \sqrt{\frac{\operatorname{Im} m + \Lambda}{M\eta}},$$

instead of (2.67). In fact, we can improve this by using the fluctuation averaging. Subtracting the average over i from the self-consistent equation (2.54) and estimating $|m|^2 \leqslant C$, we have

$$(2.80) \quad |v_i - [v]| \leqslant C\left|\sum_k s_{ik}(v_k - [v]) - (\Upsilon_i - [\Upsilon])\right| + O_\prec(\Lambda^2) \prec \tilde{\Gamma}\Lambda^2 + r(\Lambda),$$

where in the last step we used the fluctuation averaging estimate (2.75) with $s_{ik} = t_{ik}$, and $|\Upsilon_i| \prec r(\Lambda)$ from (2.56).

On the space of constant vectors, (2.53) becomes a scalar equation for the average $[v]$, which can be expanded up to second order. More precisely, assuming

$|v_i| \ll 1$, we can expand (2.53) up to second order:

$$(2.81) \qquad -\sum_k s_{ik} v_k + \Upsilon_i \; = \; -\frac{1}{m^2} v_i + \frac{1}{m^3} v_i^2 + O(\Lambda^3)\,.$$

In order to get a closed equation for $[v]$, we take the average over i:

$$(2.82) \qquad (1 - m^2)[v] - m^{-1} \frac{1}{N} \sum_i v_i^2 \; = \; -m^2 [\Upsilon] + O(\Lambda^3)\,.$$

The nonlinear term is estimated by

$$\frac{1}{N} \sum_i v_i^2 \; = \; [v]^2 + \frac{1}{N} \sum_i (v_i - [v])^2 \; = \; [v]^2 + O_\prec\Big(\big(\widetilde{\Gamma}\Lambda^2 + r(\Lambda)\big)^2 \Big),$$

where (2.80) was used. The average $[\Upsilon]$ can be estimated by $O_\prec(\Lambda_o^2)$ as in Proposition 2.14. Moreover, Λ_o is estimated by $r(\Lambda)$ as in Lemma 2.15. We thus obtain a quadratic equation for the scalar quantity $[v]$:

$$(2.83) \qquad (1 - m^2)[v] - m^{-1}[v]^2 = O_\prec\Big(\Lambda^3 + \big(\widetilde{\Gamma}\Lambda^2 + r(\Lambda)\big)^2 \Big).$$

The main control parameter in this proof is $\Theta = \|[v]\|$, and the key iterative scheme is formulated in terms of Θ. However, many intermediate estimates still involve Λ. In particular, the self-consistent equation (2.53) is effective only in the regime where v_i is already small and in the calculation above we tacitly used that $|v_i| \ll 1$. Hence we need a preparatory step to prove an apriori bound on Λ, essentially showing that $\Lambda \ll 1$, in fact we will need $\Lambda \ll \widetilde{\Gamma}^{-1}$ (compare with Proposition 2.12). This proof itself is a continuity argument similar to the proof of Proposition 2.12; now, however, we have to follow Λ and Θ in tandem. The main reason why Θ is already involved in this part is that we work in larger spectral domain $\eta \geqslant \widetilde{\eta}_E$, defined by using $\widetilde{\Gamma}$. Thus, already in this preparatory step, the self-consistent equation has to be solved separately on the subspace of constants and its orthogonal complement. We will omit here these details.

The second preparatory step is to control Λ in terms of Θ, which allows us to express the error terms in the self-consistent equation (2.83) in terms of $\Theta = \|[v]\|$. We first notice that (2.80) implies

$$|v_i| \prec \Theta + \widetilde{\Gamma}\Lambda^2 + r(\Lambda),$$

i.e.

$$\Lambda \prec \Theta + \widetilde{\Gamma}\Lambda^2 + r(\Lambda).$$

Using the apriori bound $\Lambda \ll \widetilde{\Gamma}^{-1}$, we get

$$\Lambda \prec \Theta + r(\Lambda).$$

This equation is analogous to (2.51) and can be iterated as in the application of in Proposition 2.13 leading to (2.52) to obtain

$$(2.84) \qquad \Lambda \prec \Theta + \sqrt{\frac{\operatorname{Im} m}{M\eta}} + \frac{1}{M\eta}\,.$$

Plugging this bound into (2.83), we have a self-consistent equation for the scalar quantity $[v]$ since $\Theta = \|[v]\|$. An elementary calculation, using the apriori bound

$\Lambda \ll \widetilde{\Gamma}^{-1}$, yields
(2.85)

$$(1 - m^2)[v] - m^{-1}[v]^2 = O_\prec\left(p(\Theta)^2 + M^{-\gamma/4}\Theta^2\right), \qquad p(\Theta) := \sqrt{\frac{\operatorname{Im} m + \Theta}{M\eta} + \frac{1}{M\eta}}.$$

Finally, in the main step we solve the quadratic inequality (2.85) for Θ. If we neglect the error term in (2.85), then the equation reduces to

$$(2.86) \qquad\qquad (1 - m^2)[v] = m^3[v]^2,$$

which has two solutions: either $[v] = 0$ or $[v] = (1 - m^2)/m^3$. Away from the spectral edge we have $|1 - m^2| \geqslant c$ with some positive constant c, so the two solutions are separated and they both are stable under small perturbations. The second solution would mean that $[v]$ is strictly separated away from zero. But this can be excluded by a continuity argument: for large η, say $\eta = 2$, it is easy to prove that $[v]$ is small. Since $[v]$ is a continuous function of η, we find that as η decreases continuously, $[v]$ cannot suddenly jump from a value near zero to a value near $(1 - m^2)/m^3$. Thus $[v]$ must remain in the vicinity of the zero solution to (2.86). This completes the sketch of the proof of Theorem 2.3 for the more general case $\eta \geqslant \widetilde{\eta}_E$. $\qquad\square$

3. Universality of the correlation functions for Wigner matrices

In this section we explain the sketch of the proof of Theorem 1.2.

3.1. Dyson Brownian motion and the local relaxation flow.

3.1.1. *Concept and results.* The Dyson Brownian motion (DBM) describes the evolution of the eigenvalues of a Wigner matrix as an interacting point process if each matrix element h_{ij} evolves according to independent (up to symmetry restriction) Brownian motions. We will slightly alter this definition by generating the dynamics of the matrix elements by an Ornstein-Uhlenbeck (OU) process which leaves the standard Gaussian distribution invariant. In the Hermitian case, the OU process for the rescaled matrix elements $\zeta_{ij} := N^{1/2} h_{ij}$ is given by the stochastic differential equation

$$(3.1) \qquad\qquad \mathrm{d}\zeta_{ij} = \mathrm{d}\beta_{ij} - \frac{1}{2}\zeta_{ij}\mathrm{d}t, \qquad i, j = 1, 2, \ldots N,$$

where β_{ij}, $i < j$, are independent complex Brownian motions with variance one and β_{ii} are real Brownian motions of the same variance. Denote the distribution of the eigenvalues $\boldsymbol{\lambda} = (\lambda_1, \lambda_2, \ldots, \lambda_N)$ of H_t at time t by $f_t(\boldsymbol{\lambda})\mu(\mathrm{d}\boldsymbol{\lambda})$ where μ is given by (1.10) with the Gaussian potential $V(x) = x^2/2$.

Then $f_t = f_{t,N}$ satisfies [25]

$$(3.2) \qquad\qquad \partial_t f_t = \mathscr{L} f_t,$$

where

$$(3.3) \quad \mathscr{L} = \mathscr{L}_N := \sum_{i=1}^{N} \frac{1}{2N}\partial_i^2 + \sum_{i=1}^{N}\left(-\frac{\beta}{4}\lambda_i + \frac{\beta}{2N}\sum_{j\neq i}\frac{1}{\lambda_i - \lambda_j}\right)\partial_i, \qquad \partial_i = \frac{\partial}{\partial\lambda_i}.$$

The parameter β is chosen as follows: $\beta = 2$ for complex Hermitian matrices and $\beta = 1$ for symmetric real matrices. Our formulation of the problem has already taken into account Dyson's observation that the invariant measure for this dynamics is μ. A natural question regarding the DBM is how fast the dynamics reaches equilibrium. Dyson had already posed this question in 1962:

Dyson's conjecture [25]: The global equilibrium of DBM is reached in time of order one and the local equilibrium (in the bulk) is reached in time of order $1/N$. Dyson further remarked,

> The picture of the gas coming into equilibrium in two well-separated stages, with microscopic and macroscopic time scales, is suggested with the help of physical intuition. A rigorous proof that this picture is accurate would require a much deeper mathematical analysis.

We will prove that Dyson's conjecture is correct if the initial data of the flow is a Wigner ensemble, which was Dyson's original interest. Our result in fact is valid for DBM with much more general initial data that we now survey. Briefly, it will turn out that the *global* equilibrium is indeed reached within a time of order one, but *local* equilibrium is achieved much faster if an a-priori estimate on the location of the eigenvalues (also called points) is satisfied. To formulate this estimate, let $\gamma_j = \gamma_{j,N}$ denote the location of the j-th point under the semicircle law, i.e., γ_j is defined by (2.21).

A-priori Estimate: There exists an $\mathfrak{a} > 0$ such that

$$(3.4) \qquad Q = Q_\mathfrak{a} := \sup_{t \geqslant N^{-2\mathfrak{a}}} \frac{1}{N} \int \sum_{j=1}^{N} (\lambda_j - \gamma_j)^2 f_t(\lambda)\mu(\mathrm{d}\lambda) \leqslant C N^{-1-2\mathfrak{a}}$$

with a constant C uniformly in N. This condition first appeared in [41].

The main result on the local ergodicity of Dyson Brownian motion states that if the a-priori estimate (3.4) is satisfied then the local correlation functions of the measure $f_t\mu$ are the same as the corresponding ones for the Gaussian measure, $\mu = f_\infty\mu$, provided that t is larger than $N^{-2\mathfrak{a}}$. The n-point correlation functions of the probability measure $f_t\mathrm{d}\mu$ are defined, similarly to (1.4), by
(3.5)

$$p_{t,N}^{(n)}(x_1, x_2, \ldots, x_n) = \int_{\mathbb{R}^{N-n}} f_t(\mathbf{x})\mu(\mathbf{x})\mathrm{d}x_{n+1} \ldots \mathrm{d}x_N, \qquad \mathbf{x} = (x_1, x_2, \ldots, x_N).$$

Due to the convention that one can view the locations of eigenvalues as the coordinates of particles, we have used $\mathbf{x}$, instead of $\boldsymbol{\lambda}$, in the last equation. From now on, we will use both conventions depending on which viewpoint we wish to emphasize. Notice that the probability distribution of the eigenvalues at the time t, $f_t\mu$, is the same as that of the Gaussian divisible matrix:

$$(3.6) \qquad\qquad H_t = e^{-t/2}H_0 + (1 - e^{-t})^{1/2}U,$$

where H_0 is the initial generalized Wigner matrix and U is an independent standard GUE (or GOE) matrix. This establishes the universality of the Gaussian divisible ensembles. The precise statement is the following theorem:

THEOREM 3.1. *[**42**, Theorem 2.1] Suppose that the a-priori estimate (3.4) holds for the solution f_t of the forward equation (3.2) with some exponent $\mathfrak{a} > 0$. Let $E \in (-2,2)$ and $b > 0$ such that $[E - b, E + b] \subset (-2,2)$. Then for any $s > 0$, for any integer $n \geqslant 1$ and for any compactly supported continuous test function*

$O : \mathbb{R}^n \to \mathbb{R}$, *we have*

$$
\lim_{\substack{N \to \infty \\ t \geqslant N^{-2\mathfrak{a}+s}}} \sup \int_{E-b}^{E+b} \frac{\mathrm{d}E'}{2b} \int_{\mathbb{R}^n} \mathrm{d}\alpha_1 \ldots \mathrm{d}\alpha_n \, O(\alpha_1, \ldots, \alpha_n)
$$

$$
(3.7) \qquad\qquad\qquad \times \frac{1}{\varrho_{sc}(E)^n} \left(p^{(n)}_{t,N} - p^{(n)}_{G,N} \right)
$$

$$
\times \left(E' + \frac{\alpha_1}{N \varrho_{sc}(E)}, \ldots, E' + \frac{\alpha_n}{N \varrho_{sc}(E)} \right) = 0.
$$

We can choose $b = b_N$ depending on N. In [**42**] explicit bounds on the speed of convergence and the optimal range of b were also established. In particular, thanks to the optimal rigidity estimate (2.22) which implies that (3.4) holds with any $\mathfrak{a} < 1/2$, the range of the energy averaging in (3.7) can be reduced to $b_N \geqslant N^{-1+\xi}$, $\xi > 0$, but only for $t \geqslant N^{-\xi/8}$ (Theorem 2.3 of [**48**]).

Theorem 3.1 is a consequence of the following theorem which identifies the averaged gap distribution of the eigenvalues.

THEOREM 3.2 (Universality of the Dyson Brownian motion for short time). *[**42**, Theorem 4.1]*
Suppose $\beta \geqslant 1$ and let $O : \mathbb{R} \to \mathbb{R}$ be a smooth function with compact support. Then for any sufficiently small $\varepsilon > 0$, independent of N, there exist constants $C, c > 0$, depending only on ε and O such that for any $J \subset \{1, 2, \ldots, N - 1\}$ we have

$$
(3.8) \qquad \left| \int \frac{1}{|J|} \sum_{i \in J} O(N(x_i - x_{i+1})) f_t \mathrm{d}\mu - \int \frac{1}{|J|} \sum_{i \in J} O(N(x_i - x_{i+1})) \mathrm{d}\mu \right|
$$

$$
\leqslant CN^\varepsilon \sqrt{\frac{N^2 Q}{|J| t}} + Ce^{-cN^\varepsilon}.
$$

In particular, if the a-priori estimate (3.4) holds with some $\mathfrak{a} > 0$ and $|J|$ is of order N, then for any $t > N^{-2\mathfrak{a}+3\varepsilon}$ the right hand side converges to zero as $N \to \infty$, i.e. the gap distributions for $f_t \mathrm{d}\mu$ and $\mathrm{d}\mu$ coincide.

The test functions can be generalized to

$$
(3.9) \qquad O\Big(N(x_i - x_{i+1}), N(x_{i+1} - x_{i+2}), \ldots, N(x_{i+n-1} - x_{i+n}) \Big)
$$

for any n fixed which is needed to identify higher order correlation functions. In applications, J is chosen to be the indices of the eigenvalues in the interval $[E-b, E+b]$ and thus $|J| \sim Nb$. This identifies the averaged gap distributions of eigenvalues and thus also identifies the correlation functions after energy averaging. We will not explain here in detail how to pass information from gap distribution to correlation functions (see Section 7 in [**42**]), but we note that this transfer is relatively easy if both statistics are averaged on a scale larger than the typical fluctuation of a single eigenvalue (which is smaller than $N^{-1+\varepsilon}$ in the bulk by (2.22)). This concludes Theorem 3.1. Note that the input of this theorem, the apriori estimate (3.4), identifies the location of the eigenvalues only on a scale $N^{-1/2-\mathfrak{a}}$ which is much weaker than the $1/N$ precision for the eigenvalue differences in (3.8).

By the rigidity estimate (2.22), the a-priori estimate (3.4) holds for any $\mathfrak{a} < 1/2$ if the initial data of the DBM is a generalized Wigner ensemble. Therefore, Theorem 3.2 holds for any $t \geqslant N^{-1+\varepsilon}$ for any $\varepsilon > 0$ and *this establishes Dyson's conjecture for any generalized Wigner matrices.*

3.1.2. *Main ideas behind the proof of Theorem 3.2.* The key method is to analyze the relaxation to equilibrium of the dynamics (3.2). This approach was first introduced in Section 5.1 of [**41**]; the presentation here follows [**42**].

We start with a short review of the logarithmic Sobolev inequality for a general measure. Let the probability measure μ on $\mathbb{R}^N$ be given by a general Hamiltonian $\mathcal{H}$:

$$(3.10) \qquad d\mu(\mathbf{x}) = \frac{e^{-N\mathcal{H}(\mathbf{x})}}{Z} d\mathbf{x},$$

In applications μ will be the Gaussian equilibrium measure, (1.10) with $V(x) = x^2/2$, so we use the same notation μ, but the statements in the beginning of this section hold for a general measure. Let $\mathscr{L}$ be the generator, symmetric with respect to the measure $d\mu$, defined by the associated Dirichlet form

$$(3.11) \qquad D(f) = D_\mu(f) = -\int f\mathscr{L}f\,d\mu := \frac{1}{2N}\sum_j \int (\partial_j f)^2 d\mu, \qquad \partial_j = \partial_{x_j}.$$

Recall the relative entropy of two probability measures:

$$S(\nu|\mu) := \int \frac{d\nu}{d\mu} \log\left(\frac{d\nu}{d\mu}\right) d\mu.$$

If $d\nu = f d\mu$, then we will sometimes use the notation $S_\mu(f) := S(f\mu|\mu)$. The entropy can be used to control the total variation norm via the well known inequality

$$(3.12) \qquad \int |f - 1| d\mu \leqslant \sqrt{2S_\mu(f)}.$$

Let f_t be the solution to the evolution equation

$$(3.13) \qquad \partial_t f_t = \mathscr{L}f_t, \qquad t > 0,$$

with a given initial condition f_0. The evolution of the entropy $S_\mu(f_t) = S(f_t\mu|\mu)$ satisfies

$$(3.14) \qquad \partial_t S_\mu(f_t) = -4D_\mu(\sqrt{f_t}).$$

Following Bakry and Émery [**6**], the evolution of the Dirichlet form satisfies the inequality

$$(3.15) \qquad \partial_t D_\mu(\sqrt{f_t}) \leqslant -\frac{1}{2N}\int (\nabla\sqrt{f_t})(\nabla^2\mathcal{H})\nabla\sqrt{f_t}\,d\mu.$$

If the Hamiltonian is convex, i.e.,

$$(3.16) \qquad \nabla^2\mathcal{H}(\mathbf{x}) = \operatorname{Hess}\mathcal{H}(\mathbf{x}) \geqslant \varpi \qquad \text{for all } \mathbf{x} \in \mathbb{R}^N$$

with some constant $\varpi > 0$, then we have

$$(3.17) \qquad \partial_t D_\mu(\sqrt{f_t}) \leqslant -\varpi D_\mu(\sqrt{f_t}).$$

Integrating (3.14) and (3.17) back from infinity to 0, we obtain the *logarithmic Sobolev inequality (LSI)*

$$(3.18) \qquad S_\mu(f) \leqslant \frac{4}{\varpi} D_\mu(\sqrt{f}), \quad f = f_0$$

and the *exponential relaxation of the entropy and Dirichlet form on time scale* $t \sim 1/\varpi$

$$(3.19) \qquad S_\mu(f_t) \leqslant e^{-t\varpi} S_\mu(f_0), \quad D_\mu(\sqrt{f_t}) \leqslant e^{-t\varpi} D_\mu(\sqrt{f_0}).$$

As a consequence of the logarithmic Sobolev inequality, we also have the concentration inequality for any k and $a > 0$

$$(3.20) \qquad \mathbb{P}^\mu \left(|x_k - \mathbb{E}_\mu(x_k)| > a \right) \leqslant 2 e^{-\varpi N a^2/2}.$$

We will not use this inequality in this section, but it will become important in Section 4.1.

Returning to the classical ensembles, we assume from now on that $\mathcal{H}$ is given by (1.10) with $V(x) = x^2/2$ and the equilibrium measure μ is the Gaussian one. We then have the convexity inequality

$$(3.21) \qquad \left\langle \mathbf{v}, \nabla^2 \mathcal{H}(\mathbf{x})\mathbf{v} \right\rangle \geqslant \frac{1}{2} \|\mathbf{v}\|^2 + \frac{1}{N} \sum_{i<j} \frac{(v_i - v_j)^2}{(x_i - x_j)^2} \geqslant \frac{1}{2} \|\mathbf{v}\|^2, \qquad \mathbf{v} \in \mathbb{R}^N.$$

This guarantees that μ satisfies the LSI with $\varpi = 1/2$ and the relaxation time to equilibrium is of order one.

The key idea is that the relaxation time is in fact much shorter than order one for local observables that depend only on the eigenvalue *differences*. Equation (3.21) shows that the relaxation in the direction $v_i - v_j$ is much faster than order one provided that $x_i - x_j$ are close. However, this effect is hard to exploit directly due to that all modes of different wavelengths are coupled. Our idea is to add an auxiliary strongly convex potential $W(\mathbf{x})$ to the Hamiltonian to "speed up" the convergence to local equilibrium. On the other hand, we will also show that the cost of this speeding up can be effectively controlled if the a-priori estimate (3.4) holds.

The auxiliary potential $W(\mathbf{x})$ is defined by

$$(3.22) \qquad W(\mathbf{x}) := \sum_{j=1}^{N} W_j(x_j), \qquad W_j(x) := \frac{1}{2\tau}(x_j - \gamma_j)^2,$$

i.e. it is a quadratic confinement on scale $\sqrt{\tau}$ for each eigenvalue near its classical location, where the parameter $\tau > 0$ will be chosen later. The total Hamiltonian is given by

$$(3.23) \qquad \widetilde{\mathcal{H}} := \mathcal{H} + W,$$

where $\mathcal{H}$ is the Gaussian Hamiltonian given by (1.10). The measure with Hamiltonian $\widetilde{\mathcal{H}}$,

$$(3.24) \qquad \mathrm{d}\omega := \omega(\mathbf{x})\mathrm{d}\mathbf{x}, \qquad \omega := e^{-N\widetilde{\mathcal{H}}}/\widetilde{Z},$$

will be called the *local relaxation measure*.

The *local relaxation flow* is defined to be the flow with the generator characterized by the natural Dirichlet form w.r.t. ω, explicitly, $\widetilde{\mathscr{L}}$:

$$(3.25) \qquad \widetilde{\mathscr{L}} = \mathscr{L} - \sum_{j} b_j \partial_j, \qquad b_j = W_j'(x_j) = \frac{x_j - \gamma_j}{\tau}.$$

We will choose $\tau \ll 1$ so that the additional term W substantially increases the lower bound (3.16) on the Hessian, hence speeding up the dynamics so that the relaxation time is at most τ.

The idea of adding an artificial potential W to speed up the convergence appears to be unnatural here. The current formulation is a streamlined version of a much more complicated approach that appeared in [41] and which took ideas from the earlier work [37]. Roughly speaking, in hydrodynamical limit, the short wavelength modes always have shorter relaxation times than the long wavelength modes. A direct implementation of this idea is extremely complicated due to the logarithmic interaction that couples short and long wavelength modes. Adding a strongly convex auxiliary potential $W(\mathbf{x})$ shortens the relaxation time of the long wavelength modes, but it does not affect the short modes, i.e. the local statistics, which are our main interest. The analysis of the new system is much simpler since now the relaxation is faster, uniform for all modes. Finally, we need to compare the local statistics of the original system with those of the modified one. It turns out that the difference is governed by $(\nabla W)^2$ which can be directly controlled by the a-priori estimate (3.4).

Our method for enhancing the convexity of $\mathcal{H}$ is reminiscent of a standard convexification idea concerning metastable states. To explain the similarity, consider a particle near one of the local minima of a double well potential separated by a local maximum, or energy barrier. Although the potential is not convex globally, one may still study a reference problem defined by convexifying the potential along with the well in which the particle initially resides. Before the particle reaches the energy barrier, there is no difference between these two problems. Thus questions concerning time scales shorter than the typical escape time can be conveniently answered by considering the convexified problem; in particular the escape time in the metastability problem itself can be estimated by using convex analysis. Our DBM problem is already convex, but not sufficiently convex. The modification by adding W enhances convexity without altering the local statistics. This is similar to the convexification in the metastability problem which does not alter events before the escape time.

3.1.3. *Some details on the proof of Theorem 3.2.* The core of the proof is divided into three theorems. For the flow with generator $\widetilde{\mathscr{L}}$, we have the following estimates on the entropy and Dirichlet form.

THEOREM 3.3. *Consider the forward equation*

$$(3.26) \qquad \partial_t q_t = \widetilde{\mathscr{L}} q_t, \qquad t \geqslant 0,$$

with the reversible measure ω defined in (3.24). The initial condition q_0 satisfies $\int q_0 d\omega = 1$. Then we have the following estimates

$$(3.27) \qquad \partial_t D_\omega(\sqrt{q_t}) \leqslant -\frac{1}{2\tau} D_\omega(\sqrt{q_t}) - \frac{1}{2N^2} \int \sum_{i,j=1}^{N} \frac{(\partial_i \sqrt{q_t} - \partial_j \sqrt{q_t})^2}{(x_i - x_j)^2} d\omega,$$

$$(3.28) \qquad \frac{1}{2N^2} \int_0^\infty ds \int \sum_{i,j=1}^{N} \frac{(\partial_i \sqrt{q_s} - \partial_j \sqrt{q_s})^2}{(x_i - x_j)^2} d\omega \leqslant D_\omega(\sqrt{q_0})$$

and the logarithmic Sobolev inequality

$$(3.29) \qquad S_\omega(q) \leqslant C\tau D_\omega(\sqrt{q_0})$$

with a universal constant C. Thus the relaxation time to equilibrium is of order τ:

$$(3.30) \qquad S_\omega(q_t) \leqslant e^{-Ct/\tau} S_\omega(q_0), \qquad D_\omega(q_t) \leqslant e^{-Ct/\tau} D_\omega(q_0).$$

PROOF. Denote by $h = \sqrt{q}$ and we have the equation

$$(3.31) \qquad \partial_t D_\omega(h_t) = \partial_t \frac{1}{2N} \int (\nabla h)^2 e^{-N\widetilde{\mathcal{H}}} d\mathbf{x} \leqslant -\frac{1}{2N} \int \nabla h (\nabla^2 \widetilde{\mathcal{H}}) \nabla h e^{-N\widetilde{\mathcal{H}}} d\mathbf{x}.$$

In our case, (3.21) and (3.22) imply that the Hessian of $\widetilde{\mathcal{H}}$ is bounded from below as

$$(3.32) \qquad \nabla h(\nabla^2 \widetilde{\mathcal{H}})\nabla h \geqslant \frac{C}{\tau}\sum_j (\partial_j h)^2 + \frac{1}{2N}\sum_{i,j}\frac{1}{(x_i - x_j)^2}(\partial_i h - \partial_j h)^2$$

with some positive constant C. This proves (3.27) and (3.28). The rest can be proved by straightforward arguments analogously to (3.14)–(3.19). $\qquad\square$

Notice that the estimate (3.28) is an additional information that we extracted from the Bakry-Émery argument by using the second term in the Hessian estimate (3.21). It plays a key role in the next theorem.

THEOREM 3.4 (Dirichlet form inequality). *Let q be a probability density $\int q\mathrm{d}\omega = 1$ and let $O : \mathbb{R} \to \mathbb{R}$ be a smooth function with compact support. Then for any $J \subset \{1, 2, \ldots, N-1\}$ and any $t > 0$ we have*

$$(3.33) \qquad \left|\int \frac{1}{|J|}\sum_{i\in J}O(N(x_i - x_{i+1}))q\mathrm{d}\omega - \int \frac{1}{|J|}\sum_{i\in J}O(N(x_i - x_{i+1}))\mathrm{d}\omega\right|$$

$$\leqslant C\left(t\frac{D_\omega(\sqrt{q})}{|J|}\right)^{1/2} + C\sqrt{S_\omega(q)}e^{-ct/\tau}.$$

Proof. For simplicity, we assume that $J = \{1, 2, \ldots, N-1\}$. Let q_t satisfy

$$\partial_t q_t = \widetilde{\mathscr{L}}q_t, \qquad t \geqslant 0,$$

with an initial condition $q_0 = q$. We write

$$(3.34)$$
$$\int \left[\frac{1}{|J|}\sum_{i\in J}O(N(x_i - x_{i+1}))\right](q-1)\mathrm{d}\omega$$

$$= \int \left[\frac{1}{|J|}\sum_{i\in J}O(N(x_i - x_{i+1}))\right](q - q_t)\mathrm{d}\omega + \int \left[\frac{1}{|J|}\sum_{i\in J}O(N(x_i - x_{i+1}))\right](q_t - 1)\mathrm{d}\omega.$$

The second term in (3.34) can be estimated by (3.12), the decay of the entropy (3.30) and the boundedness of O; this gives the second term in (3.33).

To estimate the first term in (3.34), by the evolution equation $\partial_t q_t = \widetilde{\mathscr{L}}q_t$ and the definition of $\widetilde{\mathscr{L}}$ we have

$$\int \frac{1}{|J|}\sum_{i\in J}O(N(x_i - x_{i+1}))q_t\mathrm{d}\omega - \int \frac{1}{|J|}\sum_{i\in J}O(N(x_i - x_{i+1}))q_0\mathrm{d}\omega$$

$$= \int_0^t \mathrm{d}s\int \frac{1}{|J|}\sum_{i\in J}O'(N(x_i - x_{i+1}))[\partial_i q_s - \partial_{i+1}q_s]\mathrm{d}\omega.$$

From the Schwarz inequality and $\partial q = 2\sqrt{q}\partial\sqrt{q}$, the last term is bounded by

$$(3.35) \qquad 2\left[\int_0^t \mathrm{d}s \int_{\mathbb{R}^N} \frac{N^2}{|J|^2} \sum_{i\in J} \Big[O'(N(x_i - x_{i+1}))\Big]^2 (x_i - x_{i+1})^2 \, q_s \mathrm{d}\omega\right]^{1/2}$$

$$\times \left[\int_0^t \mathrm{d}s \int_{\mathbb{R}^N} \frac{1}{N^2} \sum_i \frac{1}{(x_i - x_{i+1})^2} [\partial_i \sqrt{q_s} - \partial_{i+1}\sqrt{q_s}]^2 \mathrm{d}\omega\right]^{1/2}$$

$$\leqslant C\left(\frac{D_\omega(\sqrt{q_0})t}{|J|}\right)^{1/2},$$

where we have used (3.28) and that $\Big[O'(N(x_i - x_{i+1}))\Big]^2 (x_i - x_{i+1})^2 \leqslant CN^{-2}$ due to O being smooth and compactly supported. $\qquad\square$

Alternatively, we could have directly estimated the left hand side of (3.33) by using the total variation norm between $q\omega$ and ω, which in turn could be estimated by the entropy (3.12) and the Dirichlet form using the logarithmic Sobolev inequality, i.e., by

$$(3.36) \qquad C\int |q - 1|\mathrm{d}\omega \leqslant C\sqrt{S_\omega(q)} \leqslant C\sqrt{\tau D_\omega(\sqrt{q})}.$$

However, compared with this simple bound, the estimate (3.33) gains an extra factor $|J| \sim N$ in the denominator, i.e. it is in terms of Dirichlet form *per particle*. The improvement is due to the fact that the observable in (3.33) depends only on the gap, i.e. *difference* of points. This allows us to exploit the additional term (3.28) gained in the Bakry-Émery argument. This is a manifestation of the general observation that gap-observables behave much better than point-observables.

The final ingredient in proving Theorem 3.2 is the following entropy and Dirichlet form estimates.

THEOREM 3.5. *Suppose that* (3.21) *holds. Let* $\mathfrak{a} > 0$ *be fixed and recall the definition of* $Q = Q_{\mathfrak{a}}$ *from* (3.4). *Fix a constant* $\tau \geqslant N^{-2\mathfrak{a}}$ *and consider the local relaxation measure* ω *with this* τ. *Set* $\psi := \omega/\mu$ *and let* $g_t := f_t/\psi$. *Suppose there is a constant* m *such that*

$$(3.37) \qquad S(f_\tau\omega|\omega) \leqslant CN^m.$$

Then for any $t \geqslant \tau N^\varepsilon$ *the entropy and the Dirichlet form satisfy the estimates:*

$$(3.38) \qquad S(g_t\omega|\omega) \leqslant CN^2 Q\tau^{-1}, \qquad D_\omega(\sqrt{g_t}) \leqslant CN^2 Q\tau^{-2}$$

where the constants depend on ε *and* m.

Proof. The evolution of the entropy $S(f_t\mu|\omega) = S_\omega(g_t)$ can be computed explicitly by the formula [88]

$$\partial_t S(f_t\mu|\omega) = -\frac{2}{N}\sum_j \int (\partial_j\sqrt{g_t})^2\,\psi\,\mathrm{d}\mu + \int g_t \mathscr{L}\psi\,\mathrm{d}\mu.$$

Hence we have, by using (3.25),

$$\partial_t S(f_t\mu|\omega) = -\frac{2}{N}\sum_j \int (\partial_j\sqrt{g_t})^2\,\mathrm{d}\omega + \int \widetilde{\mathscr{L}}g_t\,\mathrm{d}\omega + \sum_j \int b_j\partial_j g_t\,\mathrm{d}\omega.$$

Since ω is $\widetilde{\mathscr{L}}$-invariant and time independent, the middle term on the right hand side vanishes, and from the Schwarz inequality

$$(3.39) \quad \partial_t S(f_t \mu | \omega) \leqslant -D_\omega(\sqrt{g_t}) + CN \sum_j \int b_j^2 g_t \, d\omega \leqslant -D_\omega(\sqrt{g_t}) + CN^2 Q \tau^{-2}.$$

Notice that (3.39) is reminiscent to (3.14) for the derivative of the entropy of the measure $g_t \omega = f_t \mu$ with respect to ω. The difference is, however, that g_t does not satisfy the evolution equation with the generator $\widetilde{\mathscr{L}}$. The last term in (3.39) expresses the error.

Together with the logarithmic Sobolev inequality (3.29), we have

$$(3.40) \quad \partial_t S(f_t \mu | \omega) \leqslant -D_\omega(\sqrt{g_t}) + CN^2 Q \tau^{-2} \leqslant -C\tau^{-1} S(f_t \mu | \omega) + CN^2 Q \tau^{-2}.$$

Integrating the last inequality from τ to t and using the assumption (3.37) and $t \geqslant \tau N^\varepsilon$, we have proved the first inequality of (3.38). Using this result and integrating (3.39), we have

$$\int_\tau^t D_\omega(\sqrt{g_s}) \mathrm{d}s \leqslant CN^2 Q \tau^{-1}.$$

By the convexity of the Hamiltonian, $D_\mu(\sqrt{f_t})$ is decreasing in t. Since $D_\omega(\sqrt{g_s}) \leqslant CD_\mu(\sqrt{f_s}) + CN^2 Q \tau^{-2}$, this proves the second inequality of (3.38). $\quad\square$

Finally, we complete the proof of Theorem 3.2. For any given $t > 0$ we now choose $\tau := tN^{-\varepsilon}$ and we construct the local relaxation measure ω with this τ. Set $\psi = \omega/\mu$ and let $q := g_t = f_t/\psi$ be the density q in Theorem 3.4. Then Theorem 3.5, Theorem 3.4 and an easy bound on the entropy $S_\omega(q) \leqslant CN^m$ imply that
(3.41)

$$\left| \int \frac{1}{N} \sum_{i \in J} O(N(x_i - x_{i+1}))(f_t \mathrm{d}\mu - \mathrm{d}\omega) \right| \leqslant C\left(t \frac{D_\omega(\sqrt{q})}{|J|}\right)^{1/2} + C\sqrt{S_\omega(q)} e^{-cN^\varepsilon}.$$

$$\leqslant C\left(t \frac{N^2 Q}{|J|\tau^2}\right)^{1/2} + Ce^{-cN^\varepsilon} \leqslant CN^\varepsilon \sqrt{\frac{N^2 Q}{|J|t}} + Ce^{-cN^\varepsilon},$$

i.e., the local statistics of $f_t \mu$ and ω are the same for any initial data f_τ for which (3.37) is satisfied. Applying the same argument to the Gaussian initial data, $f_0 = f_\tau = 1$, we can also compare μ and ω. We have thus proved (3.8) and hence the universality. $\quad\square$

3.2. The Green function comparison theorems and four moment matching. We now state the Green function comparison theorem, Theorem 3.6. It will quickly lead to Theorem 3.7 stating that the correlation functions of eigenvalues of two matrix ensembles are identical on a scale smaller than $1/N$ provided that the first four moments of all matrix elements of these two ensembles are almost the same. We will state a limited version for real Wigner matrices for simplicity of presentation.

THEOREM 3.6 (Green function comparison). *[46, Theorem 2.3] Suppose that we have two $N \times N$ Wigner matrices, $H^{(v)}$ and $H^{(w)}$, with matrix elements h_{ij} given by the random variables $N^{-1/2} v_{ij}$ and $N^{-1/2} w_{ij}$, respectively, with v_{ij} and w_{ij} satisfying the uniform subexponential decay condition (1.17). We assume that the first four moments of v_{ij} and w_{ij} are close to each other in the sense that*

$$(3.42) \qquad \left| \mathbb{E} v_{ij}^s - \mathbb{E} w_{ij}^s \right| \leqslant N^{-\delta-2+s/2}, \qquad 1 \leqslant s \leqslant 4,$$

holds for some $\delta > 0$. Then there are positive constants C_1 and ε, depending on ϑ and C_0 from (1.17) such that for any η with $N^{-1-\varepsilon} \leqslant \eta \leqslant N^{-1}$ and for any z_1, z_2 with $\operatorname{Im} z_j = \pm\eta$, $j = 1, 2$, we have

$$(3.43) \qquad \lim_{N \to \infty} \left[\mathbb{E} \operatorname{Tr} G^{(v)}(z_1) \operatorname{Tr} G^{(v)}(z_2) - \mathbb{E} \operatorname{Tr} G^{(w)}(z_1) \operatorname{Tr} G^{(w)}(z_2) \right] = 0,$$

where $G^{(v)}$ and $G^{(w)}$ denotes the Green functions of $H^{(v)}$ and $H^{(w)}$.

Here we formulated Theorem 3.6 for a product of two traces of the Green function, but the result holds for a large class of smooth functions depending on several individual matrix elements of the Green functions as well, see [46] for the precise statement. (The matching condition (3.42) is slightly weaker than in [46], but the proof in [46] without any change yields this slightly stronger version.) This general version of Theorem 3.6 implies the correlation functions of the two ensembles at the scale $1/N$ are identical:

THEOREM 3.7 (Correlation function comparison). *[46, Theorem 6.4] Suppose the assumptions of Theorem 3.6 hold. Let $p_{v,N}^{(n)}$ and $p_{w,N}^{(n)}$ be the $n-$point functions of the eigenvalues w.r.t. the probability law of the matrix $H^{(v)}$ and $H^{(w)}$, respectively. Then for any $|E| < 2$, any $n \geqslant 1$ and any compactly supported continuous test function $O : \mathbb{R}^n \to \mathbb{R}$ we have*
$$(3.44)$$
$$\lim_{N \to \infty} \int_{\mathbb{R}^n} \mathrm{d}\alpha_1 \ldots \mathrm{d}\alpha_n \, O(\alpha_1, \ldots, \alpha_n) \left(p_{v,N}^{(n)} - p_{w,N}^{(n)} \right) \left(E + \frac{\alpha_1}{N}, \ldots, E + \frac{\alpha_n}{N} \right) = 0.$$

Notice that these comparison theorems hold for any fixed energy E, i.e. no averaging in energy is necessary.

The basic idea for proving Theorem 3.6 is similar to Lindeberg's proof of the central limit theorem, where the random variables are replaced one by one with a Gaussian one. We will replace the matrix elements v_{ij} with w_{ij} one by one and estimate the effect of this change on the resolvent by a resolvent expansion. The idea of applying Lindeberg's method in random matrices was recently used by Chatterjee [16] for comparing the traces of the Green functions; the idea was also used by Tao and Vu [82] in the context of comparing individual eigenvalue distributions.

The four moment matching condition (3.42) with $\delta = 0$ first appeared in [82]. For comparison with Theorem 3.6, we state here the main result of [82]. Let $\lambda_1 < \lambda_2 < \ldots < \lambda_N$ and $\lambda_1' < \lambda_2' < \ldots < \lambda_N'$ denote the eigenvalues of H and H', respectively. Then the joint distribution of any k-tuple of eigenvalues on scale $1/N$ is very close to each other in the following sense:

THEOREM 3.8 (Four moment theorem for eigenvalues). *[82, Theorem 15] Let H and H' be two Wigner matrices. Assume that the first four moments of h_{ij} and h_{ij}' exactly match and the subexponential decay (1.17) holds for the single entry distributions. Then for any sufficiently small positive ε and ε' and for any function $F : \mathbb{R}^k \to \mathbb{R}$ satisfying $|\nabla^j F| \leqslant N^\varepsilon$, $j \leqslant 5$, and for any selection of k-tuple of indices $i_1, i_2, \ldots, i_k \in [\varepsilon N, (1 - \varepsilon)N]$ away from the edge, we have*

$$(3.45) \qquad \left| \mathbb{E} F\left(N\lambda_{i_1}, N\lambda_{i_2}, \ldots, N\lambda_{i_k} \right) - \mathbb{E}' F\left(N\lambda_{i_1}', N\lambda_{i_2}', \ldots, N\lambda_{i_k}' \right) \right| \leqslant N^{-c_0}$$

with some $c_0 > 0$. The exact moment matching condition can be relaxed to (3.42), but c_0 will depend on δ.

Note that the arguments in (3.45) are magnified by a factor N so the result is sufficiently precise to detect individual eigenvalue correlations. Therefore Theorem 3.6 or 3.8 can prove bulk universality for a Wigner matrix H if another H' is found, with matching four moments, for which universality is already proved. This will be explained in Section 3.3.

Both Theorem 3.6 and Theorem 3.8 rely on some version of the local semicircle law on the shortest possible scale. There are, however, three main differences between them.

(i) Theorem 3.6 compares the statistics of eigenvalues of two different ensembles near *fixed energies* while Theorem 3.8 compares the statistics of the $j_1, j_2, \ldots j_k$-th eigenvalues for *fixed labels* $j_1, j_2, \ldots j_k$.

(ii) Both theorems are of perturbative nature that require some apriori information. Theorem 3.6 uses a bound on the resolvent matrix entries, $|G_{ij}(z)|$, that has already been obtained in the local semicircle law (see, e.g. (2.19)). Theorem 3.8 needs an apriori lower bound on the gaps to exclude possible eigenvalue resonances that may render the expansion unstable. This is achieved by a level repulsion estimate that is the most complicated technical part of [82]. Previously, even more precise level repulsion estimates were obtained in [40] but only for smooth distributions.

(iii) Theorem 3.6 also compares off-diagonal Green function elements, an information that cannot be obtained from Theorem 3.8. Hence it directly provides information on the eigenvectors as well, see [59] for the development. In fact, once Theorem 3.6 is proved for all energies, it also implies Theorem 3.8. The reason is that we can integrate correlation functions in energy with a precision smaller than the typical size of the gap, hence eigenvalues with a *fixed label* can be identified. This was first done near the edge in [33] and later in the bulk in [59].

Sketch of the proof of Theorem 3.6. We fix a bijective ordering map on the index set of the independent matrix elements,

$$\phi : \{(i,j) : 1 \leqslant i \leqslant j \leqslant N\} \to \Big\{1, \ldots, \gamma(N)\Big\}, \qquad \gamma(N) := \frac{N(N+1)}{2},$$

and denote by H_γ the Wigner matrix whose matrix elements h_{ij} follow the v-distribution if $\phi(i,j) \leqslant \gamma$ and they follow the w-distribution otherwise; in particular $H^{(v)} = H_0$ and $H^{(w)} = H_{\gamma(N)}$.

Consider the telescopic sum of differences of expectations (we present only one resolvent for simplicity of the presentation):

$$(3.46)$$

$$\mathbb{E}\left(\frac{1}{N}\operatorname{Tr}\frac{1}{H^{(w)} - z}\right) - \mathbb{E}\left(\frac{1}{N}\operatorname{Tr}\frac{1}{H^{(v)} - z}\right)$$

$$= \sum_{\gamma=1}^{\gamma(N)}\left[\mathbb{E}\left(\frac{1}{N}\operatorname{Tr}\frac{1}{H_\gamma - z}\right) - \mathbb{E}\left(\frac{1}{N}\operatorname{Tr}\frac{1}{H_{\gamma-1} - z}\right)\right].$$

Let $E^{(ij)}$ denote the matrix whose matrix elements are zero everywhere except at the (i,j) position, where it is 1, i.e., $E^{(ij)}_{k\ell} = \delta_{ik}\delta_{j\ell}$. Fix a $\gamma \geqslant 1$ and let (i,j) be determined by $\phi(i,j) = \gamma$. We will compare $H_{\gamma-1}$ with H_γ. Note that these two matrices differ only in the (i,j) and (j,i) matrix elements and they can be written as

$$H_{\gamma-1} = Q + \frac{1}{\sqrt{N}} V, \qquad V := v_{ij} E^{(ij)} + v_{ji} E^{(ji)}, \qquad v_{ji} := \overline{v}_{ij},$$

$$H_{\gamma} = Q + \frac{1}{\sqrt{N}} W, \qquad W := w_{ij} E^{(ij)} + w_{ji} E^{(ji)}, \qquad w_{ji} := \overline{w}_{ij},$$

with a matrix Q that has zero matrix element at the (i,j) and (j,i) positions.

By the resolvent expansion,

$$S_{\gamma-1} = R - N^{-1/2} RVR + \cdots + N^{-2}(RV)^4 R - N^{-5/2}(RV)^5 S,$$

$$R := \frac{1}{Q - z}, \quad S_{\gamma-1} := \frac{1}{H_{\gamma-1} - z},$$

and a similar expression holds for the resolvent S_γ of by H_γ. From the local semicircle law for individual matrix elements (2.19), the matrix elements of all Green functions $R, S_{\gamma-1}, S_\gamma$ are bounded by $CN^{2\varepsilon}$. Although (2.19) is not directly applicable to $\eta \geqslant N^{-1+\varepsilon}$, it is easy to show that

$$|G_{ij}(E + i\eta)| \leqslant \max_i |G_{ii}(E + i\eta)| \leqslant \frac{\eta'}{\eta} \max_i |G_{ii}(E + i\eta')|$$

so choosing $\eta' \sim N^{-1+\varepsilon}$ we can prove a bound for η slightly below $1/N$ at the expense of a factor η/η'. The estimates of the related resolvents $R, S_{\gamma-1}, S_\gamma$ are similar.

By assumption (3.42), the difference between the expectation of matrix elements of $S_{\gamma-1}$ and S_γ is of order $N^{-2-\delta+C\varepsilon}$. Since the number of steps, $\gamma(N)$ is of order N^2, the difference in (3.46) is of order $N^2 N^{-2-\delta+C\varepsilon} \ll 1$, and this proves Theorem 3.6 for a single resolvent. It is very simple to turn this heuristic argument into a rigorous proof and to generalize it to the product of several resolvents. The real difficulty is the input that the resolvent entries can be bounded for a general class of Wigner matrices down to the almost optimal scale $\eta \sim 1/N$.

3.3. Universality for generalized Wigner matrices: putting it together. In this short section we put the previous information together to prove Theorem 1.2. We first focus on the case when b_N is independent of N. Recall that Theorem 3.1 states that the correlation functions of the Gaussian divisible ensemble,

$$(3.47) \qquad\qquad H_t = e^{-t/2} H_0 + (1 - e^{-t})^{1/2} U,$$

where H_0 is the initial Wigner matrix and U is an independent standard GUE (or GOE) matrix, are given by the corresponding GUE (or GOE) for $t \geqslant N^{-2\mathfrak{a}+\varepsilon}$ provided that the a-priori estimate (3.4) holds for the solution f_t of the forward equation (3.2) with some exponent $\mathfrak{a} > 0$. Since the rigidity of eigenvalues (2.22) holds uniformly for all generalized Wigner matrices, we have proved (3.4) for $\mathfrak{a} = 1/2 - \varepsilon$ with any $\varepsilon > 0$.

From the evolution of the OU process (3.1) for $v_{ij} = N^{1/2} h_{ij}$ we have

$$(3.48) \qquad\qquad \left| \mathbb{E} v_{ij}^s(t) - \mathbb{E} v_{ij}^s(0) \right| \leqslant Ct = CN^{-1+3\varepsilon}$$

for $s = 3, 4$ and with the choice of $t = N^{-1+3\varepsilon}$. Furthermore, $\mathbb{E} h_{ij}^s(t)$ are independent of t for $s = 1, 2$ due to $\mathbb{E} v_{ij}(0) = 0$ and $\mathbb{E} v_{ij}^2(t) = 1$. Hence (3.42) is satisfied for the matrix elements of H_t and H_0 and we can thus use Theorem 3.7 to conclude that the correlation functions of H_t and H_0 are identical at the scale $1/N$. Since the correlation functions of H_t are given by the corresponding Gaussian case, we have proved Theorem 1.2 under the condition that the probability distribution of the matrix elements decay subexponentially. Finally, we need a technical cutoff argument to relax the decay condition to (1.13) which we omit here (see Section 7 in [**33**]).

The argument for N-dependent $b = b_N$ in the range $b_N \geqslant N^{-1+\xi}$, $\xi > 0$, is slightly different. For such a small b_N, (3.7) could be established only for relatively large times, $t \geqslant N^{-\xi/8}$. We cannot therefore compare H_0 with H_t directly, since the deviation of the third moments of $v_{ij}(0)$ and $v_{ij}(t)$ in (3.48) would not satisfy (3.42). Instead, we construct an auxiliary Wigner matrix $\widehat{H}_0$ such that up to the third moment *its* time evolution $\widehat{H}_t$ under the OU flow (3.47) matches exactly the *original* matrix H_0 and the fourth moments are close even for t of order $N^{-\xi/8}$ (see Lemma 3.4 of [**47**]). Theorem 3.1 will then be applied for $\widehat{H}_t$, and Theorem 3.6 can be used to compare $\widehat{H}_t$ and H_0. This completes the sketch of the proof of Theorem 1.2.

4. Universality of the correlation functions for β-ensembles

In this section we outline the proof of Theorem 1.4. We will use the notation $\mu = \mu_{\beta,V}^{(N)}$ for the probability measure defining the general β-ensemble with a potential V on N ordered real points $\lambda_1 \leqslant \ldots \leqslant \lambda_N$, see (1.10). We let $\mathbb{P}^\mu$ and $\mathbb{E}^\mu$ denote the probability and the expectation with respect to μ. The equilibrium density is denoted by $\varrho = \varrho_V$ and its Stieltjes transform by

$$m(z) = m_V(z) = \int_{\mathbb{R}} \frac{\varrho_V(x)}{x - z} \mathrm{d}x.$$

The classical location of the k-th point will be denoted by $\gamma_k = \gamma_{k,V}$, see (1.22). Note that in this section μ, ϱ, m and γ_k refer to the quantities related to the general V and not the Gaussian one as in the previous sections. This avoids carrying the V subscripts all the time as in Section 2 we dropped the subscripts sc referring to the semicircle law.

4.1. Rigidity estimates. For simplicity of presentation we assume that the potential V is convex, i.e.,

$$(4.1) \qquad \varpi := \frac{1}{2} \inf_{x \in \mathbb{R}} V''(x) > 0,$$

the equilibrium density $\varrho(s)$ is supported on a single interval $[A, B] \subset \mathbb{R}$ and satisfies (1.20) (for the general case, see [**12**]). The Gaussian case corresponds to $V(x) = x^2/2$, in which case the equilibrium density is the semicircle law, ϱ_{sc}, given by (1.3). Our main result concerning the universality is Theorem 1.4 and similar statement holds for the universality of the averaged gap distributions directly. In fact, the proof of Theorem 1.4 goes via the averaged gap distribution as we now explain.

The first step to prove Theorem 1.4 is the following theorem which provides a rigidity estimate on the location of each individual point. The precision in the bulk is almost down to the optimal scale $1/N$, the estimate is weaker near the edges. In the following, we will denote $[\![x, y]\!] = \mathbb{N} \cap [x, y]$.

THEOREM 4.1. *[**11**, Theorem 3.1 and Lemma 3.6] Fix any $\alpha, \varepsilon > 0$ and assume that (4.1) holds. Then there are constants $\delta, c_1, c_2 > 0$ such that for any $N \geqslant 1$ and*

$k \in [\![\alpha N, (1 - \alpha)N]\!]$,

$$(4.2) \qquad \mathbb{P}^{\mu} \left(|\lambda_k - \gamma_k| > N^{-1+\varepsilon} \right) \leqslant c_1 e^{-c_2 N^{\delta}}.$$

The following weaker bound is valid close to the spectral edges:

$$(4.3) \qquad \mathbb{P}^{\mu} \left(|\lambda_k - \gamma_k| \geqslant N^{-4/15+\varepsilon} \right) \leqslant c_1 e^{-c_2 N^{\delta}}.$$

for any $k \in [\![N^{3/5+\varepsilon}, N - N^{3/5+\varepsilon}]\!]$. *Finally, the bound*

$$(4.4) \qquad \mathbb{P}^{\mu} \left(|\lambda_k - \gamma_k| \geqslant \varepsilon \right) \leqslant c_1 e^{-c_2 N^{\delta}}.$$

holds for any $k \in [\![1, N]\!]$.

We explain some ideas of the proof of (4.2), the arguments for (4.3) are similar and (4.4) follows from an easy large deviation bound, see [11]. The first ingredient to prove (4.2) is an analysis of the loop equation following Johansson [58] and Shcherbina [79]. The equilibrium density ϱ, for a convex potential V, is given by

$$(4.5) \qquad \varrho(t) = \frac{1}{\pi} r(t) \sqrt{(t - A)(B - t)} \, \mathbb{1}_{[A,B]}(t),$$

where r is a real function that can be extended to an analytic function in $\mathbb{C}$ and r has no zero in $\mathbb{R}$. Denote by $s(z) := -2r(z)\sqrt{(A - z)(B - z)}$ where the square root is defined such that its asymptotic value is z as $z \to \infty$. Recall that the density is the one-point correlation function which is characterized by

$$(4.6) \qquad \int_{\mathbb{R}} d\lambda_1 O(\lambda_1) p_N^{(1)}(\lambda_1) = \int_{\mathbb{R}^N} O(\lambda_1) d\mu_{\beta,V}^{(N)}(\lambda), \qquad \lambda = (\lambda_1, \lambda_2, \ldots, \lambda_N).$$

Let $\bar{m}_N$ and m be the Stieltjes transforms of the density $p_N^{(1)}$ and the equilibrium density ϱ, respectively. Notice that in Section 2 we have used m_N to denote the Stieltjes transform of the empirical measure (2.11); here $\bar{m}_N$ denotes the ensemble average of the analogous quantity.

Define the analytic functions

$$b_N(z) := \int_{\mathbb{R}} \frac{V'(z) - V'(t)}{z - t} (p_1^{(N)} - \varrho)(t) \mathbf{1}(t)$$

and

$$c_N(z) := \frac{1}{N^2} k_N(z) + \frac{1}{N} \left(\frac{2}{\beta} - 1 \right) \bar{m}_N'(z), \quad \text{with} \quad k_N(z) := \mathrm{Var}_{\mu} \left(\sum_{k=1}^{N} \frac{1}{z - \lambda_k} \right).$$

Here for complex random variables X we use the definition that $\mathrm{Var}(X) = \mathbb{E}(X^2) - \mathbb{E}(X)^2$.

The equation used by Johansson (which can be obtained by a change of variables in (4.6) [58] or by integration by parts [79]), is a variation of the loop equation (see, e.g., [49]) used in the physics literature and it takes the form

$$(4.7) \qquad (\bar{m}_N - m)^2 + s(\bar{m}_N - m) + b_N = c_N.$$

Equation (4.7) can be used to express the difference $\bar{m}_N - m$ in terms of $(\bar{m}_N - m)^2, b_N$ and c_N. In the regime where $|\bar{m}_N - m|$ is small, we can neglect the quadratic term. The term b_N is of the same order as $|\bar{m}_N - m|$ and is difficult to treat. As observed in [**5, 79**], for analytic V, this term vanishes when we perform a contour integration. So we have roughly the relation

$$(4.8) \qquad (\bar{m}_N - m) \sim \frac{1}{N^2} \operatorname{Var}_\mu \left(\sum_{k=1}^N \frac{1}{z - \lambda_k} \right),$$

where we dropped the less important error involving $\bar{m}'_N(z)/N$ due to the extra $1/N$ factor. In the convex setting, the variance can be estimated by the logarithmic Sobolev inequality and we immediately obtain an estimate on $\bar{m}_N - m$. We then use the Helffer-Sjöstrand formula, see (2.24), to estimate the locations of the particles. This will provide us with an accuracy of order $N^{-1/2}$ for $\mathbb{E}^\mu \lambda_k - \gamma_k$. This argument gives only an estimate on the expectation of the locations of the particles since we only have information on the averaged quantity, $\bar{m}_N$. Although it is tempting to use this new accuracy information on the particles to estimate the variance again in (4.8), the information on the expectation on λ_k alone is very difficult to use in a bootstrap argument. To estimate the variance of a non-trivial function of λ_k we need high probability estimates on λ_k.

The key idea in this section is the observation that the accuracy information on the λ's can be used to improve the local convexity of the measure μ in the direction involving the *differences* of λ's. To explain this idea, we compute the Hessian of the Hamiltonian of μ:

$$\left\langle \mathbf{v}, \nabla^2 \mathcal{H}(\lambda) \mathbf{v} \right\rangle \geqslant \varpi \|\mathbf{v}\|^2 + \frac{1}{N} \sum_{i<j} \frac{(v_i - v_j)^2}{(\lambda_i - \lambda_j)^2}.$$

The naive lower bound on $\nabla^2 \mathcal{H}$ is ϖ, but for a typical $\boldsymbol{\lambda} = (\lambda_1, \lambda_2, \ldots, \lambda_N)$ it is in fact much better in most directions. To see this effect, suppose we know $|\lambda_i - \lambda_j| \lesssim M/N$ with some M for any $i, j \in I_k^M$, where $I_k^M := [\![k - M, k + M]\!]$. Then for $\mathbf{v} = (v_{k-M}, \ldots, v_{k+M})$ with $\sum_j v_j = 0$ we have

$$(4.9) \qquad \left\langle \mathbf{v}, \nabla^2 \mathcal{H}(\boldsymbol{\lambda}) \mathbf{v} \right\rangle \geqslant \frac{N}{M^2} \sum_{i,j \in I_k^M} (v_i - v_j)^2 \geqslant C \frac{N}{M} \sum_j v_j^2.$$

This improves the convexity of the Hessian to N/M on the hyperplane $\sum_j v_j = 0$. Let

$$\lambda_k^{[M]} := \frac{1}{|I_k^M|} \sum_{j \in I_k^M} \lambda_j = \frac{1}{2M+1} \sum_{j \in I_k^M} \lambda_j$$

denote the block average of the locations of particles and rewrite

$$\lambda_k - \lambda_k^{[N^{1-\varepsilon}]} = \sum_j \left(\lambda_k^{[M_j]} - \lambda_k^{[M_{j+1}]} \right)$$

as a telescopic sum with an appropriate sequence of $M_1 = 0$, $M_2, \ldots$ with the property that $M_j/M_{j-1} \leqslant N^\varepsilon$. We can now use the improved concentration on the hyperplane $\sum_j v_j = 0$ to the variables $\lambda_k^{[M_j]} - \lambda_k^{[M_{j+1}]}$ to control the fluctuation of $\lambda_k - \lambda_k^{[N^{1-\varepsilon}]}$. Since the fluctuation of $\lambda_k^{[N^{1-\varepsilon}]}$ is very small for small ε, we finally arrive at the estimate

$$(4.10) \qquad \mathbb{P}^\mu \left(|\lambda_k - \mathbb{E}^\mu(\lambda_k)| > a \right) \leqslant C e^{-CN^2 a^2/M}.$$

From (4.10) we thus have that $|\lambda_k - \mathbb{E}^\mu \lambda_k| \lesssim \sqrt{M}/N$ with high probability. This improves the starting accuracy $|\lambda_i - \lambda_j| \lesssim M/N$ for $i, j \in I_k^M$ to $|\lambda_i - \lambda_j| \lesssim M'/N$

with some $M' \ll M$, provided that we can prove that $|\mathbb{E}^\mu(\lambda_i - \lambda_j)| \ll M'/N$. But the last inequality involves only expectations and it will follow from the analysis of the loop equation (4.7) we just mentioned above. Starting from $M = N$, this procedure can be repeated by decreasing M step by step until we get the optimal accuracy, $M \sim O(1)$. The implementation of this argument in [11] is somewhat different from this sketch due to various technical issues, but it follows the same basic idea.

4.2. The local equilibrium measure. Having completed the first step, the rigidity estimate, we now focus on the second step, i.e. on the uniqueness of the local Gibbs measure. Let $0 < \kappa < 1/2$. Choose $q \in [\kappa, 1 - \kappa]$ and set $L = [Nq]$ (the integer part). Fix an integer $K = N^k$ with $k < 1$. We will study the local spacing statistics of K consecutive particles

$$\{\lambda_j \, : \, j \in I\}, \qquad I = I_L := [\![L+1, L+K]\!].$$

These particles are typically located near E_q determined by the relation

$$\int_{-\infty}^{E_q} \varrho(t)\mathrm{d}t = q.$$

Note that $|\gamma_L - E_q| \leqslant C/N$.

We will distinguish the inside and outside particles by renaming them as

$$(4.11) \quad (\lambda_1, \lambda_2, \ldots, \lambda_N) := (y_1, \ldots y_L, x_{L+1}, \ldots, x_{L+K}, y_{L+K+1}, \ldots y_N) \in \Xi^{(N)},$$

but note that they keep their original indices. The notation $\Xi^{(N)}$ refers to the simplex $\{\mathbf{z} \, : \, z_1 < z_2 < \ldots < z_N\}$ in $\mathbb{R}^N$. In short we will write

$$\mathbf{x} = (x_{L+1}, \ldots, x_{L+K}), \qquad \text{and} \qquad \mathbf{y} = (y_1, \ldots, y_L, y_{L+K+1}, \ldots, y_N),$$

all in increasing order, i.e. $\mathbf{x} \in \Xi^{(K)}$ and $\mathbf{y} \in \Xi^{(N-K)}$. We will refer to the y's as *external points* and to the x's as *internal points*.

We will fix the external points (also called as boundary conditions) and study conditional measures on the internal points. We define the *local equilibrium measure* on $\mathbf{x}$ with fixed boundary condition $\mathbf{y}$ by

$$(4.12) \qquad \mu_{\mathbf{y}}(\mathrm{d}\mathbf{x}) = \mu_{\mathbf{y}}(\mathbf{x})\mathrm{d}\mathbf{x}, \qquad \mu_{\mathbf{y}}(\mathbf{x}) := \mu(\mathbf{y}, \mathbf{x}) \left[\int \mu(\mathbf{y}, \mathbf{x})\mathrm{d}\mathbf{x} \right]^{-1}.$$

Note that for any fixed $\mathbf{y} \in \Xi^{(N-K)}$, the measure $\mu_{\mathbf{y}}$ is supported on configurations of K points $\mathbf{x} = \{x_j\}_{j \in I}$ located in the interval $[y_L, y_{L+K+1}]$.

The Hamiltonian $\mathcal{H}_{\mathbf{y}}$ of the measure $\mu_{\mathbf{y}}(\mathrm{d}\mathbf{x}) \sim \exp(-\beta N \mathcal{H}_{\mathbf{y}}(\mathbf{x}))\mathrm{d}\mathbf{x}$ is given by

$$(4.13)$$
$$\mathcal{H}_{\mathbf{y}}(\mathbf{x}) := \sum_{i \in I} \frac{1}{2} V_{\mathbf{y}}(x_i) - \frac{1}{N} \sum_{\substack{i,j \in I \\ i<j}} \log|x_j - x_i| \quad \text{with} \quad V_{\mathbf{y}}(x) := V(x) - \frac{2}{N} \sum_{j \notin I} \log|x - y_j|.$$

We now define the set of *good boundary configurations* with a parameter $\delta = \delta(N) > 0$

$$(4.14)$$
$$\mathcal{G}_\delta = \mathcal{G} := \left\{ \mathbf{y} \in \Xi^{(N-K)} : |y_j - \gamma_j| \leqslant \delta, \, \forall j \in [\![N\kappa/2, L]\!] \cup [\![L + K + 1, N(1 - \kappa/2)]\!] \right\},$$

where κ is a small constant to cutoff points near the spectral edges. Some rather weak additional conditions for $\mathbf{y}$ near the spectral edges will also be needed. They can be built in the definition of $\mathcal{G}$ based upon the bounds (4.3) and (4.4) but we will neglect this issue here.

Let σ and μ be two measures of the form (1.10) with potentials W and V and densities $\varrho = \varrho_W$ and ϱ_V, respectively. For our purpose $W(x) = x^2/2$, i.e., σ is the Gaussian β-ensemble and its density $\varrho_W(t) = \frac{1}{2\pi}(4 - t^2)_+^{1/2}$ is the Wigner semicircle law. Let the sequence γ_j be the classical locations for μ and the sequence θ_j be the classical locations for σ. Similarly to the construction of the measure $\mu_{\mathbf{y}}$, for any positive integer $L' \in [\![1, N - K]\!]$ we can construct the measure $\sigma_{\boldsymbol{\theta}}$ conditioned that the particles outside are given by the classical locations θ_j for $j \notin [\![L', L'+K]\!]$. More precisely, we define a *reference local Gaussian measure* $\sigma_{\boldsymbol{\theta}} \sim \exp(-\beta N \mathcal{H}_{\boldsymbol{\theta}}(\mathbf{x}))\mathrm{d}\mathbf{x}$ on the set $[\theta_{L'}, \theta_{L'+K+1}]$ via the Hamiltonian

$$(4.15) \qquad \mathcal{H}_{\boldsymbol{\theta}}(\mathbf{x}) = \sum_{i \in I'} \left[\frac{1}{4}x_i^2 - \frac{1}{N}\sum_{j \notin I'} \log |x_i - \theta_j| \right] - \frac{1}{N} \sum_{\substack{i,j \in I' \\ i < j}} \log |x_j - x_i|,$$

where $I' := [\![L' + 1, L' + K]\!]$. Since L' will not play an active role, we will abuse the notation and set $L' = L$.

The measure $\mu_{\mathbf{y}}$ lives on the interval $[y_L, y_{L+K+1}]$ while the measure $\sigma_{\boldsymbol{\theta}}$ lives on the interval $[\theta_L, \theta_{L+K+1}]$ and it is difficult to compare them. But after an appropriate translation and dilation, they will live on the same interval and from now on we assume that $[y_L, y_{L+K+1}] = [\theta_L, \theta_{L+K+1}]$. The parameter $K = N^k$ has to be sufficiently small since ϱ_V and ϱ_W are not constant functions and we have to match these two densities quite precisely in the whole interval. There are some other subtle issues related to the rescaling, but we will neglect them here to concentrate on the main ideas. Our main result is the following theorem which is essentially a combination of Proposition 4.2 and Theorem 4.4 from [11].

THEOREM 4.2. *Let* $0 < \varphi \leqslant \frac{1}{38}$. *Fix* $K = N^k$, $\delta = N^{-1+\varphi}$ *and* $k = \frac{39}{2}\varphi$. *Then for* $\mathbf{y} \in \mathcal{G}_\delta$ *we have*

$$(4.16) \qquad \left| \mathbb{E}^{\mu_{\mathbf{y}}} \frac{1}{K} \sum_{i \in I} O\big(N(x_i - x_{i+1})\big) - \mathbb{E}^{\sigma_{\boldsymbol{\theta}}} \frac{1}{K} \sum_{i \in I} O\big(N(x_i - x_{i+1})\big) \right| \to 0$$

as $N \to \infty$ *for any smooth and compactly supported test function* O. *A similar formula holds for more complicated observables of the form* (3.9).

From the rigidity estimate, Theorem 4.1, it follows that $\mathcal{G}_\delta$ has an overwhelming probability, so the expectation $\mathbb{E}^{\mu_{\mathbf{y}}}$ can be changed to $\mathbb{E}^{\mu}$ and similarly for the reference measure. Once (4.16) is proven for all observables of the form (3.9), we get that the locally averaged gap statistics for μ coincide with those of the reference Gaussian case, hence they are universal. Since averaged gap statistics identifies locally averaged correlation functions, we obtain Theorem 1.4.

It remains to prove Theorem 4.2. The basic idea is to use the Dirichlet form inequality (3.33). Although (3.33) was stated for an infinite volume measure, it holds for any measure with repulsive logarithmic interactions in a finite volume and with the parameter τ^{-1} being the lower bound on the Hessian of the Hamiltonian. In our setting, we denote by τ_σ^{-1} the lower bound for $\nabla^2 \mathcal{H}_{\boldsymbol{\theta}}$, and the Dirichlet form

inequality becomes
(4.17)
$$\left| [\mathbb{E}^{\mu_{\mathbf{y}}} - \mathbb{E}^{\sigma_{\theta}}] \frac{1}{K} \sum_{i \in I} O\big(N(x_i - x_{i+1})\big) \right| \leqslant C \left(\frac{\tau_\sigma N^\varepsilon}{K} D(\mu_{\mathbf{y}} | \sigma_{\theta}) \right)^{1/2} + C e^{-cN^\varepsilon} \sqrt{S(\mu_{\mathbf{y}} | \sigma_{\theta})},$$

where
(4.18)
$$D(\mu_{\mathbf{y}} \mid \sigma_{\theta}) := \frac{1}{2N} \int \left| \nabla \sqrt{\frac{\mathrm{d}\mu_{\mathbf{y}}}{\mathrm{d}\sigma_{\theta}}} \right|^2 \mathrm{d}\sigma_{\theta}.$$

Thus our task is to prove that

(4.19)
$$\tau_\sigma N^\varepsilon \frac{D(\mu_{\mathbf{y}} \mid \sigma_{\theta})}{K} \to 0.$$

By definition,
$$\frac{\tau_\sigma}{K} D(\mu_{\mathbf{y}} \mid \sigma_{\theta}) \leqslant \frac{\tau_\sigma N}{K} \int \sum_{L+1 \leqslant j \leqslant L+K} Z_j^2 \mathrm{d}\mu_{\mathbf{y}},$$

where Z_j is defined as

(4.20)
$$Z_j := \frac{\beta}{2} V'(x_j) - \frac{\beta}{N} \sum_{\substack{k < L \\ k > L+K}} \frac{1}{x_j - y_k} - \frac{\beta}{2} W'(x_j) + \frac{\beta}{N} \sum_{\substack{k < L \\ k > L+K}} \frac{1}{x_j - \theta_k}.$$

Using the equilibrium relation (1.20) between the potentials V, W and the densities ϱ_V, ϱ_W, we have

$$Z_j = \beta \int_{\mathbb{R}} \frac{\varrho_V(y)}{x_j - y} \mathrm{d}y - \frac{\beta}{N} \sum_{\substack{k < L \\ k > L+K}} \frac{1}{x_j - y_k} - \beta \int_{\mathbb{R}} \frac{\varrho_W(y)}{x_j - y} \mathrm{d}y + \frac{\beta}{N} \sum_{\substack{k < L \\ k > L+K}} \frac{1}{x_j - \theta_k}.$$

Hence Z_j is the sum of the error terms,

(4.21)
$$A_j := \int_{y \notin [y_L, y_{L+K+1}]} \frac{\varrho_V(y)}{x_j - y} \mathrm{d}y - \frac{1}{N} \sum_{\substack{k < L \\ k > L+K}} \frac{1}{x_j - y_k},$$

(4.22)
$$B_j := \int_{y_L}^{y_{L+K+1}} \frac{\varrho_V(y) - \varrho_W(y)}{x_j - y} \mathrm{d}y,$$

and there is a term similar to A_j with y_j replaced by θ_j and ϱ_V replaced by ϱ_W.

With our convention, the total numbers of particles in the interval $[y_{L+K+1}, y_L]$ are equal and thus
$$\int_{y_L}^{y_{L+K+1}} \varrho_V(y) \mathrm{d}y = \int_{y_L}^{y_{L+K+1}} \varrho_W(y) \mathrm{d}y.$$

Since the densities ρ_V and ρ_W are C^1 functions away from the endpoints A and B and $y_{L+K+1} - y_L$ is small, $|\rho_V - \rho_W|$ is small in the interval $[y_{L+K+1}, y_L]$ and thus B_j is small. For estimating A_j, we can replace the integral

$$\int_{-\infty}^{y_L} \frac{\varrho_V(y)}{x_j - y} \mathrm{d}y \quad \text{by} \quad \frac{1}{N} \sum_{k < L} \frac{1}{x_j - \gamma_k}$$

with negligible errors, at least for j's away from the edges, $j \in [\![L+N^\varepsilon, L+K-N^\varepsilon]\!]$. Thus

(4.23)
$$|A_j| \leqslant \frac{C}{N} \left| \sum_{\substack{k < L \\ k > L+K}} T_j^k \right|, \qquad T_j^k := \frac{1}{x_j - y_k} - \frac{1}{x_j - \gamma_k},$$

and T_j^k can be estimated by the assumption $|y_k - \gamma_k| \leqslant \delta$ from $\mathbf{y} \in \mathcal{G}_\delta$. The same argument works if j is close to the edge, but k is away from the edges, i.e. $k \leqslant L - N^\varepsilon$ or $k \geqslant L + K + N^\varepsilon$. The edge terms, T_j^k for $|j - k| \leqslant N^\varepsilon$, are difficult to estimate due to the singularity in the denominator and the event that many y_k's with $k < L$ may pile up near y_L. To resolve this difficulty, we show that the averaged local statistics of the measure $\mu_{\mathbf{y}}$ are insensitive to the change of the boundary conditions for $\mathbf{y}$ near the edges. This can be achieved by the simple inequality

(4.24)
$$\left| \frac{1}{K} \sum_{i \in I} \int O\big(N(x_i - x_{i+1})\big) [\mathrm{d}\mu_{\mathbf{y}'} - \mathrm{d}\mu_{\mathbf{y}}] \right| \leqslant C \int |\mathrm{d}\mu_{\mathbf{y}'} - \mathrm{d}\mu_{\mathbf{y}}| \leqslant C\sqrt{S(\mu_{\mathbf{y}'}|\mu_{\mathbf{y}})}$$

for any two boundary conditions $\mathbf{y}$ and $\mathbf{y}'$. Although we still have to estimate the entropy that includes a logarithmic singularity, this can be done much more easily since entropy is less sensitive to singularities than Dirichlet form. Therefore, we can replace the boundary condition y_k with $y_k' = \theta_k$ for $|j - k| \leqslant N^\varepsilon$ and then the most singular edge terms in (4.20) cancel out.

We note that we can perform this replacement only for a small number of index pairs (j, k), since estimating the gap distribution by the total entropy, as noted in (3.36) in Section 3.1, is not as efficient as the estimate using the Dirichlet form per particle. Thus we can afford to use this argument only for the edge terms, $|j - k| \leqslant N^\varepsilon$. For all other index pairs (j, k) we still have to estimate T_j^k by exploiting that $\mathbf{y}$ is a good configuration, i.e. $y_k - \gamma_k$ is small.

Unfortunately, even with the optimal accuracy $\delta \sim N^{-1+\varepsilon'}$ in (4.14) as an input, the relation (4.19) still cannot be satisfied for any choice of $N^{c\varepsilon'} \leqslant K \leqslant N^{1-c\varepsilon'}$. To understand this problem, we remark that while the edge terms become a smaller percentage of the total terms in (4.24) as K gets bigger, the relaxation time to equilibrium for σ_θ, determined by the convexity of $\mathcal{H}_\theta''$, increases at the same time. At the end of our calculation, there is no good regime for the choice of K. Fortunately, this can be resolved by using the idea of the local relaxation measure as in (3.22), i.e., we add a quadratic term $\frac{1}{2\tau}(x_j - \gamma_j)^2$ to the Hamiltonian of the measure $\mu_{\mathbf{y}}$ and $\frac{1}{2\tau_\sigma}(x_j - \theta_j)^2$ for the measure σ_θ. With these ideas, we can complete the proof of Theorem 4.2.

5. Single gap universality

In this section we outline the proofs of Theorem 1.3 and 1.5 following closely [**45**]. Both proofs rely on the single gap universality for the locally conditioned measure $\mu_{\mathbf{y}}$ introduced already in (4.12). This will be stated in Theorem 5.1, whose proof takes up most of this section. At the end, in Section 5.4 we complete the proofs of Theorem 1.3 and 1.5.

5.1. Statements on local equilibrium measures.

5.1.1. *Definition.* We work in the bulk spectrum and we consider the local equilibrium measure on $\mathcal{K} := 2K + 1$ points which is the conditional measure after fixing all other points. To define it precisely, we fix two small positive numbers, $\alpha, \delta > 0$ and choose two positive integer parameters L, K such that

(5.1)
$$L \in [\![\alpha N, (1-\alpha)N]\!], \qquad N^\delta \leqslant K \leqslant N^{1/4}.$$

All results will hold for any sufficiently small α, δ and for any sufficiently large $N \geqslant N_0$, where the threshold N_0 depends on α, δ and maybe on other parameters of the model.

Denote $I = I_{L,K} := [\![L - K, L + K]\!]$ the set of $\mathcal{K}$ consecutive indices in the bulk. As in Section 4.2, we will distinguish external and internal points by renaming them as
$$(5.2)$$
$$(\lambda_1, \lambda_2, \ldots, \lambda_N) := (y_1, \ldots y_{L-K-1}, x_{L-K}, \ldots, x_{L+K}, y_{L+K+1}, \ldots y_N) \in \Xi^{(N)},$$

the only difference is that here the internal particles are labelled symmetrically to L. This discrepancy is only notational, but we prefer to follow the notations of the original papers. In short we will write

$$\mathbf{x} = (x_{L-K}, \ldots x_{L+K}) \in \Xi^{(\mathcal{K})}, \quad \text{and} \quad \mathbf{y} = (y_1, \ldots y_{L-K-1}, y_{L+K+1}, \ldots y_N) \in \Xi^{(N-\mathcal{K})}.$$

As in (4.12) we again define the local equilibrium measure on $\mathbf{x}$ with boundary condition $\mathbf{y}$ by

$$(5.3) \qquad \mu_{\mathbf{y}}(\mathrm{d}\mathbf{x}) := \mu_{\mathbf{y}}(\mathbf{x})\mathrm{d}\mathbf{x}, \qquad \mu_{\mathbf{y}}(\mathbf{x}) := \mu(\mathbf{y}, \mathbf{x}) \left[\int \mu(\mathbf{y}, \mathbf{x})\mathrm{d}\mathbf{x} \right]^{-1},$$

where $\mu = \mu(\mathbf{y}, \mathbf{x})$ is the (global) equilibrium measure (1.10). For a fixed $\mathbf{y}$, this measure can also be written as a Gibbs measure,

$$(5.4) \qquad \mu_{\mathbf{y}} = \mu_{\mathbf{y}, \beta, V} = Z_{\mathbf{y}}^{-1} e^{-N\beta \mathcal{H}_{\mathbf{y}}},$$

with Hamiltonian
$$(5.5)$$
$$\mathcal{H}_{\mathbf{y}}(\mathbf{x}) := \sum_{i \in I} \frac{1}{2} V_{\mathbf{y}}(x_i) - \frac{1}{N} \sum_{\substack{i,j \in I \\ i < j}} \log|x_j - x_i|, \qquad V_{\mathbf{y}}(x) := V(x) - \frac{2}{N} \sum_{k \notin I} \log|x - y_k|.$$

Here $V_{\mathbf{y}}(x)$ can be viewed as the external potential of a β-log-gas of the points $\{x_i : i \in I\}$ in the configuration interval $J = J_{\mathbf{y}} := (y_{L-K-1}, y_{L+K+1})$.

5.1.2. *Universality of the local gap statistics for $\mu_{\mathbf{y}}$.* Our main technical result, Theorem 5.1 below, asserts that the local gap statistics is essentially independent of V and $\mathbf{y}$ as long as the boundary conditions $\mathbf{y}$ are regular. This property is expressed by defining the following set of "good" boundary conditions with some given positive parameters ξ, α (thet set $\mathcal{G}$ in (4.14) played exactly the same role)

$$(5.6)$$
$$\mathcal{R} = \mathcal{R}_{L,K}(\xi, \alpha) := \{\mathbf{y} : |y_k - \gamma_k| \leqslant N^{-1} K^{\xi}, \quad k \in [\![\alpha N, (1 - \alpha)N]\!] \setminus I_{L,K}\}$$
$$\cap \{\mathbf{y} : |y_k - \gamma_k| \leqslant N^{-4/15} K^{\xi}, \quad k \in [\![N^{3/5} K^{\xi}, N - N^{3/5} K^{\xi}]\!]\}$$
$$\cap \{\mathbf{y} : |y_k - \gamma_k| \leqslant 1, \quad k \in [\![1, N]\!] \setminus I_{L,K}\}.$$

This definition is taylored to the rigidity bounds for the β-ensemble, see Theorem 4.1. Note that $\mathcal{R}$ has a key parameter, the exponent ξ, which will be chosen as an arbitrary small positive number in the applications. We will not follow its dependence precisely and we will often neglect it from the notation, i.e. we will talk about "good" boundary conditions $\mathbf{y} \in \mathcal{R}$.

Good boundary conditions give rise to a regular potential $V_{\mathbf{y}}$. More precisely, if $\mathbf{y} \in \mathcal{R}$, then

$$(5.7) \qquad |J_{\mathbf{y}}| = \frac{\mathcal{K}}{N\varrho(\bar{y})} + O\left(\frac{K^{\xi}}{N}\right),$$

$$(5.8) \qquad V_{\mathbf{y}}'(x) = \varrho(\bar{y}) \log \frac{d_+(x)}{d_-(x)} + O\left(\frac{K^{\xi}}{Nd(x)}\right), \qquad x \in J_{\mathbf{y}},$$

$$(5.9) \qquad V_{\mathbf{y}}''(x) \geqslant \inf V'' + \frac{c}{d(x)}, \qquad x \in J_{\mathbf{y}}.$$

Here

$$\bar{y} := \frac{1}{2}(y_{L-K-1} + y_{L+K+1})$$

denotes the midpoint of the configuration interval, $d(x) := \min\{|x - y_{L-K-1}|, |x - y_{L+K+1}|\}$ is the distance of x to the boundary of the configuration interval $J = (y_{L-K-1}, y_{L+K+1})$ and $d_-(x)$ and $d_+(x)$ are regularized versions of the distances of x to the closest and to the farthest endpoints of J, respectively. The key point is that the leading term of $V_{\mathbf{y}}'(x)$ depends on the boundary conditions only through the density in the center, $\varrho(\bar{y})$. We also introduce

$$(5.10) \qquad \alpha_j := \bar{y} + \frac{j - L}{K + 1}|J|, \qquad j \in I_{L,K},$$

to denote the $\mathcal{K}$ equidistant points within the interval J.

THEOREM 5.1 (Gap universality for local measures). *Fix $L, \widetilde{L}$ and $\mathcal{K} = 2K + 1$ satisfying (5.1) with an exponent $\delta > 0$. Consider two boundary conditions $\mathbf{y}, \widetilde{\mathbf{y}}$ such that the configuration intervals coincide,*

$$(5.11) \qquad J = (y_{L-K-1}, y_{L+K+1}) = (\widetilde{y}_{\widetilde{L}-K-1}, \widetilde{y}_{\widetilde{L}+K+1}).$$

We consider the measures $\mu = \mu_{\mathbf{y},\beta,V}$ and $\widetilde{\mu} = \mu_{\widetilde{\mathbf{y}},\beta,\widetilde{V}}$ defined as in (5.4), with possibly two different external potentials V and $\widetilde{V}$. Let $\xi > 0$ be a small constant. Assume that $|J|$ satisfies

$$(5.12) \qquad |J| = \frac{\mathcal{K}}{N\varrho(\bar{y})} + O\left(\frac{K^{\xi}}{N}\right).$$

Suppose that $\mathbf{y}, \widetilde{\mathbf{y}} \in \mathcal{R}$ and that

$$(5.13) \qquad \max_{j \in I_{L,K}} \left|\mathbb{E}^{\mu_{\mathbf{y}}} x_j - \alpha_j\right| + \max_{j \in I_{\widetilde{L},K}} \left|\mathbb{E}^{\widetilde{\mu}_{\widetilde{\mathbf{y}}}} x_j - \alpha_j\right| \leqslant CN^{-1}K^{\xi}$$

holds. Let the integer number p satisfy $|p| \leqslant K - K^{1-\xi^*}$ for some small $\xi^* > 0$. Then there exists $\xi_0 > 0$, depending on δ, such that if $\xi, \xi^* \leqslant \xi_0$ then for any n fixed and any bounded smooth observable $O : \mathbb{R}^n \to \mathbb{R}$ with compact support we have

$$(5.14) \qquad \left| \mathbb{E}^{\mu_{\mathbf{y}}} O\big(N(x_{L+p} - x_{L+p+1}), \ldots N(x_{L+p} - x_{L+p+n})\big) \right.$$

$$\left. - \mathbb{E}^{\widetilde{\mu}_{\widetilde{\mathbf{y}}}} O\big(N(x_{\widetilde{L}+p} - x_{\widetilde{L}+p+1}), \ldots N(x_{\widetilde{L}+p} - x_{\widetilde{L}+p+n})\big) \right| \leqslant CK^{-\varepsilon}$$

for some $\varepsilon > 0$ depending on δ, α and for some C depending on O. This holds for any $N \geqslant N_0$ sufficiently large, where N_0 depends on the parameters ξ, ξ^*, α, and C in (5.13).

5.1.3. *Rigidity and level repulsion of* $\mu_{\mathbf{y}}$. In the following two theorems we establish rigidity and level repulsion estimates for the local log-gas $\mu_{\mathbf{y}}$ with good boundary conditions $\mathbf{y}$. While both rigidity and level repulsion are basic questions for log gases and are interesting in themselves, our main motivation to prove these theorems is to use them in the proof of Theorem 5.1.

We remark that an almost optimal rigidity estimate in the bulk was given in Theorem 4.1 and some level repulsion bound was given in (4.11) of [**11**], these results hold with respect to the global measure μ. For the proof of Theorem 5.1 we need their local versions with respect to $\mu_{\mathbf{y}}$, at least for most $\mathbf{y}$. Naively, this looks as a simple conditioning argument, but there is a subtle point. From the estimates w.r.t. μ, one can conclude that $\mu_{\mathbf{y}}$ has a good rigidity bound for a set of boundary conditions with high probability w.r.t. the global measure μ. This will be sufficient for the proof of Theorem 1.5, but not for Theorem 1.3. In the proof for the gap universality of Wigner matrices we will need a rigidity estimate for $\mu_{\mathbf{y}}$ for a set of $\mathbf{y}$'s with high probability with respect to by the time evolved measure $f_t\mu$ which may be asymptotically singular to μ for large N. The following result asserts that a rigidity estimate holds for $\mu_{\mathbf{y}}$ provided that $\mathbf{y}$ itself satisfies a rigidity bound and an extra condition, (5.15), holds. This condition will have to be verified with different methods in the Wigner case.

THEOREM 5.2 (Rigidity estimate for local measures). *For* $\mathbf{y} \in \mathcal{R}$ *consider the local equilibrium measure* $\mu_{\mathbf{y}}$ *defined in* (5.4) *and assume that*

$$(5.15) \qquad \left| \mathbb{E}^{\mu_{\mathbf{y}}} x_k - \alpha_k \right| \leqslant CN^{-1}K^{\xi}, \quad k \in I = I_{L,K},$$

is satisfied. Then there are positive constants C, c, *depending on* ξ, *such that for any* $k \in I$ *and* $u > 0$,

$$(5.16) \qquad \mathbb{P}^{\mu_{\mathbf{y}}} \left(N \left| x_k - \alpha_k \right| \geqslant uK^{\xi} \right) \leqslant Ce^{-cu^2}.$$

The proof of this result is similar to that of the concentration estimate (4.10) in Theorem 4.1. To estimate $x_k - \mathbb{E}^{\mu_{\mathbf{y}}} x_k$, we again use a multiscale argument of local averages for which stronger convexity bounds are available. The analogue of the accuracy estimate controlling $\mathbb{E}^{\mu_{\mathbf{y}}} x_k - \gamma_k$ in Theorem 4.1 is replaced by the assumption (5.15). Notice that, unlike for the global measure μ, a direct accuracy control via the loop equation is not available for $\mu_{\mathbf{y}}$ since the potential $V_{\mathbf{y}}$ is not analytic.

Now we state the level repulsion estimates.

THEOREM 5.3 (Level repulsion estimate for local measures). *For* $\mathbf{y} \in \mathcal{R}$ *we have the following estimates:*
i) [Weak form of level repulsion] For any $s > 0$ *we have*

$$(5.17) \qquad \mathbb{P}^{\mu_{\mathbf{y}}} [N(x_{i+1} - x_i) \leqslant s] \leqslant C(Ns)^{\beta+1}, \qquad i \in [\![L - K - 1, L + K]\!].$$

ii) [Strong form of level repulsion] Suppose that there exist positive constants C, c *such that the following rigidity estimate holds for any* $k \in I$:

$$(5.18) \qquad \mathbb{P}^{\mu_{\mathbf{y}}} \left(N \left| x_k - \alpha_k \right| \geqslant CK^{\xi^2} \right) \leqslant C\exp\left(-K^c \right).$$

Then there exists small a constant θ, *depending on* C, c *in* (5.18), *such that for any* $s \geqslant \exp(-K^{\theta})$. *we have*

$$(5.19) \qquad \mathbb{P}^{\mu_{\mathbf{y}}} [N(x_{i+1} - x_i) \leqslant s] \leqslant C \left(K^{\xi} s \log N \right)^{\beta+1}, \qquad i \in [\![L - K - 1, L + K]\!].$$

The level repulsion bounds will mostly be used in the following estimate which trivially follows from Theorem 5.3

COROLLARY 5.4. *Let* $\mathbf{y} \in \mathcal{R}$, *then for any* $p < \beta + 1$ *we have*

$$(5.20) \qquad \mathbb{E}^{\mu_{\mathbf{y}}} \frac{1}{[N|x_i - x_{i+1}|]^p} \leqslant C_p K^{C_3 \xi}, \qquad i \in [\![L - K - 1, L + K]\!].$$

$\square$

5.1.4. *Sketch of the proof of the level repulsion.* The proof of part (ii) of Theorem 5.3 goes in three steps. For simplicity, we consider (5.19) only for the first gap, i.e. $i = L - K - 1$ and we also assume that $\bar{y} = 0$ by a simple shift.

Step 1. In this step we prove

$$(5.21) \qquad \mathbb{P}^{\mu_{\mathbf{y}}} \left(x_{L-K} - y_{L-K-1} \leqslant s/N \right) \leqslant CKs \log N,$$

which is essentially (5.19) but with factor K instead of K^ξ and with the exponent $\beta + 1$ replaced with one. The proof of (5.21) is dilation argument. For a nonnegative parameter φ, we define

$$Z_\varphi := \int \dots \int_{-a+a\varphi}^{a-a\varphi} \mathrm{d}\mathbf{x} \prod_{\substack{i,j \in I \\ i<j}} (x_i - x_j)^\beta e^{-N\frac{\beta}{2} \sum_j V_{\mathbf{y}}(x_j)}$$

$$(5.22) \qquad = (1-\varphi)^{K+\beta K(K-1)/2} \int \dots \int_{-a}^{a} \mathrm{d}\mathbf{w} \prod_{i<j} (w_i - w_j)^\beta e^{-N\frac{\beta}{2} \sum_j V_{\mathbf{y}}((1-\varphi)w_j)},$$

where we set

$$a := -y_{L-K-1}, \qquad w_j := (1-\varphi)^{-1} x_{L+j}, \qquad \mathrm{d}\mathbf{x} = \prod_{|j| \leqslant K} \mathrm{d}x_{L+j} \qquad \mathrm{d}\mathbf{w} = \prod_{|j| \leqslant K} \mathrm{d}w_j.$$

Clearly $Z_{\varphi=0}$ is the normalization constant of the measure $\mu_{\mathbf{y}}$ and we have

$$(5.23) \qquad \mathbb{P}^{\mu_{\mathbf{y}}} \left(x_{L-K} - (-a) \geqslant a\varphi \right) \geqslant \frac{Z_\varphi}{Z_0}.$$

The multiple integral on the r.h.s of (5.22) is almost the same as Z_0, except that the argument of $V_{\mathbf{y}}$ is rescaled by $1 - \varphi$. This effect can be estimated from the explicit formula (5.5) for $V_{\mathbf{y}}$. The external potential V in (5.5) is unproblematic since it is smooth. Due to $\mathbf{y} \in \mathcal{R}$, the points y_j are regularly spaced on scales at least K^ξ/N, thus the sum of the interaction terms $\log|x - y_k|$ for k's away from the edges of I^c, i.e. $k \leqslant L - 2K$ or $k \geqslant L + 2K$, is a regular function of x and the effect of dilation can be well approximated by Taylor expansion. For nearby k's right below the lower edge, i.e. $L - 2K \leqslant k \leqslant L - K$, we use the trivial bound $(1-\varphi)x - y_k \geqslant (1-\varphi)(x - y_k)$. From these estimates it follows that

$$(5.24) \qquad \frac{Z_\varphi}{Z_0} \geqslant 1 - CK^2 \varphi \log N.$$

Since $a \sim K/N$, together with (5.23) it implies (5.21).

Step 2. Now we consider an auxiliary measure which are slightly modified version of the local equilibrium measures:

$$(5.25) \qquad \mu^{(0)} := Z^{(0)} (x_{L-K} - y_{L-K-1})^{-\beta} \mu_{\mathbf{y}};$$

where $Z^{(0)}$ are chosen for normalization. In other words, we drop the term $(x_{L-K} - y_{L-K-1})^\beta$ from the measure $\mu_{\mathbf{y}}$. Setting $X := x_{L-K} - y_{L-K-1}$ for brevity, we have

$$(5.26) \qquad \mathbb{P}^{\mu_{\mathbf{y}}}[X \leqslant s/N] = \frac{\mathbb{E}^{\mu^{(0)}}[\mathbf{1}(X \leqslant s/N)X^\beta]}{\mathbb{E}^{\mu^{(0)}}[X^\beta]}.$$

The estimate (5.21) also holds for $\mu^{(0)}$ and thus

$$\mathbb{E}^{\mu^{(0)}}[\mathbf{1}(X \leqslant s/N)X^\beta] \leqslant C(s/N)^\beta Ks \log N$$

and with the choice $s = cK^{-1}(\log N)^{-1}$ in (5.21) we also have

$$\mathbb{P}^{\mu^{(0)}}\left(X \geqslant \frac{c}{NK \log N}\right) \geqslant 1/2$$

with some positive constant c. This implies that

$$\mathbb{E}^{\mu^{(0)}}[X^\beta] \geqslant \frac{1}{2}\left(\frac{c}{NK \log N}\right)^\beta.$$

Combining with (5.26), we have thus proved that

$$(5.27) \qquad \mathbb{P}^{\mu_{\mathbf{y}}}[X \leqslant s/N] \leqslant C\left(Ks \log N\right)^{\beta+1},$$

i.e. we obtained (5.21) but with an exponent $\beta + 1$ in the r.h.s.

Step 3. We now improve the constant K to K^ξ in the r.h.s of (5.21). The factor K originated from the number of particles in $\mu_{\mathbf{y}}$. We can further condition the measure $\mu_{\mathbf{y}}$ on the points

$$z_j := x_j \qquad j \geqslant L - K + K^\xi,$$

and we let $\mu_{\mathbf{y},\mathbf{z}}$ denote the conditional measure on the remaining x variables $\{x_j : L - K \leqslant j \leqslant L - K + K^\xi\}$. From the rigidity estimate (5.18) we have $(\mathbf{y}, \mathbf{z}) \in \mathcal{R}$ with a very high probability w.r.t. $\mu_{\mathbf{y}}$. We will now apply (5.27) to the measure $\mu_{\mathbf{y},\mathbf{z}}$ to obtain

$$(5.28) \qquad \mathbb{P}^{\mu_{\mathbf{y},\mathbf{z}}}[X \leqslant s/N] \leqslant C\left(K^\xi s \log N\right)^{\beta+1}.$$

This holds for all z with a high $\mu_{\mathbf{y}}$-probability. The subexponential lower bound on s, assumed in part ii) of Theorem 5.3, allows us to include the probability of the complement of $\mathcal{R}$ in the estimate, we thus have proved (5.19).

The proofs of the weaker bound (5.17) for any $s > 0$ use similar arguments that have led to (5.21), but without assuming $\mathbf{y} \in \mathcal{R}$ which yields that one factor of K has to be replaced with N in (5.24). The assumption that the boundary conditions are good needs to be dropped since in Step 3 of the above argument, (5.21) is also used after additional conditioning on $\mathbf{z}$, distributed according to $\mu_{\mathbf{y}}$, and without (5.18) there is no rigidity result available for $\mu_{\mathbf{y}}$.

5.2. Proof of Theorem 5.1. In this section, we start to compare gap distributions of two local log-gases on the same configuration interval but with different external potential and boundary conditions. For simplicity, we consider only an observable of a single gap; a few consecutive gaps can be handled similarly. From now on, we use microscopic coordinates, i.e. we replace x_j with x_j/N, and we also relabel the indices so that the coordinates of x_j are $j \in I = \{-K, \ldots, 0, 1, \ldots K\}$. This will have the advantage that K remains the only large parameter; N disappears.

The local equilibrium measures and their Hamiltonians will be denoted by the same symbols, $\mu_{\mathbf{y}}$ and $\mathcal{H}_{\mathbf{y}}$, as before, but with a slight abuse of notations we redefine them now to the microscopic scaling, i.e.

(5.29)

$$\mathcal{H}_{\mathbf{y}}(\mathbf{x}) := \sum_{i \in I} \frac{1}{2} V_{\mathbf{y}}(x_i) - \sum_{\substack{i,j \in I \\ i<j}} \log |x_j - x_i|, \qquad V_{\mathbf{y}}(x) := NV(x/N) - 2\sum_{j \notin I} \log |x - y_j|,$$

The other Hamiltonian $\widetilde{H}_{\widetilde{\mathbf{y}}}$ is defined in a similar way with V in (5.29) replaced with another external potential $\widetilde{V}$. We also rewrite (5.13) in the microscopic coordinate as

(5.30)

$$|\mathbb{E}^{\mu_{\mathbf{y}}} x_j - \alpha_j| + |\mathbb{E}^{\widetilde{\mu}_{\widetilde{\mathbf{y}}}} x_j - \alpha_j| \leqslant CK^\xi,$$

where $\alpha_j := \frac{j}{\mathcal{K}+1}|J|$ is the rescaled version of the definition given in (5.10), but we keep the same notation. The concept of "good" set $\mathcal{R}$ is also rescaled accordingly.

Suppose that $\mathbf{y}, \widetilde{\mathbf{y}} \in \mathcal{R}$ and define the interpolating measures

(5.31)

$$\omega_{\mathbf{y},\widetilde{\mathbf{y}}}^r = Z_r e^{-\beta r(\widetilde{V}_{\widetilde{\mathbf{y}}}(\mathbf{x}) - V_{\mathbf{y}}(\mathbf{x}))} \mu_{\mathbf{y}}, \qquad r \in [0,1],$$

so that $\omega_{\mathbf{y},\widetilde{\mathbf{y}}}^1 = \widetilde{\mu}_{\widetilde{\mathbf{y}}}$ and $\omega_{\mathbf{y},\widetilde{\mathbf{y}}}^0 = \mu_{\mathbf{y}}$ (Z_r is a normalization constant). This is again a local log-gas with Hamiltonian

(5.32)

$$\mathcal{H}_{\mathbf{y},\widetilde{\mathbf{y}}}^r(\mathbf{x}) = \frac{1}{2}\sum_{i \in I} V_{\mathbf{y},\widetilde{\mathbf{y}}}^r(x_i) - \frac{1}{N}\sum_{i<j} \log |x_i - x_j|, \qquad V_{\mathbf{y},\widetilde{\mathbf{y}}}^r(x) := (1-r)V_{\mathbf{y}}(x) + r\widetilde{V}_{\widetilde{\mathbf{y}}}(x).$$

For any fixed r, the measure $\omega_{\mathbf{y},\widetilde{\mathbf{y}}}^r$ inherits all relevant properties of $\mu_{\mathbf{y}}$. In particular the rigidity bound in the form

(5.33)

$$\mathbb{P}^\omega\left(|x_i - \alpha_i| \geqslant CK^{C\xi}\right) \leqslant Ce^{-K^\theta}, \quad i \in I,$$

the level repulsion bounds (5.17)–(5.19) and their consequence in (5.20) hold w.r.t. the measure $\omega = \omega_{\mathbf{y},\widetilde{\mathbf{y}}}^r$ as well (in the new microscopic coordinates there are no N factors in the left hand sides of these inequalities). The proofs are basically parallel with the arguments for $\mu_{\mathbf{y}}$; the only nontrivial step is to show that (5.30) implies the analogous bound

$$|\mathbb{E}^\omega x_k - \alpha_k| \leqslant CK^\xi$$

w.r.t. $\omega = \omega_{\mathbf{y},\widetilde{\mathbf{y}}}^r$ as well. Although ω appears to be some easy combination of $\mu_{\mathbf{y}}$ and $\widetilde{\mu}_{\widetilde{\mathbf{y}}}$, this conclusion is nontrivial. It requires comparing ω and $\mu_{\mathbf{y}}$ via the entropy inequality, which involves controlling the exponential moment of $|x_k - \alpha_k|$ w.r.t. $\mu_{\mathbf{y}}$. At this point the Gaussian tail proven in (5.16) is necessary.

The right hand side of (5.14) with $n = 1$, in the rescaled coordinates and with $L = \widetilde{L} = 0$, is estimated by

(5.34)

$$\left|[\mathbb{E}^{\mu_{\mathbf{y}}} - \mathbb{E}^{\widetilde{\mu}_{\widetilde{\mathbf{y}}}}]O(x_p - x_{p+1})\right| \leqslant \int_0^1 dr \frac{d}{dr}\mathbb{E}^{\omega_{\mathbf{y},\widetilde{\mathbf{y}}}^r} O(x_p - x_{p+1}).$$

For any bounded smooth function O with compact support

(5.35)

$$\frac{d}{dr}\mathbb{E}^{\omega_{\mathbf{y},\widetilde{\mathbf{y}}}^r} O(x_p - x_{p+1}) = \beta\langle h_0; O(x_p - x_{p+1})\rangle_{\omega_{\mathbf{y},\widetilde{\mathbf{y}}}^r},$$

where

(5.36)

$$h_0 = h_0(\mathbf{x}) = \sum_{i \in I}(V_{\mathbf{y}}(x_i) - \widetilde{V}_{\widetilde{\mathbf{y}}}(x_i))$$

and $\langle f; g\rangle_\omega := \mathbb{E}^\omega fg - (\mathbb{E}^\omega f)(\mathbb{E}^\omega g)$ denotes the covariance. Thus Theorem 5.1 follows immediately from the following estimate on the gap covariance function. $\square$

THEOREM 5.5. *Consider two smooth potentials* $V, \widetilde{V}$ *and two good boundary conditions,* $\mathbf{y}, \widetilde{\mathbf{y}} \in \mathcal{R}$, *such that the configuration intervals coincide,* $J_\mathbf{y} = J_{\widetilde{\mathbf{y}}}$. *For any* $r \in [0, 1]$ *let* $\omega = \omega^r_{\mathbf{y}, \widetilde{\mathbf{y}}}$ *be the interpolating measure defined in* (5.31). *Assume that* (5.30) *holds for both boundary conditions* $\mathbf{y}, \widetilde{\mathbf{y}}$. *Fix* $\xi^* > 0$. *Then there exist* $\varepsilon > 0$ *and* $C > 0$, *depending on* ξ^*, *such that for any sufficiently small* ξ, *for* $|p| \leqslant K^{1-\xi^*}$ *we have*

$$(5.37) \qquad |\langle h_0(\mathbf{x}); O(x_p - x_{p+1})\rangle_\omega| \leqslant K^{C\xi} K^{-\varepsilon}$$

for any smooth function $O : \mathbb{R} \to \mathbb{R}$ *with compact support provided that* K *is large enough.*

Theorem 5.5 is our key technical result. The main difficulty behind it is due to the fact that the covariance function of two points, $\langle x_i; x_j\rangle_\omega$, decays only logarithmically. In fact, for the GUE, Gustavsson proved that (Theorem 1.3 in [**55**])

$$(5.38) \qquad \langle x_i; x_j\rangle_{GUE} \sim \log \frac{N}{[|i-j|+1]},$$

and a similar formula is expected for ω. Although $h_0(\mathbf{x})$ depends strongly only on points near the boundary and x_p is away from the boundary, it is still very difficult to prove Theorem 5.5 based on this slow logarithmic decay. However, the covariance function of the type

$$(5.39) \qquad \langle g_1(x_i); g_2(x_j - x_{j+1})\rangle_\omega$$

decays much faster in $|i-j|$. Since the second factor $g_2(x_j - x_{j+1})$ depends only on the difference of two neighboring points, it is expected that the decay is the (discrete) derivative in j of the covariance (5.38), i.e. it is $|i-j|^{-1}$. The actual result (5.37) is much weaker, but it still provides a power-law decay in K instead of a logarithmic decay. Covariances of the form $\langle g_1(x_i - x_{i+1}); g_2(x_j - x_{j+1})\rangle_\omega$ are expected to decay even faster but we have not pursued this direction further.

We point out that the fact that observables of differences of particles behave much nicer was a basic observation in DBM analysis (Theorem 3.4), see the explanation around (3.36).

5.3. Decay of correlation functions: Proof of Theorem 5.5. We will express the difference of gap distributions between two measures in terms of random walks in time dependent random environments. The decay of correlation functions will be translated into a partial regularity property of the corresponding parabolic equation. This partial regularity is a discrete version of the De Giorgi-Nash-Moser theory but with a long range elliptic part.

5.3.1. *Random Walk Representation.* In this section we derive a random walk representation for the gap correlation function on the left hand side of (5.37). We will apply it for the interpolating measure $\omega = \omega^r_{\mathbf{y}, \widetilde{\mathbf{y}}}$ (5.31) and for the function h_0 given in (5.36), but the representation formula (Proposition 5.6 below) is valid for any ω and h_0.

Let $\mathscr{L}^\omega$ be the reversible generator given by the Dirichlet form

$$(5.40) \qquad D^\omega(f) = -\int f\mathscr{L}^\omega f\,\mathrm{d}\omega = \sum_{|j|\leqslant K} \int (\partial_j f)^2 \mathrm{d}\omega.$$

This process can also be characterized by the following SDE

$$(5.41) \qquad \mathrm{d}x_i = \mathrm{d}B_i + \beta\Big[-\frac{1}{2}(V^r_{\mathbf{y},\widetilde{\mathbf{y}}})'(x_i) + \frac{1}{2}\sum_{j\neq i}\frac{1}{(x_i - x_j)}\Big]\mathrm{d}t,$$

where $\{B_i \ : \ |i| \leqslant K\}$ is a family of independent standard real Brownian motions. Let $\mathbb{E}_{\mathbf{x}}$ denote the expectation for this process with initial point $\mathbf{x}(0) = \mathbf{x}$. The expectation with respect to the process starting from equilibrium is $\mathbb{E}^{\omega}[\cdot] = \int \mathbb{E}_{\mathbf{x}}[\cdot]\omega(\mathrm{d}\mathbf{x})$. With a slight abuse of notations, when we talk about the process, we will use $\mathbb{P}^{\omega}$ and $\mathbb{E}^{\omega}$ also to denote the probability and expectation w.r.t. this dynamics with initial data distributed w.r.t. ω, i.e., in equilibrium.

Suppose $h(t) = h(t,\mathbf{x})$ is the solution of the equation $\partial_t h = \mathscr{L}^{\omega} h$ with an initial condition h_0. Introduce the notation

$$(5.42) \qquad \mathbf{v}(t,\mathbf{x}) = \nabla_{\mathbf{x}}h(t,\mathbf{x}), \quad \text{i.e.} \quad v_j(t,\mathbf{x}) := \partial_{x_j}h(t,\mathbf{x}).$$

By integrating the time derivative of $\langle h(t,\mathbf{x}); O(x_p - x_{p+1})\rangle_{\omega}$ and using the equation $h(t,\mathbf{x})$ satisfies, we have
$$(5.43)$$

$$\langle h_0(\mathbf{x}); O(x_p - x_{p+1})\rangle_{\omega} = \int_0^{\infty} \mathrm{d}\sigma \int O'(x_p - x_{p+1})[v_p(\sigma,\mathbf{x}) - v_{p+1}(\sigma,\mathbf{x})]\mathrm{d}\omega(\mathbf{x}).$$

For any fixed σ, the inner integral on the right hand side can be expressed by a random walk representation. Fix a path $\{\mathbf{x}(s) \ : \ s \in [0,\sigma]\}$. Define the following operators on $\mathbb{R}^{\mathcal{K}}$

$$(5.44) \qquad \mathcal{A}(s) := \mathcal{B}(s) + \mathcal{W}(s)$$

$$(5.45) \qquad [\mathcal{B}(s)\mathbf{v}]_j = -\sum_k B_{jk}(\mathbf{x}(\sigma - s))(v_k - v_j), \quad B_{jk}(\mathbf{x}) = \frac{1}{(x_j - x_k)^2} \geqslant 0$$

$$[\mathcal{W}(s)\mathbf{v}]_j = \mathcal{W}_j(s)v_j, \qquad \mathcal{W}_j(s) := [V^r_{\mathbf{y},\widetilde{\mathbf{y}}}]''(x_j(\sigma - s)).$$

Clearly $\mathcal{B}(s)$ is diffusion operator with random rates and $\mathcal{W}(s)$ is a potential representing a random environment. These operators depend on the whole path $\mathbf{x}(s)$, but we omit this fact from the notation.

With these notations we have the following representation:

PROPOSITION 5.6. *For any smooth function* $h_0 \ : \ J^{\mathcal{K}} \to \mathbb{R}$, *for any* $p \in I$, $-K \leqslant p \leqslant K - 1$, *we have*

$$(5.46) \quad \langle h_0; O(x_p - x_{p+1})\rangle_{\omega}$$

$$= \int_0^{\infty} \mathrm{d}\sigma \int O'(x_p - x_{p+1})\mathbb{E}_{\mathbf{x}}[w_p(\sigma,\mathbf{x}(\cdot);\sigma) - w_{p+1}(\sigma,\mathbf{x}(\cdot);\sigma)]\omega(\mathrm{d}\mathbf{x}).$$

Here, for any $\sigma > 0$ *and for any fixed path* $\{\mathbf{x}(s) \ : \ s \in [0,\sigma]\}$ *we let* $\mathbf{w}$ *denote the solution of the evolution equation*

$$(5.47) \qquad \partial_s \mathbf{w}(s;\mathbf{x}(\cdot),\sigma) = -\mathcal{A}(s)\mathbf{w}(s;\mathbf{x}(\cdot),\sigma),$$

with initial data $\mathbf{w}(0;\mathbf{x}(\cdot),\sigma) := \nabla h_0(\mathbf{x}(\sigma))$.

This representation in a slightly different setting already appeared in Proposition 2.2 of [17] (see also Proposition 3.1 in [54]), which was a probabilistic formulation of the idea of Helffer and Sjöstrand [56] and Naddaf and Spencer [70]. The proof relies on taking the gradient of the equation $\partial_t h = \mathscr{L}^\omega h$. A direct computation of the commutator $[\nabla, \mathscr{L}^\omega]$ yields that

$$(5.48) \qquad \partial_t \mathbf{v}(t,\mathbf{x}) = \mathscr{L}^\omega \mathbf{v}(t,\mathbf{x}) - \widetilde{\mathcal{A}}(\mathbf{x})\mathbf{v}(t,\mathbf{x}),$$

with initial condition $\mathbf{v}_0(\mathbf{x}) = \mathbf{v}(0,\mathbf{x}) = \nabla h_0(\mathbf{x})$. Here $\widetilde{\mathcal{A}}(\mathbf{x}) = \mathcal{B}(\mathbf{x}) + \mathcal{W}(\mathbf{x})$, where $\mathcal{B}(\mathbf{x})$ is the operator given by the matrix B_{jk} in (5.45) and $\mathcal{W}(\mathbf{x})$ is the diagonal multiplication operator by $[V^r_{\mathbf{y},\widetilde{\mathbf{y}}}]''(x_j)$. Since $\mathscr{L}^\omega$ generates the process $\mathbf{x}(t)$, we can represent the solution to (5.48) by the Feynman-Kac formula which can be written in the form (5.46).

When applying this proposition to our case, we will choose the initial condition h_0 be given by (5.36). The initial condition for the random walk (5.47) is given by ∇h_0. Notice that the leading term in $\partial_j h_0(\mathbf{x}) = V'_{\mathbf{y}}(x_j) - [\widetilde{V}_{\widetilde{\mathbf{y}}}]'(x_j)$ cancel; this is because the leading term in (5.8) depends only on the density $\varrho(\bar{y})$ which is matched for $\mathbf{y}$ and $\widetilde{\mathbf{y}}$ by $J_{\mathbf{y}} = J_{\widetilde{\mathbf{y}}}$, see (5.7). We thus have

$$(5.49) \qquad |\partial_j h_0(x)| \leqslant \frac{CK^\xi}{d(x_j)},$$

i.e. initially $\mathbf{w}$ is small away from the boundary and for the small σ regime the inner integral in the r.h.s. of (5.46) is small. After very long time $\mathbf{w}$ becomes constant, but then the right hand side of (5.46) is zero. The analysis of (5.46) requires to monitor what happens to $\mathbf{w}$ for coordinates p away from the boundary at intermediate times.

In the following sections we make a few preparations that exclude irrelevant regimes. First, it is easy to see that the regular spacing of $\mathbf{y}, \widetilde{\mathbf{y}} \in \mathcal{R}$ implies that $W_j(s) \geqslant cK^{-1}$, which means that the L^1-norm of the solution to (5.47) decays at a rate of order K. Thus the integral in (5.46) can be truncated at $\sigma \leqslant CK \log K$.

5.3.2. *Preparation for the De Giorgi-Nash-Moser bound: Restriction to the good paths.* The representation (5.46) expresses the covariance function in terms of the discrete spatial derivative of the solution to (5.47). To estimate $w_p(\sigma, \mathbf{x}(\cdot); \sigma) - w_{p+1}(\sigma, \mathbf{x}(\cdot); \sigma)$ in (5.46), we will now study the Hölder continuity of the solution $\mathbf{w}(s, \mathbf{x}(\cdot); \sigma)$ to (5.47) at time $s = \sigma$ and at the spatial point p. We will do it for each fixed path $\mathbf{x}(\cdot)$, with the exception of a set of "bad" paths that will have a small probability.

Notice that if all points x_i were approximately regularly spaced in the interval J, then the operator $\mathcal{B}$ had a kernel $\mathcal{B}_{ij} \sim (i-j)^{-2}$, i.e. it were essentially a discrete version of the operator $|p| = \sqrt{-\Delta}$ (in one dimension). Hölder continuity will thus be the consequence of the De Giorgi-Nash-Moser bound for the parabolic equation (5.47). However, we need to control the coefficients in this equation, which depend on the random walk $\mathbf{x}(\cdot)$.

For the De Giorgi-Nash-Moser theory we need both upper and lower bounds on the time dependent kernel $\mathcal{B}_{ij}(s)$. The rigidity bound (5.33) guarantees a lower bound on $\mathcal{B}_{ij}$, up to a factor $K^{-C\xi}$. Since the subexponential probabilistic estimate in (5.33) is very strong, one can easily guarantee a very similar estimate uniformly in time, i.e.

$$(5.50) \qquad \mathbb{P}^\omega \left\{ \mathbf{x}(s) : \sup_{0 \leqslant s \leqslant CK \log K} \sup_{|j| \leqslant K} |x_j(s) - \alpha_j| \leqslant K^{C\xi} \right\} \geqslant 1 - e^{-K^\theta}$$

(maybe after reducing θ from (5.33)). This follows from the fact that ω is invariant under the dynamics and $\mathbf{x}(t)$ has some stochastic continuity.

The level repulsion estimate implies certain upper bounds on $\mathcal{B}_{ij}$, but these estimates not particularly strong. Even in the $\beta > 1$ case, the bound (5.20) implies only that

$$\mathbb{E}^{\omega} \mathcal{B}_{i,i+1}(s) = \mathbb{E}^{\omega} \frac{1}{(x_{i+1} - x_i)^2} \leqslant K^{C\xi}$$

is finite. In the $\beta = 1$ borderline case even the expectation of $\mathcal{B}_{i,i+1}$ is infinite. Such a weak control does not allow us to guarantee an effective simultaneous bound on $\mathcal{B}_{i,i+1}$ for all i and for all time. Instead of supremum bounds, we control these coefficients only in an average sense and we can show that for any fixed index $Z \in I$, time s and parameter M, we have

$$(5.51) \quad \mathbb{P}^{\omega} \left\{ \mathbf{x}(s) \ : \ \frac{1}{1+s} \int_0^s da \, \frac{1}{M} \sum_{|i-Z| \leqslant M} \sum_j B_{ij}(\mathbf{x}(\sigma-a)) \leqslant K^{\rho} \right\} \geqslant 1 - K^{C\xi - \rho}.$$

Here ρ will be chosen as large constant times ξ. The summation over j is harmless since for $|i - j| \geqslant K^{\xi}$ the rigidity estimate can be used to bound $\mathcal{B}_{ij}$. By a dyadic choice of the parameters s, M, it is easy to upgrade (5.51) to hold for any $M \leqslant K$ and $s \leqslant CK \log K$. But it is essential that a reference point Z be fixed, one cannot guarantee that none of the gaps closes.

The expectation over the paths, $\int \mathbb{E}_{\mathbf{x}}[\,\cdot\,]\omega(d\mathbf{x})$, in (5.46) will be restricted to the sets given in (5.50) and (5.51). Due to the strong subexponential bound, the restriction to the set in (5.50) is unproblematic. However, the estimate (5.51) is quite weak; the probability of the "bad" paths is bounded only by a small negative power of K. This is not sufficient to compensate the time integration in (5.46) even after the upper cutoff $\sigma \leqslant CK \log K$. We will need to use that the heat kernel of the equation (5.47) has an $L^1 \to L^{\infty}$ decay of order $1/s$ after time s. Thus the solution $w_p(\sigma, \mathbf{x}(\cdot); \sigma)$ decays as $1/\sigma$ which renders the $d\sigma$ integration in (5.46) harmless.

For completeness, we state the $L^p \to L^q$ heat kernel decay estimate in a general form. Notice that we only assume a lower bound in $\mathcal{B}_{ij}$ to guarantee sufficient ellipticity; there is no upper bound required for these bounds.

PROPOSITION 5.7. *Consider the evolution equation*

$$(5.52) \quad \partial_s \mathbf{u}(s) = -\mathcal{A}(s)\mathbf{u}(s), \qquad \mathbf{u}(s) \in \mathbb{R}^{\mathcal{K}}$$

and fix $\sigma > 0$. Suppose that for some constant b we have

$$(5.53) \quad \mathcal{B}_{jk}(s) \geqslant \frac{b}{(j-k)^2}, \quad 0 \leqslant s \leqslant \sigma, \quad j \neq k,$$

and

$$(5.54) \quad \mathcal{W}_j(s) \geqslant \frac{b}{d_j}, \qquad d_j := \big|\, |j| - K \,\big| + 1, \quad 0 \leqslant s \leqslant \sigma.$$

Then for any $1 \leqslant p \leqslant q \leqslant \infty$ we have the decay estimate

$$(5.55) \quad \|\mathbf{u}(s)\|_q \leqslant (sb)^{-\left(\frac{1}{p} - \frac{1}{q}\right)} \|\mathbf{u}(0)\|_p, \qquad 0 < s \leqslant \sigma.$$

The proof relies on the usual Nash argument and uses the following critical Gagliardo-Nirenberg-type inequality for the discrete version of the operator $\sqrt{-\Delta}$:

PROPOSITION 5.8. *There exists a positive constant C such that*

$$(5.56) \quad \|f\|_{L^4(\mathbb{Z})}^4 \leqslant C\|f\|_{L^2(\mathbb{Z})}^2 \sum_{i \neq j \in \mathbb{Z}} \frac{|f_i - f_j|^2}{|i - j|^2}$$

holds for any function $f : \mathbb{Z} \to \mathbb{R}$.

The continuous version of this inequality, $\|\phi\|_4^4 \leqslant C\|\phi\|_2^2\langle\phi,|p|\phi\rangle$, was first proven in [**71**].

5.3.3. *Preparation for the De Giorgi-Nash-Moser bound: Finite speed of propagation.* The Hölder continuity of the parabolic equation (5.47) emerges only after a certain time, thus for the small σ regime in the integral (5.46) we need a different argument. Since we are interested in the Hölder continuity around the middle of the interval I (note that $|p| \leqslant K^{1-\xi^*}$ in Theorem 5.5), and the initial condition ∇h_0 is small in this region (see (5.49)), a finite speed of propagation estimate guarantees that $w_p(\sigma; \mathbf{x}(0), \sigma)$ is small if σ is not too large.

Since (5.47) is linear, for the finite speed of propagation it is sufficient to consider the fundamental solution. For a fixed, let $\mathbf{u}^a(s)$ denote the solution

$$(5.57) \qquad \partial_s \mathbf{u}^a(s) = -\mathcal{A}(s)\mathbf{u}^a(s), \quad u_j^a(0) = \delta_{aj}.$$

with a delta function as initial data. We will assume that the coefficients of $\mathcal{A}$ satisfy, for some fixed $|Z| \leqslant K/2$ and $\rho > 0$, the bound

$$(5.58) \quad \sup_{0\leqslant s\leqslant\sigma} \sup_{0\leqslant M\leqslant K} \frac{1}{1+s} \int_0^s \frac{1}{M} \sum_{i\in I\,:\,|i-Z|\leqslant M} \sum_{j\in I\,:\,|j-Z|\leqslant M} \mathcal{B}_{ij}(\sigma-a)\mathrm{d}a \leqslant CK^\rho.$$

Notice that (5.58) is satisfied on the set of good path given by (5.50) and (5.51). The following lemma provides a finite speed of propagation estimate for the equation (5.57) under the condition of (5.58). This estimate is not optimal, but it is sufficient for our purpose.

LEMMA 5.9. *[Finite Speed of Propagation Estimate] Fix $a \in I$ and $\sigma \leqslant CK\log K$. We assume that the coefficients of $\mathcal{A}$ satisfy* (5.53) *and* (5.54) *with $b = K^{-\xi}$. Assume that* (5.58) *is satisfied for some fixed Z, $|Z| \leqslant K/2$. Then for the fundamental solution* (5.57) *we have the estimate for any $s \leqslant \sigma$ and $p \in I$*

$$(5.59) \qquad |u_p^a(s)| \leqslant \frac{CK^{\rho+2\xi+1/2}\sqrt{s+1}}{|p-a|}.$$

For the proof, we split the operator $\mathcal{A} = \mathcal{S} + \mathcal{R}$ into a short range and a long range part, where the short range part $\mathcal{S}(s)$ is defined by

$$(5.60) \qquad (\mathcal{S}(s)\mathbf{v})_j := -\sum_{k\,:\,|j-k|\leqslant\ell} \mathcal{B}_{jk}(s)(v_k - v_j) + \mathcal{W}_j(s)v_j$$

with some cutoff parameter ℓ. The norm of the long range part in any L^p is bounded by ℓ^{-1} and it is treated as a perturbation via Duhamel formula. For the short range part, we control the exponentially weighted norm of the solution of $\partial_s\mathbf{r}(s) = -\mathcal{S}(s)\mathbf{r}(s)$, i.e. we derive a Gronwall bound for

$$f(s) = \sum_{j\in I} e^{|j-a|/\theta}r_j^2(s).$$

The result is

$$f(s) \leqslant \exp\left[C\theta^{-2}\ell^2 \int_0^s \sum_{k,j:|j-k|\leqslant\ell} \mathcal{B}_{kj}(s')\mathrm{d}s'\right] f(0).$$

The exponent is estimated by (5.58) with $M = K$. The optimization of the length-scale θ together with the cutoff parameter ℓ yields (5.59). $\qquad\square$

Inserting the estimate (5.59) into (5.46) and using the estimate (5.49) on the initial data ∇h_0, we obtain that the contribution of the short time regime, $\sigma \leqslant K^{1/4}$, is negligible if p is away from edge, $|p| \leqslant K^{1-\xi^*}$ for some $\xi^* > 0$. This allows us to disregard the $\sigma \leqslant K^{1/4}$ regime in (5.46) and focus on $\sigma \in [K^{1/4}, CK \log K]$.

5.3.4. *A discrete De Giorgi-Nash-Moser bound.* We will now treat the main part of the integral (5.46) by parabolic regularity. The preparations in the previous sections ensure that is is sufficient to consider the integration regime $\sigma \in [K^{1/4}, CK \log K]$ and we can assume that the path $\mathbf{x}(\cdot)$ is good in the sense of the estimates (5.50) and (5.51). In particular, the rigidity estimate implies not only lower bounds but also upper bounds for distant indices; more precisely we have

$$(5.61) \qquad \mathcal{B}_{ij}(s) \leqslant \frac{C}{(i-j)^2}$$

for any $|i-j| \geqslant CK^\xi$ and $0 \leqslant s \leqslant CK \log K$; and similarly

$$(5.62) \qquad \mathcal{W}_i(s) \leqslant \frac{K^\xi}{d_i}, \quad \text{if} \quad d_i \geqslant K^{C\xi}.$$

The following regularity theorem combined with (5.49) completes the estimate of (5.46) and completes the proof of Theorem 5.5. $\qquad\qquad\qquad\qquad\square$

THEOREM 5.10 (Parabolic partial regularity with singular coefficients). *Let $\mathbf{u}$ be a solution to (5.57), where $\mathbf{u} = \mathbf{u}^a$ for any choice of a. Suppose that the coefficients of $\mathcal{A}$ satisfy the lower bounds (5.53) and (5.54) with $b = K^{-\xi}$, the upper bounds (5.61), (5.62) for distant indices and the upper bound in (5.58) in average sense for all indices. Let $\sigma \in [K^{c_1}, C_1 K \log K]$ be fixed, where $c_1 > 0$ is an arbitrary positive constant. Then for any $0 < q' < 1$ there exists $q > 0$ so that for any $|Z| \leqslant K/2$*

$$(5.63) \qquad \sup_{\max(|j-Z|,|j'-Z|)\leqslant\sigma^{1-q'}} |u_j(\sigma) - u_{j'}(\sigma)| \leqslant C\sigma^{-1-q},$$

where $\mathbf{u} = \mathbf{u}^a$ for any choice of a.

Notice that this result is deterministic, all probability estimates are comprised in verifying the conditions. We also remark that if we define the rescaled function $v(j/K, t) := t u_j(t)$, then (5.63) can be interpreted as a type of Hölder regularity of v on scale $\sigma^{1-q'} K^{-1} \ll 1$ at the point Z/K:

$$(5.64) \qquad |v(x,\sigma) - v(y,\sigma)| \leqslant \sigma^{-q} \leqslant |x-y|^{c_1 q}$$

for $1/K \leqslant |x-y| \leqslant \sigma^{1-q'}/K$ and x, y near Z/K. The Hölder exponent is thus at least $c_1 q$.

Although the statement of Theorem 5.10 seems to be complicated, the underlying mechanism is that there is a positive exponent q in (5.63), which to a great degree is an universal constant. This exponent provides an extra smallness factor in addition to the natural size of $u_j(\sigma)$, which is σ^{-1} from the $L^1 \to L^\infty$ decay. As (5.64) indicates, this gain comes from a Hölder regularity on the relevant scale.

Our equation (5.57) is of the type considered in [15], but it is discrete and in a finite interval. The key difference, however, is that the coefficient $\mathcal{B}_{ij} = (x_i - x_j)^{-2}$ in the elliptic part of (5.57) can be singular if gaps close, even temporarily, while [15] assumed the uniform bound $\mathcal{B}_{ij} \leqslant C/|i-j|^2$. The only control we have for the singular behavior of $\mathcal{B}_{ij}$ is the estimate (5.58) which is very weak. This estimate essentially says that the space-time maximum function of $\mathcal{B}_{i,i+1}(t)$ at a fixed space-time point $(Z,0)$ is bounded by K^ρ. Our main task is to show that this condition is sufficient for proving Hölder continuity at the same point. Our strategy follows the approach of Caffarelli-Chan-Vasseur [**15**]. The main new feature of our argument is the derivation of a local energy dissipation estimate for parabolic equation with singular coefficients satisfying (5.53), (5.54) as lower bounds and only (5.58) as an upper bound. The analogous result in [**15**], called the *first De Giorgi lemma*, is proved under uniform bounds on the coefficients. For our proof, roughly, we have to run the argument of the first De Giorgi lemma twice; first we get a bound only in $L^2(\mathbb{Z})$ then using this information we upgrade it to an $L^\infty(\mathbb{Z})$ bound. This concludes the sketch of the proof of Theorem 5.10. $\qquad\square$

5.4. From local measures to Wigner matrices and β-ensembles. Given Theorem 5.1, the proofs of Theorem 1.3 and 1.5 follow relatively standard ideas from previous results, some of them were reviewed in Sections 3 and 4. The key inputs are to verify the condition (5.13) and to ensure that the configuration intervals coincide (5.11).

For the β-ensemble, (5.13) simply follows from conditioning the global rigidity estimate in Theorem 4.1. For matching the configuration intervals, first we match the local density by scaling and translation that guarantees that $|J_{\mathbf{y}}| \sim |J_{\widetilde{\mathbf{y}}}|$, see (5.12). Then, with a second scaling, we fine tune the slight discrepancy between the lengths of $J_{\mathbf{y}}$ and $J_{\widetilde{\mathbf{y}}}$. This finishes the proof of Theorem 1.5. $\qquad\square$

In the Wigner case, we always work on the same configuration interval, so matching of J is automatic. The proof of (5.13), however, requires a bit more effort than for the β-ensemble, but it will relatively easily follow from other information we already collected along the three step strategy described in Section 1.5. As we explained in the proof of Theorem 1.2, the averaging over the energy was really needed only in the second step, where the closeness of the local statistics of $f_t\mu$ and μ was shown for small t, where f_t is the evolution of the DBM. As a byproduct of this step, we obtain bounds on the global entropy and Dirichlet form, see (3.38). In particular, the local Dirichlet form w.r.t $\mu_{\mathbf{y}}$ can be estimated by the global one, which then can be used to compare expectations w.r.t. the conditional measures $f_{t,\mathbf{y}}\mu_{\mathbf{y}}$ and $\mu_{\mathbf{y}}$;

$$(5.65) \qquad \mathbb{E}^{f_{t,\mathbf{y}}\mu_{\mathbf{y}}} O(\mathbf{x}) - \mathbb{E}^{\mu_{\mathbf{y}}} O(\mathbf{x}).$$

We are especially interested in controlling the difference

$$(5.66) \qquad \left| \mathbb{E}^{f_{t,\mathbf{y}}\mu_{\mathbf{y}}} x_j - \mathbb{E}^{\mu_{\mathbf{y}}} x_j \right| \leqslant CK^\xi N^{-1}.$$

Since $\mathbb{E}^{f_t\mu} x_j$ is close to its classical location γ_j by rigidity (5.16) for Wigner matrices, after conditioning, we obtain that $\mathbb{E}^{f_{t,\mathbf{y}}\mu_{\mathbf{y}}} x_j$ is also close to γ_j, at least for most $\mathbf{y}$ w.r.t. $f_t\mu$. Combining this information with (5.66) yields (5.13). Therefore Theorem 5.1 applies and we will use it for a Gaussian case, $V(x) = \widetilde{V}(x) = x^2/2$ but with two different boundary conditions $\mathbf{y}, \widetilde{\mathbf{y}} \in \mathbb{R}$. Since this holds for most $\widetilde{\mathbf{y}}$ w.r.t the measure μ, it also holds for μ itself, i.e. the gap statistics of $\mu_{\mathbf{y}}$ and μ coincide. On the other hand, the estimate (5.65) applied to the observable $O(x_j - x_{j+1})$ implies directly that the single gap distribution w.r.t. $f_{t,\mathbf{y}}\mu_{\mathbf{y}}$ and $\mu_{\mathbf{y}}$ coincide for most of the $\mathbf{y}$ w.r.t. $f_t\mu$. Finally, the gap statistics of $f_{t,\mathbf{y}}\mu_{\mathbf{y}}$ and $f_t\mu$ coincide for most $\mathbf{y}$ w.r.t. $f_t\mu$ by conditioning. Putting these relations together we obtain that the gap statistics of $f_t\mu$ and μ coincide, i.e. the local measures, that played an important auxiliary role, are eliminated.

Finally, the small Gaussian component present in $f_t\mu$ for small but non-zero t can be removed by the Green function comparison theorem, Theorem 3.6. Although the direct application of the Green functions give information only on eigenvalues around a fixed energy and not on an eigenvalue with a fixed label, the estimates are strong enough to transfer fixed energy information to fixed label. The main reason for this flexibility is that Theorem 3.6 allows for very small $\eta \sim N^{-1-\varepsilon}$, i.e. well below the typical spacing. Indeed, Theorem 1.10 from [59] implies that if the first four moments of two generalized Wigner ensembles, $H^{\mathbf{v}}$ and $H^{\mathbf{w}}$, are the same, then we have

$$(5.67) \qquad \lim_{N\to\infty} \left[\mathbb{E}^{\mathbf{v}} - \mathbb{E}^{\mathbf{w}}\right] O\big(N(x_j - x_{j+1}), N(x_j - x_{j+2}), \ldots, N(x_j - x_{j+n})\big) = 0.$$

Roughly speaking, the proof of (5.67) in [59] was based on Theorem 3.6. In order to convert fixed energy to a fixed eigenvalue index, one needs to know that the total number of eigenvalues up to a fixed energy is the same for the two ensembles. The total number of eigenvalues up to a fixed energy E can be expressed in terms of integration of imaginary part of the trace of Green functions, i.e.,

$$\int_{-\infty}^{E} dy \; \mathrm{Im} \; \mathrm{Tr} \; \frac{1}{H - (y + i\eta)}$$

with an η slightly smaller than $1/N$. Thus the basic idea of the Green function comparison theorem can be employed and this leads to (5.67). This completes the proof of Theorem 1.3. $\qquad\square$

References

[1] Aizenman, M., and Molchanov, S.: Localization at large disorder and at extreme energies: an elementary derivation, *Commun. Math. Phys.* **157**, 245–278 (1993)

[2] Anantharaman, N., Nonnenmacher, S.: Half-delocalization of eigenfunctions for the Laplacian on an Anosov manifold. *Annales de l'Institut Fourier* **57**, no. 7, 2465–2523 (2007)

[3] Anderson, G., Guionnet, A., Zeitouni, O.: An Introduction to Random Matrices. Studies in advanced mathematics, **118**, Cambridge University Press, 2009.

[4] Anderson, P.: Absences of diffusion in certain random lattices, *Phys. Rev.* **109**, 1492–1505 (1958)

[5] Albeverio, S., Pastur, L., Shcherbina, M.: On the $1/n$ expansion for some unitary invariant ensembles of random matrices, *Commun. Math. Phys.* **224**, 271–305 (2001).

[6] Bakry, D., Émery, M.: Diffusions hypercontractives. In: *Séminaire de probabilités, XIX, 1983/ 84, vol. 1123 of Lecture Notes in Math.* Springer, Berlin, 1985, pp. 177–206.

[7] Ben Arous, G., Péché, S.: Universality of local eigenvalue statistics for some sample covariance matrices. *Comm. Pure Appl. Math.* **LVIII.** (2005), 1–42.

[8] Berry, M.V., Tabor, M.: Level clustering in the regular spectrum, *Proc. Roy. Soc.* **A 356** (1977) 375-394

[9] Bleher, P., Its, A.: Semiclassical asymptotics of orthogonal polynomials, Riemann-Hilbert problem, and universality in the matrix model. *Ann. of Math.* **150** (1999), 185–266.

[10] Bohigas, O.; Giannoni, M.-J.; Schmit, C.: Characterization of chaotic quantum spectra and universality of level fluctuation laws. *Phys. Rev. Lett.* **52**, no. 1, 1–4, (1984)

[11] Bourgade, P., Erdős, Yau, H.-T.: Universality of General β-Ensembles, arXiv:1104.2272

[12] Bourgade, P., Erdős, Yau, H.-T.: Bulk Universality of General β-Ensembles with Non-convex Potential, *J. Math. Phys.* **53**, 095221 (2012)

[13] Bourgade, P., Erdős, Yau, H.-T.: Edge Universality of General β-Ensembles. In preparation.

[14] Brézin, E., Hikami, S.: Correlations of nearby levels induced by a random potential. *Nucl. Phys. B* **479** (1996), 697–706, and Spectral form factor in a random matrix theory. *Phys. Rev. E* **55**, 4067–4083 (1997)

[15] Caffarelli, L., Chan, C.H., Vasseur, A.: Regularity theory for parabolic nonlinear integral operators, *J. Amer. Math. Soc.* **24**, no. 3, 849–889 (2011)

[16] Chatterjee, S.: A generalization of the Lindeberg principle. *Ann. Probab.* **34**, no. 6, 2061–2076 (2006)

[17] Deuschel, J.-D., Giacomin, G., Ioffe, D.: Large deviations and concentration properties for $\nabla \varphi$ interface models. *Probab. Theor. Relat. Fields.* **117**, 49–111 (2000)

[18] Deift, P.: Orthogonal polynomials and random matrices: a Riemann-Hilbert approach. *Courant Lecture Notes in Mathematics* **3**, American Mathematical Society, Providence, RI, 1999

[19] Deift, P., Gioev, D.: Universality in random matrix theory for orthogonal and symplectic ensembles. *Int. Math. Res. Pap. IMRP* 2007, no. 2, Art. ID rpm004, 116 pp

[20] Deift, P., Gioev, D.: Random Matrix Theory: Invariant Ensembles and Universality. *Courant Lecture Notes in Mathematics* **18**, American Mathematical Society, Providence, RI, 2009

[21] Deift, P., Kriecherbauer, T., McLaughlin, K.T-R, Venakides, S., Zhou, X.: Uniform asymptotics for polynomials orthogonal with respect to varying exponential weights and applications to universality questions in random matrix theory. *Comm. Pure Appl. Math.* **52**, 1335–1425 (1999)

[22] Deift, P., Kriecherbauer, T., McLaughlin, K.T-R, Venakides, S., Zhou, X.: Strong asymptotics of orthogonal polynomials with respect to exponential weights. *Comm. Pure Appl. Math.* **52**, 1491–1552 (1999)

[23] Dumitriu, I., Edelman, A.: Matrix Models for Beta Ensembles, *Journal of Mathematical Physics* **43** (11), 5830–5847 (2002)

[24] Dyson, F.J.: Statistical theory of energy levels of complex systems, I, II, and III. *J. Math. Phys.* **3**, 140-156, 157-165, 166-175 (1962)

[25] Dyson, F.J.: A Brownian-motion model for the eigenvalues of a random matrix. *J. Math. Phys.* **3**, 1191-1198 (1962)

[26] Dyson, F.J.: Correlations between eigenvalues of a random matrix. *Commun. Math. Phys.* **19**, 235-250 (1970)

[27] Erdős, L.: Universality of Wigner Random Matrices: a Survey of Recent Results. *Russian Math. Surveys* **66** (3) 67–198.

[28] Erdős, L., Knowles, A.: Quantum Diffusion and Eigenfunction Delocalization in a Random Band Matrix Model. *Commun. Math. Phys.* **303** no. 2, 509–554 (2011)

[29] Erdős, L., A. Knowles, A.: Quantum Diffusion and Delocalization for Band Matrices with General Distribution. *Annales Inst. H. Poincaré*, **12** (7), 1227-1319 (2011)

[30] Erdős, L., A. Knowles, A., Yau, H.-T.: Averaging Fluctuations in Resolvents of Random Band Matrices. Preprint. arXiv:1205.5664.

[31] Erdős, L., A. Knowles, A., Yau, H.-T., J. Yin.: The local semicircle law for a general class of random matrices. Preprint. arXiv:1212.0164.

[32] Erdős, L., Knowles, A., Yau, H.-T., Yin, J.: Spectral Statistics of Erdős-Rényi Graphs I: Local Semicircle Law. To appear in *Annals Probab.* Preprint: arXiv:1103.1919

[33] Erdős, L., Knowles, A., Yau, H.-T., Yin, J.: Spectral Statistics of Erdős-Rényi Graphs II: Eigenvalue Spacing and the Extreme Eigenvalues. *Comm. Math. Phys.* **314** no. 3. 587–640 (2012)

[34] Erdős, L., Knowles, A., Yau, H.-T., Yin, J.: Delocalization and Diffusion Profile for Random Band Matrices. Preprint: arXiv:1205.5669

[35] Erdős, L., Péché, G., Ramírez, J., Schlein, B., and Yau, H.-T., Bulk universality for Wigner matrices. *Commun. Pure Appl. Math.* **63**, No. 7, 895–925 (2010)

[36] Erdős, L., Ramirez, J., Schlein, B., Tao, T., Vu, V., Yau, H.-T.: Bulk Universality for Wigner Hermitian matrices with subexponential decay. *Math. Res. Lett.* **17** (2010), no. 4, 667–674.

[37] Erdős, L., Ramirez, J., Schlein, B., Yau, H.-T.: Universality of sine-kernel for Wigner matrices with a small Gaussian perturbation. *Electron. J. Prob.* **15**, Paper 18, 526–604 (2010)

[38] Erdős, L., Schlein, B., Yau, H.-T.: Semicircle law on short scales and delocalization of eigenvectors for Wigner random matrices. *Ann. Probab.* **37**, No. 3, 815–852 (2009)

[39] Erdős, L., Schlein, B., Yau, H.-T.: Local semicircle law and complete delocalization for Wigner random matrices. *Commun. Math. Phys.* **287**, 641–655 (2009)

[40] Erdős, L., Schlein, B., Yau, H.-T.: Wegner estimate and level repulsion for Wigner random matrices. *Int. Math. Res. Notices.* **2010**, No. 3, 436-479 (2010)

[41] Erdős, L., Schlein, B., Yau, H.-T.: Universality of random matrices and local relaxation flow. *Invent. Math.* **185** (2011), no.1, 75–119.

[42] Erdős, L., Schlein, B., Yau, H.-T., Yin, J.: The local relaxation flow approach to universality of the local statistics for random matrices. *Annales Inst. H. Poincaré (B), Probability and Statistics* **48**, no. 1, 1–46 (2012)

[43] Erdős, L., Yau, H.-T.: Universality of local spectral statistics of random matrices. *Bull. Amer. Math. Soc.* **49**, no.3 (2012), 377–414.

[44] Erdős, L., Yau, H.-T.: A comment on the Wigner-Dyson-Mehta bulk universality conjecture for Wigner matrices. *Electron. J. Probab.* **17**, no 28. 1–5 (2012)

[45] Erdős, L., Yau, H.-T.: Gap universality of generalized Wigner and β-ensembles. Preprint arXiv:1211.3786

[46] Erdős, L., Yau, H.-T., Yin, J.: Bulk universality for generalized Wigner matrices. To appear in *Prob. Theor. Rel. Fields.* Preprint arXiv:1001.3453

[47] Erdős, L., Yau, H.-T., Yin, J.: Universality for generalized Wigner matrices with Bernoulli distribution. *J. of Combinatorics,* **1**, no. 2, 15–85 (2011)

[48] Erdős, L., Yau, H.-T., Yin, J.: Rigidity of Eigenvalues of Generalized Wigner Matrices. *Adv. Math.* **229**, no. 3, 1435–1515 (2012)

[49] Eynard, B.: Master loop equations, free energy and correlations for the chain of matrices. *J. High Energy Phys.* **11**, 018 (2003)

[50] Fokas, A. S., Its, A. R., Kitaev, A. V.: The isomonodromy approach to matrix models in 2D quantum gravity. *Comm. Math. Phys.* **147**, 395–430 (1992)

[51] Fröhlich, J., Spencer, T.: Absence of diffusion in the Anderson tight binding model for large disorder or low energy, *Commun. Math. Phys.* **88**, 151–184 (1983)

[52] Fyodorov, Y.V. and Mirlin, A.D.: Scaling properties of localization in random band matrices: a σ-model approach. *Phys. Rev. Lett.* **67**, 2405–2409 (1991)

[53] Gaudin, M.: Sur la loi limit de l'espacement des valeurs propres d'une matrice aléatoire. *Nucl. Phys.* **25**, 447-458.

[54] Giacomin, G., Olla, S., Spohn, H.: Equilibrium fluctuations for $\nabla\varphi$ interface model. *Ann. Probab.* **29**, no.3., 1138–1172 (2001)

[55] Gustavsson, J. : Gaussian fluctuations of eigenvalues in the GUE. *Ann. Inst. H. Poincaré Probab. Statist.* **41**, no. 2, 151–178 (2005)

[56] Helffer, B., Sjöstrand, J.: On the correlation for Kac-like models in the convex case. *J. Statis. Phys.* **74**, no.1-2, 349–409 (1994)

[57] Johansson, K.: Universality of the local spacing distribution in certain ensembles of Hermitian Wigner matrices. *Comm. Math. Phys.* **215**, no.3. 683–705 (2001)

[58] Johansson, K.: On the fluctuations of eigenvalues of random Hermitian matrices. *Duke Math. J.* **91**, 151–204 (1998)

[59] Knowles, A., Yin, J.: Eigenvector distribution of Wigner matrices. Preprint arXiv:1102.0057.

[60] Kriecherbauer, T., Shcherbina, M.: Fluctuations of eigenvalues of matrix models and their applications. Preprint arXiv:1003.6121

[61] Lee, J. O., Yin, J.: A Necessary and Sufficient Condition for Edge Universality of Wigner matrices. Preprint. arXiv:1206.2251

[62] Lindenstrauss, E.: Invariant measures and arithmetic quantum ergodicity *Ann. Math.* **163**, 165–219 (2006)

[63] Lubinsky, D.S.: A New Approach to Universality Limits Involving Orthogonal Polynomials, *Ann. Math.,* **170**, 915-939 (2009)

[64] Marklof, J.: Energy level statistics, lattice point problems and almost modular functions, *in* Cartier, Julia, Moussa, Vanhove (Herausgeber): Frontiers in Number Theory, Physics and Geometry (Les Houches Lectures 2003), Band 1, Springer Verlag 2006, S. 163–181

[65] Mehta, M.L.: *Random Matrices.* Third Edition, Academic Press, New York, 1991.

[66] Mehta, M.L.: A note on correlations between eigenvalues of a random matrix. *Commun. Math. Phys.* **20** no.3. 245–250 (1971)

[67] Mehta, M.L., Gaudin, M.: On the density of eigenvalues of a random matrix. *Nuclear Phys.* **18**, 420-427 (1960).

[68] Minami, N.: Local fluctuation of the spectrum of a multidimensional Anderson tight binding model. *Commun. Math. Phys.* **177**, 709–725 (1996)

[69] Montgomery, H.L.: The pair correlation of zeros of the zeta function. Analytic number theory, Proc. of Sympos. in Pure Math. **24**), Amer. Math. Soc. Providence, R.I., 181–193 (1973).

[70] Naddaf, A., Spencer, T.: On homogenization and scaling limit of some gradient perturbations of a massless free field, *Commun. Math. Phys.* **183**, no.1., 55–84 (1997)

[71] Ogawa, T.; Ozawa, T.: Trudinger type inequalities and uniqueness of weak solutions for the nonlinear Schrödinger mixed problem. *J. Math. Anal. Appl.* **155**, no. 2, 531–540 (1991)

[72] Pastur, L., Shcherbina, M.: Universality of the local eigenvalue statistics for a class of unitary invariant random matrix ensembles. *J. Stat. Phys.* **86**, 109-147 (1997)

[73] Pastur, L., Shcherbina M.: Bulk universality and related properties of Hermitian matrix models. *J. Stat. Phys.* **130**, no.2., 205-250 (2008)

[74] Pillai, N.S. and Yin, J.: Universality of covariance matrices. Preprint arXiv:1110.2501

[75] Pillai, N.S. and Yin, J.: Edge universality of covariance matrices. Preprint arXiv:1112.2381

[76] Ramirez, J., Rider, B., Virág, B.: Beta ensembles, stochastic Airy spectrum, and a diffusion. *J. Amer. Math. Soc.* **24**, 919-944 (2011)

[77] Rudnick, Z. and Sarnak, P.: The pair correlation function of fractional parts of polynomials, *Comm. Math. Phys.* **194**, 61–70 (1998)

[78] Schenker, J.: Eigenvector localization for random band matrices with power law band width. *Commun. Math. Phys.* **290**, 1065–1097 (2009)

[79] Shcherbina, M.: Orthogonal and symplectic matrix models: universality and other properties. *Comm. Math. Phys.* **307**, no.3., 761–790 (2011)

[80] Sinai, Y.: Poisson distribution in a geometrical problem, *Adv. Sov. Math.* AMS Publ. **3**, 199–215 (1991)

[81] Spencer, T.: Random banded and sparse matrices (Chapter 23), to appear in "Oxford Handbook of Random Matrix Theory" edited by G. Akemann, J. Baik, and P. Di Francesco

[82] Tao, T. and Vu, V.: Random matrices: Universality of the local eigenvalue statistics. *Acta Math.*, **206**, no. 1, 127–204 (2011)

[83] Tao, T. and Vu, V.: The Wigner-Dyson-Mehta bulk universality conjecture for Wigner matrices. *Electron. J. Probab.* **16**, no 77, 2104–2121 (2011)

[84] Tao, T.: The asymptotic distribution of a single eigenvalue gap of a Wigner matrix. Preprint. arxiv:1203.1605

[85] Valkó, B.; Virág, B.: Continuum limits of random matrices and the Brownian carousel. *Invent. Math.* **177**, no. 3, 463-508 (2009)

[86] Widom H.: On the relation between orthogonal, symplectic and unitary matrix ensembles. *J. Statist. Phys.* **94**, no. 3-4, 347–363 (1999)

[87] Wigner, E.: Characteristic vectors of bordered matrices with infinite dimensions. *Ann. of Math.* **62**, 548-564 (1955)

[88] Yau, H. T.: Relative entropy and the hydrodynamics of Ginzburg-Landau models, *Lett. Math. Phys.* **22**, 63–80 (1991)

INSTITUTE OF MATHEMATICS, UNIVERSITY OF MUNICH, THERESIENSTR. 39, D-80333 MUNICH, GERMANY

E-mail address: lerdos@math.lmu.de

Mirror symmetry and the Strominger-Yau-Zaslow conjecture

Mark Gross

ABSTRACT. We trace progress and thinking about the Strominger-Yau-Zaslow conjecture since its introduction in 1996. We begin with the original differential geometric conjecture and its refinements, and explain how insights gained in this context led to the algebro-geometric program developed by the author and Siebert. The objective of this program is to explain mirror symmetry by studying degenerations of Calabi-Yau manifolds. This introduces logarithmic and tropical geometry into the mirror symmetry story, and gives a clear path towards a conceptual understanding of mirror symmetry within an algebro-geometric context. After explaining the overall philosophy, we explain how recent results fit into this program.

CONTENTS

Introduction

Mirror symmetry got its start in 1989 with work of Greene and Plesser [17] and Candelas, Lynker and Schimmrigk [9]. These two works first observed the existence

2010 *Mathematics Subject Classification.* 14J32.

of pairs of Calabi-Yau manifolds exhibiting an exchange of Hodge numbers. Recall that by Yau's proof of the Calabi conjecture [77], a Calabi-Yau manifold is an n-dimensional complex manifold X with a nowhere vanishing holomorphic n-form Ω and a Ricci-flat Kähler metric with Kähler form ω. Ricci-flatness is equivalent to $\omega^n = C\Omega \wedge \bar{\Omega}$ for a constant C.

The most famous example of a Calabi-Yau manifold is a smooth quintic three-fold $X \subseteq \mathbb{P}^4$. The Hodge numbers of X are $h^{1,1}(X) = 1$ and $h^{1,2}(X) = 101$, with topological Euler characteristic -200. The original construction of Greene and Plesser gave a mirror to X, as follows. Let $Y \subseteq \mathbb{P}^4$ be given by the equation

$$x_0^5 + \cdots + x_4^5 = 0,$$

and let

$$G = \{(a_0, \ldots, a_4) \in \mathbb{Z}_5^5 \mid \sum_i a_i = 0\}.$$

An element $(a_0, \ldots, a_4) \in G$ acts on Y by

$$(x_0, \ldots, x_4) \mapsto (\xi^{a_0} x_0, \ldots, \xi^{a_4} x_4)$$

for ξ a primitive fifth root of unity. The quotient Y/G is highly singular, but there is a resolution of singularities $\check{X} \to Y/G$ such that $\check{X}$ is also Calabi-Yau, and one finds $h^{1,1}(\check{X}) = 101$ and $h^{1,2}(\check{X}) = 1$, so that $\check{X}$ has topological Euler characteristic 200.

The relationship between these two Calabi-Yau manifolds proved to be much deeper than just this exchange of Hodge numbers. Pioneering work of Candelas, de la Ossa, Greene and Parkes [10] performed an amazing calculation, following string-theoretic predictions which suggested that certain enumerative calculations on X should give the same answer as certain period calculations on $\check{X}$. The calculations on $\check{X}$, though subtle, could be carried out: these involved integrals of the holomorphic form on $\check{X}$ over three-cycles as the complex structure on $\check{X}$ is varied. On the other hand, the corresponding calculations on X involved numbers of rational curves on X of each degree. For example, the number of lines on a generic quintic threefold is 2875 and the number of conics is 609250. String theory thus gave predictions for these numbers for every degree, an astonishing feat given that most of these numbers seemed far beyond the reach of algebraic geometry at the time.

More generally, string theory introduced the concepts of the *A-model* and *B-model*. The *A*-model involves the symplectic geometry of Calabi-Yau manifolds. Properly defined, the counts of rational curves are in fact symplectic invariants, now known as Gromov-Witten invariants. The *B*-model, on the other hand, involves the complex geometry of Calabi-Yau manifolds. Holomorphic forms of course depend on the complex structure, so the period calculations mentioned above can be thought of as *B*-model calculations. Ultimately, string theory predicts an isomorphism between the *A*-model of a Calabi-Yau manifold X and the *B*-model of its mirror, $\check{X}$. The equality of numerical invariants is then a consequence of this isomorphism.

Proofs of these string-theoretic predictions of curve-counting invariants were given by Givental [15] and Lian, Liu and Yau [59], with successively simpler proofs by many other researchers. However, all the proofs relied on the geometry of the ambient space $\mathbb{P}^4$ in which the quintic is contained. Roughly speaking, one considers all rational curves in $\mathbb{P}^4$, and tries to understand how to compute how many of these are contained in a given quintic hypersurface.

This raised the question: *is there some underlying intrinsic geometry to mirror symmetry?*

Historically the first approach to an intrinsic formulation of mirror symmetry is Kontsevich's Homological Mirror Symmetry conjecture, stated in 1994 in [52]. This made mathematically precise the notion of an isomorphism between the A- and B-models. The homological mirror symmetry conjecture posits an isomorphism between two categories, the Fukaya category of Lagrangian submanifolds of X (the A-model) and the derived category of coherent sheaves on the mirror $\check{X}$ (the B-model). Morally, this states that the symplectic geometry of X is the same as the complex geometry of $\check{X}$. At the time this conjecture was made, however, there was no clear idea as to how such an isomorphism might be realised, nor did this conjecture state how to construct mirror pairs.

The second approach is due to Strominger, Yau and Zaslow in their 1996 paper [75]. They made a remarkable proposal, based on new ideas in string theory, which gave a very concrete geometric interpretation for mirror symmetry.

Let me summarize, very roughly, the physical argument they used here. Developments in string theory in the mid-1990s introduced the notion of *Dirichlet branes*, or *D*-branes. These are submanifolds of space-time, with some additional data, which serve as boundary conditions for open strings, i.e., we allow open strings to propagate with their endpoints constrained to lie on a *D*-brane. Remembering that space-time, according to string theory, looks like $\mathbb{R}^{1,3} \times X$, where $\mathbb{R}^{1,3}$ is ordinary space-time and X is a Calabi-Yau three-fold, we can split a *D*-brane into a product of a submanifold of $\mathbb{R}^{1,3}$ and one on X. It turned out, simplifying a great deal, that there are two particular types of submanifolds on X of interest: *holomorphic D*-branes, i.e., holomorphic submanifolds with a holomorphic line bundle, and *special Lagrangian D*-branes, which are *special Lagrangian submanifolds* with flat $U(1)$-bundle:

DEFINITION 0.1. Let X be an n-dimensional Calabi-Yau manifold with ω the Kähler form of a Ricci-flat metric on X and Ω a nowhere vanishing holomorphic n-form. Then a submanifold $M \subseteq X$ is *special Lagrangian* if it is Lagrangian, i.e., $\dim_{\mathbb{R}} M = \dim_{\mathbb{C}} X$ and $\omega|_M = 0$, and in addition, $\operatorname{Im} \Omega|_M = 0$.

Holomorphic D-branes can be viewed as B-model objects, and special Lagrangian D-branes as A-model objects. The isomorphism between the B-model on X and the A-model on $\check{X}$ then suggests that the moduli space of holomorphic D-branes on X should be isomorphic to the moduli space of special Lagrangian D-branes on $\check{X}$. (This is now seen as a physical manifestation of the homological mirror symmetry conjecture). Now X itself is the moduli space of points on X. So each point on X should correspond to a pair (M, ∇), where $M \subseteq \check{X}$ is a special Lagrangian submanifold and ∇ is a flat $U(1)$-connection on M.

A theorem of McLean [61] tells us that the tangent space to the moduli space of special Lagrangian deformations of a special Lagrangian submanifold $M \subseteq \check{X}$ is $H^1(M, \mathbb{R})$. Of course, the moduli space of flat $U(1)$-connections modulo gauge equivalence on M is the torus $H^1(M, \mathbb{R})/H^1(M, \mathbb{Z})$. In order for this moduli space to be of the correct dimension, we need $\dim H^1(M, \mathbb{R}) = n$, the complex dimension of X. This suggests that X consists of a family of tori which are dual to a family of special Lagrangian tori on $\check{X}$. An elaboration of this argument yields the following conjecture:

CONJECTURE 0.2. The Strominger-Yau-Zaslow conjecture. *If X and $\check{X}$ are a mirror pair of Calabi-Yau n-folds, then there exists fibrations $f : X \to B$ and $\check{f} : \check{X} \to B$ whose fibres are special Lagrangian, with general fibre an n-torus. Furthermore, these fibrations are dual, in the sense that canonically $X_b = H^1(\check{X}_b, \mathbb{R}/\mathbb{Z})$ and $\check{X}_b = H^1(X_b, \mathbb{R}/\mathbb{Z})$ whenever X_b and $\check{X}_b$ are non-singular tori.*

This conjecture motivated a great deal of work in the five years following its introduction in 1996, some of which will be summarized in the following sections. There was a certain amount of success, as we shall see, with the conjecture proved for some cases, including the quintic three-fold, at the topological level. Further, the conjecture gave a solid framework for thinking about mirror symmetry at an intuitive level. However, work of Dominic Joyce demonstrated that the conjecture was unlikely to be literally true. Nevertheless, it is possible that weaker limiting forms of the conjecture still hold.

In the first several sections of this survey, I will clarify the conjecture, review what is known about it, and state a weaker form which seems accessible. Most importantly, I will explain how the SYZ conjecture leads to the study of affine manifolds (manifolds with transition functions being affine linear) and hence to an algebro-geometric interpretation of the conjecture, developed by me and Bernd Siebert. This removes the hard analysis, and gives a powerful framework for understanding mirror symmetry at a conceptual level.

The bulk of the paper is devoted to outlining this framework as developed over the last ten years. I explain how affine manifolds are related to degenerations of Calabi-Yau manifolds. Once one begins to consider degenerations, log geometry of K. Kato and Fontaine–Illusie comes into the picture. Conjecturally, the base of the SYZ fibration incorporates key combinatorial information about log structures on degenerations of Calabi-Yau manifolds. Log geometry then gives a connection with tropical geometry and log Gromov-Witten theory, which theoretically allows a description of A-model curve counting using tropical geometry. On the mirror side, we explain how again tropical geometry is used to describe complex structures. This identifies tropical geometry as the geometry underlying both sides of mirror symmetry, and guides us towards a conceptual understanding of mirror symmetry. We end with a description of recent work with Pandharipande and Siebert [**28**] which provides a snapshot of the relationship between the two sides of mirror symmetry.

Acknowledgments. I would like to thank the organizers of Current Developments in Mathematics 2012 for inviting me to take part in the conference, and Bernd Siebert, my collaborator on much of the work described here. Some of the material appearing in this article was first published in my article "The Strominger-Yau-Zaslow conjecture: From torus fibrations to degenerations," in *Algebraic Geometry: Seattle 2005*, edited by D. Abramovich, et al., Proceedings of Symposia in Pure Mathematics Vol. 80, part 1, 149-192, published by the American Mathematical Society. (c) 2009 by the American Mathematical Society. Finally, I would like to thank Lori Lejeune and the Clay Institute for Figure 3.

This work was partially supported by NSF grant 1105871.

1. Moduli of special Lagrangian submanifolds

The first step in understanding the SYZ conjecture is to examine the structures which arise on the base of a special Lagrangian fibration. These structures arise from

McLean's theorem on the moduli space of special Lagrangian submanifolds [**61**], and these structures and their relationships were explained by Hitchin in [**42**]. We outline some of these ideas here. McLean's theorem says that the moduli space of deformations of a compact special Lagrangian submanifold of a compact Calabi-Yau manifold X is unobstructed. Further, the tangent space at the point of moduli space corresponding to a special Lagrangian $M \subseteq X$ is canonically isomorphic to the space of harmonic 1-forms on M. This isomorphism is seen explicitly as follows. Let $\nu \in \Gamma(M, N_{M/X})$ be a normal vector field to M in X. Then the restriction of the contractions $(\iota(\nu)\omega)|_M$ and $(\iota(\nu)\operatorname{Im}\Omega)|_M$ are both seen to be well-defined forms on M: one needs to lift ν to a vector field but the choice is irrelevant because ω and $\operatorname{Im}\Omega$ restrict to zero on M. McLean shows that if M is special Lagrangian then

$$\iota(\nu)\operatorname{Im}\Omega = - * \iota(\nu)\omega,$$

where $*$ denotes the Hodge star operator on M. Furthermore, ν corresponds to an infinitesimal deformation preserving the special Lagrangian condition if and only if $d(\iota(\nu)\omega) = d(\iota(\nu)\operatorname{Im}\Omega) = 0$. This gives the correspondence between harmonic 1-forms and infinitesimal special Lagrangian deformations.

Let $f : X \to B$ be a special Lagrangian fibration with torus fibres, and assume for now that all fibres of f are non-singular. Then we obtain three structures on B, namely two affine structures and a metric, as we shall now see.

DEFINITION 1.1. Let B be an n-dimensional manifold. An *affine structure* on B is given by an atlas $\{(U_i, \psi_i)\}$ of coordinate charts $\psi_i : U_i \to \mathbb{R}^n$, whose transition functions $\psi_i \circ \psi_j^{-1}$ lie in $\operatorname{Aff}(\mathbb{R}^n)$. We say the affine structure is *tropical* if the transition functions lie in $\mathbb{R}^n \rtimes GL(\mathbb{Z}^n)$, i.e., have integral linear part. We say the affine structure is *integral* if the transition functions lie in $\operatorname{Aff}(\mathbb{Z}^n)$.

If an affine manifold B carries a Riemannian metric g, then we say the metric is *affine Kähler* or *Hessian* if g is locally given by $g_{ij} = \partial^2 K / \partial y_i \partial y_j$ for some convex function K and $y_1, \ldots, y_n$ affine coordinates.

We obtain the three structures as follows:

Affine structure 1. For a normal vector field ν to a fibre X_b of f, $(\iota(\nu)\omega)|_{X_b}$ is a well-defined 1-form on X_b, and we can compute its periods as follows. Let $U \subseteq B$ be a small open set, and suppose we have submanifolds $\gamma_1, \ldots, \gamma_n \subseteq f^{-1}(U)$ which are families of 1-cycles over U and such that $\gamma_1 \cap X_b, \ldots, \gamma_n \cap X_b$ form a basis for $H_1(X_b, \mathbb{Z})$ for each $b \in U$. Consider the 1-forms $\omega_1, \ldots, \omega_n$ on U defined by fibrewise integration:

$$\omega_i(\nu) = \int_{X_b \cap \gamma_i} \iota(\nu)\omega,$$

for ν a tangent vector on B at b, which we can lift to a normal vector field of X_b. We have $\omega_i = f_*(\omega|_{\gamma_i})$, and since ω is closed, so is ω_i. Thus there are locally defined functions $y_1, \ldots, y_n$ on U with $dy_i = \omega_i$. Furthermore, these functions are well-defined up to the choice of basis of $H_1(X_b, \mathbb{Z})$ and constants. Finally, they give well-defined coordinates, as follows from the fact that $\nu \mapsto \iota(\nu)\omega$ yields an isomorphism of $T_{B,b}$ with $H^1(X_b, \mathbb{R})$ by McLean's theorem. Thus $y_1, \ldots, y_n$ define local coordinates of a tropical affine structure on B.

Affine structure 2. We can play the same trick with $\operatorname{Im}\Omega$: choose submanifolds

$$\Gamma_1, \ldots, \Gamma_n \subseteq f^{-1}(U)$$

which are families of $n-1$-cycles over U and such that $\Gamma_1 \cap X_b, \ldots, \Gamma_n \cap X_b$ form a basis for $H_{n-1}(X_b, \mathbb{Z})$. We define λ_i by $\lambda_i = -f_*(\operatorname{Im}\Omega|_{\Gamma_i})$, or equivalently,

$$\lambda_i(\nu) = -\int_{X_b \cap \Gamma_i} \iota(\nu)\operatorname{Im}\Omega.$$

Again $\lambda_1, \ldots, \lambda_n$ are closed 1-forms, with $\lambda_i = d\check{y}_i$ locally, and again $\check{y}_1, \ldots, \check{y}_n$ are affine coordinates for a tropical affine structure on B.

The McLean metric. The Hodge metric on $H^1(X_b, \mathbb{R})$ is given by

$$g(\alpha, \beta) = \int_{X_b} \alpha \wedge *\beta$$

for α, β harmonic 1-forms, and hence induces a metric on B, which can be written as

$$g(\nu_1, \nu_2) = -\int_{X_b} \iota(\nu_1)\omega \wedge \iota(\nu_2)\operatorname{Im}\Omega.$$

A crucial observation of Hitchin [**42**] is that these structures are related by the Legendre transform:

PROPOSITION 1.2. *Let $y_1, \ldots, y_n$ be local affine coordinates on B with respect to the affine structure induced by ω. Then locally there is a function K on B such that*

$$g(\partial/\partial y_i, \partial/\partial y_j) = \partial^2 K/\partial y_i \partial y_j.$$

Furthermore, $\check{y}_i = \partial K/\partial y_i$ form a system of affine coordinates with respect to the affine structure induced by $\operatorname{Im}\Omega$, and if

$$\check{K}(\check{y}_1, \ldots, \check{y}_n) = \sum \check{y}_i y_i - K(y_1, \ldots, y_n)$$

is the Legendre transform of K, then

$$y_i = \partial\check{K}/\partial\check{y}_i$$

and

$$\partial^2 \check{K}/\partial\check{y}_i \partial\check{y}_j = g(\partial/\partial\check{y}_i, \partial/\partial\check{y}_j).$$

PROOF. Take families $\gamma_1, \ldots, \gamma_n, \Gamma_1, \ldots, \Gamma_n$ as above over an open neighbourhood U with the two bases being Poincaré dual, i.e., $(\gamma_i \cap X_b) \cdot (\Gamma_j \cap X_b) = \delta_{ij}$ for $b \in U$. Let $\gamma_1^*, \ldots, \gamma_n^*$ and $\Gamma_1^*, \ldots, \Gamma_n^*$ be the dual bases for $\Gamma(U, R^1 f_* \mathbb{Z})$ and $\Gamma(U, R^{n-1} f_* \mathbb{Z})$ respectively. From the choice of γ_i's, we get local coordinates $y_1, \ldots, y_n$ with $dy_i = \omega_i$, so in particular

$$\delta_{ij} = \omega_i(\partial/\partial y_j) = \int_{\gamma_i \cap X_b} \iota(\partial/\partial y_j)\omega,$$

hence $\iota(\partial/\partial y_j)\omega$ defines the cohomology class γ_j^* in $H^1(X_b, \mathbb{R})$. Similarly, let

$$g_{ij} = -\int_{\Gamma_i \cap X_b} \iota(\partial/\partial y_j)\operatorname{Im}\Omega;$$

then $-\iota(\partial/\partial y_j)\operatorname{Im}\Omega$ defines the cohomology class $\sum_i g_{ij}\Gamma_i^*$ in $H^{n-1}(X_b, \mathbb{R})$, and $\lambda_i = \sum_j g_{ij} dy_j$. Thus

$$\begin{aligned} g(\partial/\partial y_j, \partial/\partial y_k) &= -\int_{X_b} \iota(\partial/\partial y_j)\omega \wedge \iota(\partial/\partial y_k)\operatorname{Im}\Omega \\ &= g_{jk}. \end{aligned}$$

On the other hand, let $\check{y}_1, \ldots, \check{y}_n$ be coordinates with $d\check{y}_i = \lambda_i$. Then

$$\partial \check{y}_i / \partial y_j = g_{ij} = g_{ji} = \partial \check{y}_j / \partial y_i,$$

so $\sum \check{y}_i dy_i$ is a closed 1-form. Thus there exists locally a function K such that $\partial K / \partial y_i = \check{y}_i$ and $\partial^2 K / \partial y_i \partial y_j = g(\partial/\partial y_i, \partial/\partial y_j)$. A simple calculation then confirms that $\partial \check{K} / \partial \check{y}_i = y_i$. On the other hand,

$$
\begin{aligned}
g(\partial/\partial \check{y}_i, \partial/\partial \check{y}_j) &= g\left(\sum_k \frac{\partial y_k}{\partial \check{y}_i} \frac{\partial}{\partial y_k}, \sum_\ell \frac{\partial y_\ell}{\partial \check{y}_j} \frac{\partial}{\partial y_\ell} \right) \\
&= \sum_{k,\ell} \frac{\partial y_k}{\partial \check{y}_i} \frac{\partial y_\ell}{\partial \check{y}_j} g(\partial/\partial y_k, \partial/\partial y_\ell) \\
&= \sum_{k,\ell} \frac{\partial y_k}{\partial \check{y}_i} \frac{\partial y_\ell}{\partial \check{y}_j} \frac{\partial \check{y}_k}{\partial y_\ell} \\
&= \frac{\partial y_j}{\partial \check{y}_i} = \frac{\partial^2 \check{K}}{\partial \check{y}_i \partial \check{y}_j}.
\end{aligned}
$$

$\square$

Thus we introduce the notion of the *Legendre transform* of an affine manifold with a multi-valued convex function.

DEFINITION 1.3. Let B be an affine manifold. A *multi-valued* function K on B is a collection of functions on an open cover $\{(U_i, K_i)\}$ such that on $U_i \cap U_j$, $K_i - K_j$ is affine linear. We say K is *convex* if the Hessian $(\partial^2 K_i / \partial y_j \partial y_k)$ is positive definite for all i, in any, or equivalently all, affine coordinate systems $y_1, \ldots, y_n$.

Given a pair (B, K) of affine manifold and convex multi-valued function, the *Legendre transform* of (B, K) is a pair $(\check{B}, \check{K})$ where $\check{B}$ is an affine structure on the underlying manifold of B with coordinates given locally by $\check{y}_i = \partial K / \partial y_i$, and $\check{K}$ is defined by

$$\check{K}_i(\check{y}_1, \ldots, \check{y}_n) = \sum \check{y}_j y_j - K_i(y_1, \ldots, y_n).$$

EXERCISE 1.4. *Check that $\check{K}$ is also convex, and that the Legendre transform of $(\check{B}, \check{K})$ is (B, K).*

Curiously, this Legendre transform between affine manifolds with Hessian metric seems to have first appeared in a work in statistics predating mirror symmetry, see [**2**].

2. Semi-flat mirror symmetry

Let's forget about special Lagrangian fibrations for the moment. Instead, we will look at how the structures found on B in the previous section give a toy version of mirror symmetry.

DEFINITION 2.1. Let B be a tropical affine manifold.

(1) Denote by $\Lambda \subseteq \mathcal{T}_B$ the local system of lattices generated locally by $\partial/\partial y_1, \ldots, \partial/\partial y_n$, where $y_1, \ldots, y_n$ are local affine coordinates. This is well-defined because transition maps are in $\mathbb{R}^n \rtimes GL_n(\mathbb{Z})$. Set

$$X(B) := \mathcal{T}_B / \Lambda.$$

This is a torus bundle over B. In addition, $X(B)$ carries a complex structure defined locally as follows. Let $U \subseteq B$ be an open set with affine coordinates $y_1, \ldots, y_n$, so $\mathcal{T}_U$ has coordinate functions $y_1, \ldots, y_n$, $x_1 = dy_1, \ldots, x_n = dy_n$. Then

$$q_j = e^{2\pi i (x_j + i y_j)}$$

gives a system of holomorphic coordinates on $\mathcal{T}_U / \Lambda|_U$, and the induced complex structure is independent of the choice of affine coordinates. This is called the *semi-flat* complex structure on $X(B)$.

Later we will need a variant of this: for $\epsilon > 0$, set

$$X_\epsilon(B) := \mathcal{T}_B / \epsilon \Lambda.$$

This has a complex structure with coordinates given by

$$q_j = e^{2\pi i (x_j + i y_j)/\epsilon}.$$

(As we shall see later, the limit $\epsilon \to 0$ corresponds to a "large complex structure limit.")

(2) Define $\check{\Lambda} \subseteq \mathcal{T}_B^*$ to be the local system of lattices generated locally by $dy_1, \ldots, dy_n$, with $y_1, \ldots, y_n$ local affine coordinates. Set

$$\check{X}(B) := \mathcal{T}_B^* / \check{\Lambda}.$$

Of course $\mathcal{T}_B^*$ carries a canonical symplectic structure, and this symplectic structure descends to $\check{X}(B)$.

$\square$

We write $f : X(B) \to B$ and $\check{f} : \check{X}(B) \to B$ for these torus fibrations; these are clearly dual.

Now suppose in addition we have a Hessian metric g on B, with local potential function K. Then the following propositions show that in fact both $X(B)$ and $\check{X}(B)$ become Kähler manifolds.

PROPOSITION 2.2. *$K \circ f$ is a (local) Kähler potential on $X(B)$, defining a Kähler form $\omega = 2i\partial\bar\partial(K \circ f)$. This metric is Ricci-flat if and only if K satisfies the real Monge-Ampère equation*

$$\det \frac{\partial^2 K}{\partial y_i \partial y_j} = constant.$$

PROOF. Working locally with affine coordinates (y_j) and complex coordinates

$$z_j = \frac{1}{2\pi i} \log q_j = x_j + i y_j,$$

we compute $\omega = 2i\partial\bar\partial(K \circ f) = \frac{i}{2} \sum \frac{\partial^2 K}{\partial y_j \partial y_k} dz_j \wedge d\bar{z}_k$ which is clearly positive. Furthermore, if $\Omega = dz_1 \wedge \cdots \wedge dz_n$, then ω^n is proportional to $\Omega \wedge \bar\Omega$ if and only if $\det(\partial^2 K / \partial y_j \partial y_k)$ is constant. $\square$

We write this Kähler manifold as $X(B, K)$.

Dually we have

PROPOSITION 2.3. *In local canonical coordinates $y_i, \check{x}_i$ on $\mathcal{T}_B^*$, the complex coordinate functions $z_j = \check{x}_j + i \partial K / \partial y_j$ on $\mathcal{T}_B^*$ induce a well-defined complex structure on $\check{X}(B)$, with respect to which the canonical symplectic form ω is the Kähler form* *of a metric. Furthermore this metric is Ricci-flat if and only if K satisfies the real Monge-Ampère equation*

$$\det \frac{\partial^2 K}{\partial y_j \partial y_k} = constant.$$

PROOF. It is easy to see that an affine linear change in the coordinates y_j (and hence an appropriate change in the coordinates $\check{x}_j$) results in a linear change of the coordinates z_j, so they induce a well-defined complex structure invariant under $\check{x}_j \mapsto \check{x}_j + 1$, and hence a complex structure on $\check{X}(B)$. Then one computes that

$$\omega = \sum d\check{x}_j \wedge dy_j = \frac{i}{2} \sum g^{jk} dz_j \wedge d\bar{z}_k$$

where $g_{ij} = \partial^2 K/\partial y_j \partial y_k$. Then the metric is Ricci-flat if and only if $\det(g^{jk}) = constant$, if and only if $\det(g_{jk}) = constant$. $\qquad\square$

As before, we call this Kähler manifold $\check{X}(B, K)$.

This motivates the definition

DEFINITION 2.4. An affine manifold with metric of Hessian form is a *Monge-Ampère manifold* if the local potential function K satisfies the Monge-Ampère equation $\det(\partial^2 K/\partial y_i \partial y_j) = constant$.

Hessian and Monge-Ampère manifolds were first studied by Cheng and Yau in [**12**].

EXERCISE 2.5. *Show that the identification of T_B and T_B^* given by a Hessian metric induces a canonical isomorphism $X(B, K) \cong \check{X}(\check{B}, \check{K})$ of Kähler manifolds, where $(\check{B}, \check{K})$ is the Legendre transform of (B, K).*

There is a key extra parameter which appears in mirror symmetry known as the B-field. This appears as a field in the non-linear sigma model with Calabi-Yau target space, and is required mathematically to make sense of mirror symmetry. Mirror symmetry roughly posits an isomorphism between the complex moduli space of a Calabi-Yau manifold X and the Kähler moduli space of $\check{X}$. If one interprets the Kähler moduli space to mean the space of all Ricci-flat Kähler forms on $\check{X}$, then one obtains only a real manifold as moduli space, and one needs a complex manifold to match up with the complex moduli space of X. The B-field is interpreted as an element $\mathbf{B} \in H^2(\check{X}, \mathbb{R}/\mathbb{Z})$, and one views $\mathbf{B} + i\omega$ as a complexified Kähler class on $\check{X}$ for ω a Kähler class on $\check{X}$.

In the context of our toy version of mirror symmetry, we view the B-field as an element $\mathbf{B} \in H^1(B, \Lambda_{\mathbb{R}}/\Lambda)$, where $\Lambda_{\mathbb{R}} = \Lambda \otimes_{\mathbb{Z}} \mathbb{R}$. This does not quite agree with the above definition of the B-field, as this group does not necessarily coincide with $H^2(\check{X}, \mathbb{R}/\mathbb{Z})$. However, in many important cases, such as for simply connected Calabi-Yau threefolds with torsion-free integral cohomology, these two groups do coincide. More generally, including the case of K3 surfaces and abelian varieties, one would need to pass to generalized complex structures [**43**], [**39**], [**7**], [**3**], [**44**], which we do not wish to do here.

Noting that a section of $\Lambda_{\mathbb{R}}/\Lambda$ over an open set U can be viewed as a section of $T_U/\Lambda|_U$, such a section acts on $T_U/\Lambda|_U$ via translation, and this action is in fact holomorphic with respect to the semi-flat complex structure. Thus a Čech 1-cocycle (U_{ij}, β_{ij}) representing $\mathbf{B}$ allows us to reglue $X(B)$ via translations over the intersections U_{ij}. This is done by identifying the open subsets $f^{-1}(U_{ij}) \subseteq f^{-1}(U_i)$

and $f^{-1}(U_{ij}) \subseteq f^{-1}(U_j)$ via the automorphism of $f^{-1}(U_{ij})$ given by translation by the section β_{ij}. This gives a new complex manifold $X(B, \mathbf{B})$. If in addition there is a multi-valued potential function K defining a metric, these translations preserve the metric and yield a Kähler manifold $X(B, \mathbf{B}, K)$.

Thus the full toy version of mirror symmetry is as follows:

CONSTRUCTION 2.6 (The toy mirror symmetry construction). Suppose given an affine manifold B with potential K and B-fields $\mathbf{B} \in H^1(B, \Lambda_{\mathbb{R}}/\Lambda)$, $\check{\mathbf{B}} \in H^1(B, \check{\Lambda}_{\mathbb{R}}/\check{\Lambda})$. It is not difficult to see, and you will have seen this already if you've done Exercise 2.5, that the local system $\check{\Lambda}$ defined using the affine structure on B is the same as the local system Λ defined using the affine stucture on $\check{B}$. So we say the pair

$$(X(B, \mathbf{B}, K), \check{\mathbf{B}})$$

is mirror to

$$(X(\check{B}, \check{\mathbf{B}}, \check{K}), \mathbf{B}).$$

This provides a reasonably fulfilling picture of mirror symmetry in a simple context. Many more aspects of mirror symmetry can be worked out in this semi-flat context, see [55] and [3], Chapter 6. This semi-flat case is an ideal testing ground for concepts in mirror symmetry. However, ultimately this only sheds limited insight into the general case. The only compact Calabi-Yau manifolds with semi-flat Ricci-flat metric which arise in this way are complex tori (shown by Cheng and Yau in [12]). To deal with more interesting cases, we need to allow singular fibres, and hence, singularities in the affine structure of B. The existence of singular fibres are fundamental for the most interesting aspects of mirror symmetry.

3. Affine manifolds with singularities

To deal with singular fibres, we define

DEFINITION 3.1. A *(tropical, integral) affine manifold with singularities* is a (C^0) manifold B with an open subset $B_0 \subseteq B$ which carries a (tropical, integral) affine structure, and such that $\Gamma := B \setminus B_0$ is a locally finite union of locally closed submanifolds of codimension ≥ 2.

Here we will give a relatively simple construction of such affine manifolds with singularities; a broader class of examples is given in [22]; see also [40] and [41].

Let Δ be a reflexive polytope in $M_{\mathbb{R}} = M \otimes_{\mathbb{Z}} \mathbb{R}$, where $M = \mathbb{Z}^n$. This means that Δ is a lattice polytope with a unique interior integral point $0 \in \Delta$, and the polar dual polytope

$$\nabla := \{n \in N_{\mathbb{R}} | \langle m, n \rangle \geq -1 \text{ for all } m \in \Delta\}$$

is also a lattice polytope.

Let $B = \partial \Delta$, and let $\mathscr{P}$ be a decomposition of B into lattice polytopes, i.e., $\mathscr{P}$ is a set of lattice polytopes contained in B such that (1) $B = \bigcup_{\sigma \in \mathscr{P}} \sigma$; (2) $\sigma_1, \sigma_2 \in \mathscr{P}$ implies $\sigma_1 \cap \sigma_2$ lies in $\mathscr{P}$ and is a face of both σ_1 and σ_2; (3) if $\sigma \in \mathscr{P}$, any face of σ lies in $\mathscr{P}$.

We now define a structure of integral affine manifold with singularities on B, with discriminant locus $\Gamma \subseteq B$ defined as follows. Let $\mathrm{Bar}(\mathscr{P})$ denote the first barycentric subdivision of $\mathscr{P}$ and let $\Gamma \subseteq B$ be the union of all simplices of $\mathrm{Bar}(\mathscr{P})$ not containing a vertex of $\mathscr{P}$ (a zero-dimensional cell) or intersecting the interior of a maximal cell of $\mathscr{P}$. Setting $B_0 := B \setminus \Gamma$, we define an affine structure on B_0 as follows. B_0 has an open cover

$$\{W_\sigma | \sigma \in \mathscr{P} \text{ maximal}\} \cup \{W_v | v \in \mathscr{P} \text{ a vertex}\}$$

where $W_\sigma = \operatorname{Int}(\sigma)$, the interior of σ, and

$$W_v = \bigcup_{\substack{\tau \in \operatorname{Bar}(\mathscr{P}) \\ v \in \tau}} \operatorname{Int}(\tau)$$

is the (open) star of v in $\operatorname{Bar}(\mathscr{P})$. We define an affine chart

$$\psi_\sigma : W_\sigma \hookrightarrow \mathbb{A}_\sigma \subseteq N_\mathbb{R}$$

given by the inclusion of W_σ in $\mathbb{A}_\sigma$, which denotes the unique $(n-1)$-dimensional affine hyperplane in $N_\mathbb{R}$ containing σ. Also, take $\psi_v : W_v \to M_\mathbb{R}/\mathbb{R}v$ to be the projection. One checks easily that for $v \in \sigma$, $\psi_\sigma \circ \psi_v^{-1}$ is integral affine linear (integrality follows from reflexivity of Δ!) so B is an integral affine manifold with singularities.

EXAMPLE 3.2. Let $\Delta \subseteq \mathbb{R}^4$ be the convex hull of the points

$$(-1,-1,-1,-1),$$
$$(4,-1,-1,-1),$$
$$(-1,4,-1,-1),$$
$$(-1,-1,4,-1),$$
$$(-1,-1,-1,4).$$

Choose a triangulation $\mathscr{P}$ of $B = \partial\Delta$ into standard simplices; this can be done in a regular way so that the restriction of $\mathscr{P}$ to each two-dimensional face of Δ is as given by the light lines in Figure 1. This gives a discriminant locus Γ depicted by the dark lines in the figure; the line segments coming out of the boundary of the two-face are meant to illustrate the pieces of discriminant locus contained in adjacent two-faces. The discriminant locus there is not contained in the plane of the two-face. In particular, the discriminant locus is not planar at the vertices of Γ on the edges of Ξ with respect to the affine structure we define. Note Γ is a trivalent graph, with two types of trivalent vertices, the non-planar ones just mentioned and the planar vertices contained in the interior of two-faces.

For an affine manifold, the monodromy of the local system Λ is an important feature of the affine structure. In this example, it is very useful to analyze this monodromy around loops about the discriminant locus. If v is a vertex of Γ contained in the interior of a two-face of Δ, one can consider loops based near v in B_0 around the three line segments of Γ adjacent to v. It is an enjoyable exercise to calculate that these monodromy matrices take the form, in a suitable basis,

$$T_1 = \begin{pmatrix} 1 & 0 & 0 \\ 1 & 1 & 0 \\ 0 & 0 & 1 \end{pmatrix}, \qquad T_2 = \begin{pmatrix} 1 & 0 & 0 \\ 0 & 1 & 0 \\ 1 & 0 & 1 \end{pmatrix}, \qquad T_3 = \begin{pmatrix} 1 & 0 & 0 \\ -1 & 1 & 0 \\ -1 & 0 & 1 \end{pmatrix}.$$

They are computed by studying the composition of transition maps between charts that a loop passes through. These matrices can be viewed as specifying the obstruction to extending the affine structure across a neighbourhood of v in Γ. Of course,

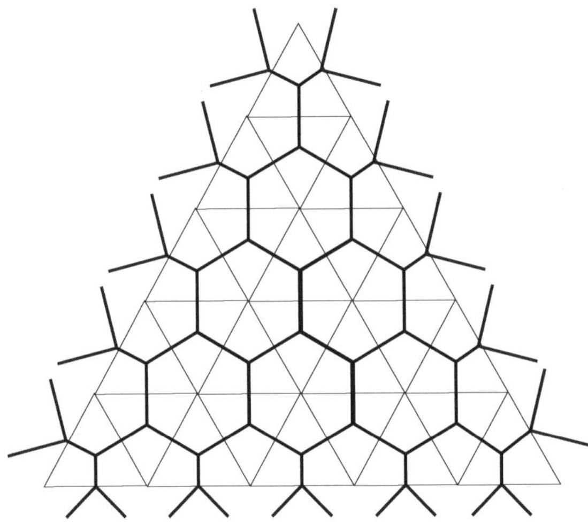

FIGURE 1

the monodromy of $\check{\Lambda}$ is the transpose inverse of these matrices. Similarly, if v is a vertex of Γ contained in an edge of Δ, then the monodromy will take the form

$$T_1 = \begin{pmatrix} 1 & -1 & 0 \\ 0 & 1 & 0 \\ 0 & 0 & 1 \end{pmatrix}, \qquad T_2 = \begin{pmatrix} 1 & 0 & -1 \\ 0 & 1 & 0 \\ 0 & 0 & 1 \end{pmatrix}, \qquad T_3 = \begin{pmatrix} 1 & 1 & 1 \\ 0 & 1 & 0 \\ 0 & 0 & 1 \end{pmatrix}.$$

So we see that the monodromy of the two types of vertices are interchanged between Λ and $\check{\Lambda}$.

One main result of [20] is

THEOREM 3.3. *If B is a three-dimensional tropical affine manifold with singularities such that Γ is trivalent and the monodromy of Λ at each vertex is one of the above two types, then $f_0 : X(B_0) \to B_0$ can be compactified to a topological fibration $f : X(B) \to B$. Dually, $\check{f}_0 : \check{X}(B_0) \to B_0$ can be compactified to a topological fibration $\check{f} : \check{X}(B) \to B$. Both $X(B)$ and $\check{X}(B)$ are topological manifolds.*

We won't give any details here of how this is carried out, but it is not particularly difficult, as long as one restricts to the category of topological (not C^∞) manifolds. However, it is interesting to look at the singular fibres we need to add.

If $b \in \Gamma$ is a point which is not a vertex of Γ, then $f^{-1}(b)$ is homeomorphic to $I_1 \times S^1$, where I_1 denotes a Kodaira type I_1 elliptic curve, i.e., a pinched torus.

If v is a vertex of Γ, with monodromy of the first type, then $f^{-1}(v) = S^1 \times S^1 \times S^1 / \sim$, with $(a,b,c) \sim (a',b',c')$ if $(a,b,c) = (a',b',c')$ or $a = a' = 1$, where S^1 is identified with the unit circle in $\mathbb{C}$. This is the three-dimensional analogue of a pinched torus, and $\chi(f^{-1}(v)) = +1$. We call this a *positive* fibre.

If v is a vertex of Γ, with monodromy of the second type, then $f^{-1}(v)$ can be described as $S^1 \times S^1 \times S^1 / \sim$, with $(a,b,c) \sim (a',b',c')$ if $(a,b,c) = (a',b',c')$ or $a = a' = 1, b = b'$, or $a = a', b = b' = 1$. The singular locus of this fibre is a figure eight, and $\chi(f^{-1}(v)) = -1$. We call this a *negative* fibre.

So we see a very concrete local consequence of SYZ duality: in the compactifications $X(B)$ and $\check{X}(B)$, the positive and negative fibres are interchanged. Of course, this results in the observation that the Euler characteristic changes sign under mirror symmetry for Calabi-Yau threefolds.

EXAMPLE 3.4. Continuing with Example 3.2, it was proved in [20] that $\check{X}(B)$ is homeomorphic to the quintic and $X(B)$ is homeomorphic to the mirror quintic. Modulo a paper [30] whose appearance has been long-delayed because of other, more pressing, projects, the results of [22] imply that the SYZ conjecture holds for all complete intersections in toric varieties at a topological level.

W.-D. Ruan in [70] gave a description of *Lagrangian* torus fibrations for hypersurfaces in toric varieties using a symplectic flow argument, and his construction should coincide with a *symplectic* compactification of the symplectic manifolds $\check{X}(B_0)$. In the three-dimensional case, such a symplectic compactification has been constructed by Ricardo Castaño-Bernard and Diego Matessi [8]. If this compactification is applied to the affine manifolds with singularities described here, the resulting symplectic manifolds should be symplectomorphic to the corresponding toric hypersurface, but this has not yet been shown.

4. Tropical geometry

Recalling that mirror symmetry is supposed to allow us to count curves, let us discuss at an intuitive level how the picture so far gives us insight into this question. Let B be a tropical affine manifold. Then as we saw, $X(B)$ carries the semi-flat complex structure, and it is easy to describe some complex submanifolds of $X(B)$ as follows. Let $L \subseteq B$ be a linear subspace with rational slope, i.e., the tangent space $\mathcal{T}_{L,b}$ to L at any $b \in L$ can be written as $M \otimes_{\mathbb{Z}} \mathbb{R}$ for some sublattice $M \subseteq \Lambda_b$. Then we obtain a submanifold

$$X(L) := \mathcal{T}_L/(\mathcal{T}_L \cap \Lambda) \subseteq X(B).$$

One checks easily that this is a complex submanifold. For example, if $B = \mathbb{R}^n$, so that $X(B)$ is just an algebraic torus $(\mathbb{C}^*)^n$ with coordinates $q_1, \ldots, q_n$, and $L \subseteq B$ is a codimension p affine linear subspace defined by equations

$$\sum_j c_{ij} y_j = d_i, \quad 1 \leq i \leq p,$$

with $c_{ij} \in \mathbb{Z}$, $d_i \in \mathbb{R}$, then the corresponding submanifold of $X(B)$ is the subtorus given by the equations

$$\prod_j q_j^{c_{ij}} = e^{-2\pi d_j}, \quad 1 \leq i \leq p.$$

Of course, subtori of tori are not particularly interesting. How might we build more complicated submanifolds? Let us focus on curves, where we take the linear submanifolds of B to be of dimension one. Then if we take L to be a line segment, ray, or line, $X(L)$ is a cylinder, with or without boundary in the various cases. We can then try to glue such cylinders together to obtain more complicated curves. For example, imagine we are given rays meeting at a point $b \in B = \mathbb{R}^2$ as pictured in Figure 2. Take primitive integral tangent vectors $v_1, v_2, v_3 \in \mathbb{R}^2$ to L_1, L_2 and L_3 pointing outwards from the point b where the three segments intersect. Now we have the three cylinders $X(L_i)$ which do not match up over b: the fibre $f^{-1}(b) = \mathbb{R}^2/\mathbb{Z}^2$ intersects $X(L_i)$ in a circle $\mathbb{R}v_i/\mathbb{Z}v_i$. These circles are represented in $H_1(f^{-1}(b), \mathbb{Z}) = \Lambda_b$ precisely by the vectors v_1, v_2, v_3, and so the

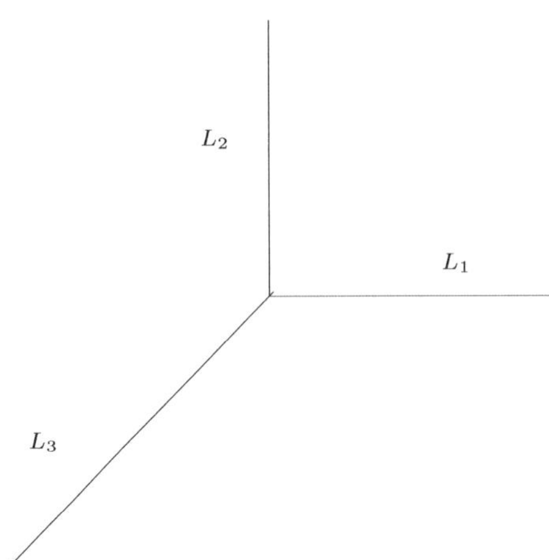

FIGURE 2

condition that the circles bound a surface in $f^{-1}(b)$ is that $v_1 + v_2 + v_3 = 0$. Thus, if this condition holds, we can glue in a surface S contained in $f^{-1}(b)$ so that $X(L_1) \cup X(L_2) \cup X(L_3) \cup S$ now has no boundary at b. Of course, it is very far from being a holomorphic submanifold. The expectation, however, is that this sort of object can be deformed to a nearby holomorphic curve.

Precisely, continuing with the above example, suppose $b = 0$ and $v_1 = (1,0)$, $v_2 = (0,1)$ and $v_3 = (-1,-1)$. With holomorphic coordinates q_1, q_2 on $X(B) = (\mathbb{C}^*)^2$, consider the curve $C \subseteq (\mathbb{C}^*)^2$ defined by $1 + q_1 + q_2 = 0$. Look at the image of this curve under the map $f : X(B) \to B$, which here can be written explicitly as $(q_1, q_2) \mapsto \frac{-1}{2\pi}(\log|q_1|, \log|q_2|)$. One finds that one obtains a thickening of the trivalent graph above, typically known as an amoeba. Further, if one considers not $X(B)$ but $X_\epsilon(B)$, where now holomorphic coordinates are given by $q_j = e^{2\pi i(x_j + iy_j)/\epsilon}$ and $f_\epsilon : X_\epsilon(B) \to B$ is given by $(q_1, q_2) \mapsto -\frac{\epsilon}{2\pi}(\log|q_1|, \log|q_2|)$, one finds that as $\epsilon \to 0$, $f_\epsilon(C)$ converges to the trivalent graph in the above figure. In this sense the trivalent graph on B is a limiting version of curves on a family of varieties tending towards a large complex structure limit.

This basic picture for curves in algebraic tori is now very well studied. In particular, this study spawned the subject of *tropical geometry*. The word *tropical* is motivated by the role that the *tropical semiring* plays. This is the semiring $(\mathbb{R}, \oplus, \odot)$ where addition and multiplication are given by

$$a \oplus b := \min(a, b)$$
$$a \odot b := a + b.$$

The word "tropical" is used in honor of the Brazilian mathematician Imre Simon, who pioneered use of this semi-ring.

We now consider polynomials over the tropical semiring, as follows. Let $S \subseteq \mathbb{Z}^n$ be a finite subset, and consider tropical polynomials on $\mathbb{R}^n$ of the form

$$g := \sum_{(p_1,\ldots,p_n) \in S} c_{p_1 \ldots p_n} x_1^{p_1} \cdots x_n^{p_n}$$

where the coefficients lie in $\mathbb{R}$ and the operations are in the tropical semiring. Then g is a convex piecewise linear function on $\mathbb{R}^n$, and the locus where g is not linear is called a *tropical hypersurface*. In particular, in the case $n = 2$, we obtain a tropical curve. In the example of Figure 2, the relevant tropical polynomial could be taken to be $0 \oplus x_1 \oplus x_2$.

While the tropical semiring has been used extensively in tropical geometry, it is not so convenient for us to view our tropical curves on B as being defined by equations, since typically these curves will be of high codimension. Instead, it is better to follow Mikhalkin [62] and use parameterized tropical curves.

The domain of a parameterized tropical curve will be a weighted graph. In what follows, $\overline{\Gamma}$ will denote a connected graph. Such a graph can be viewed in two different ways. First, it can be viewed as a purely combinatorial object, i.e., a set $\overline{\Gamma}^{[0]}$ of vertices and a set $\overline{\Gamma}^{[1]}$ of edges consisting of unordered pairs of elements of $\overline{\Gamma}^{[0]}$, indicating the endpoints of an edge. We can also view $\overline{\Gamma}$ as the topological realization of the graph, i.e., a topological space which is the union of line segments corresponding to the edges. We shall confuse these two viewpoints at will. We will then denote by Γ the topological space obtained from $\overline{\Gamma}$ by deleting the univalent vertices of $\overline{\Gamma}$, so that Γ may have some non-compact edges.

We also take $\overline{\Gamma}$ to come with a *weight function*, a map

$$w : \overline{\Gamma}^{[1]} \to \mathbb{N} = \{0, 1, 2, \ldots\}.$$

Replacing $\mathbb{R}^n$ with a general tropical affine manifold B, we now arrive at the following definition:

DEFINITION 4.1. A parameterized tropical curve in B is a continuous map

$$h : \Gamma \to B$$

where Γ is obtained from a graph $\overline{\Gamma}$ as above, satisfying the following two properties:

(1) If $E \in \Gamma^{[1]}$ and $w(E) = 0$, then $h|_E$ is constant; otherwise $h|_E$ is a proper embedding of E into B as a line segment, ray or line of rational slope.

(2) *The balancing condition.* Let $V \in \overline{\Gamma}^{[0]}$ be a vertex with valency larger than 1, with adjacent edges $E_1, \ldots, E_\ell$. Let $v_i \in \Lambda_{h(V)}$ be a primitive tangent vector to $h(E_i)$ at $h(V)$, pointing away from $h(V)$. Then

$$\sum_{i=1}^{\ell} w(E_i) v_i = 0.$$

Here the balancing condition is just expressing the topological requirement that the boundaries of the various cylinders $X(h(E_i)) \subseteq X(B)$ can be connected up with a surface contained in the fibre of $X(B) \to B$ over $h(V)$. The weights can be interpreted as taking the cylinders $X(h(E_i))$ with multiplicity.

An important question then arises:

QUESTION 4.2. *When can a given parameterized tropical curve be viewed as a limit of holomorphic curves in $X_\epsilon(B)$ as $\epsilon \to 0$?*

This question has attracted a great deal of attention when $B = \mathbb{R}^n$, with completely satisfactory results in the case $n = 2$ (Answer: always), and less complete results when $n \geq 3$. The $n = 2$ case was first treated by Mikhalkin [62], and results in all dimensions were first obtained by Nishinou and Siebert [64]. In particular,

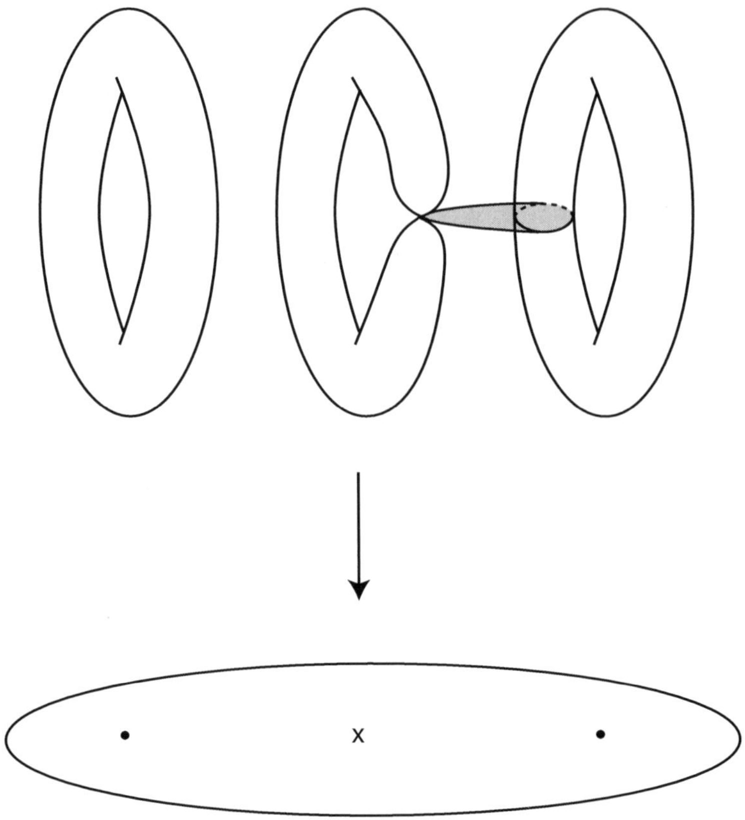

FIGURE 3

Mikhalkin proved that in this two-dimensional case, one can calculate numbers of curves of a given degree and genus passing through a fixed set of points, showing that difficult holomorphic enumerative problems can be solved by a purely combinatorial approach. This work gives hope that one can really count curves combinatorially in much more general settings. In the two-dimensional case, again, my own work [**24**] showed that the mirror side (for mirror symmetry for $\mathbb{P}^2$) could also be interpreted tropically, giving a completely tropical interpretation of mirror symmetry for $\mathbb{P}^2$.

So far we have not considered the case that B has singularities. In case B has singularities, we expect that one should be able to relax the balancing condition when a vertex falls inside of a point of the singular locus, and in particular one can allow univalent vertices which map to the singular locus. The reason for this is that once we compactify $X(B_0)$ to $X(B)$, one expects to find holomorphic disks fibering over line segments emanating from singular points: see Figure 3 for a depiction of this when B is two-dimensional, having isolated singularities.

We will avoid giving a precise definition of what a tropical curve should mean in the case that B has singularities, largely because it is not clear yet what the precise definition should be. Hopefully, though, this discussion makes it clear that at an intuitive level, counting curves should be something which can be done on B.

5. The problems with the SYZ conjecture, and how to get around them

The discussion of §3 demonstrates that the SYZ conjecture gives a beautiful description of mirror symmetry at a purely topological level. This, by itself, can often be useful, but fails to get at the original hard differential geometric conjecture and fails to give insight into why mirror symmetry counts curves.

In order for the full-strength version of the SYZ conjecture to hold, the strong version of duality for topological torus fibrations we saw in §3 should continue to hold at the special Lagrangian level. This would mean that a mirror pair $X, \check{X}$ would possess special Lagrangian torus fibrations $f : X \to B$ and $\check{f} : \check{X} \to B$ with codimension two discriminant loci, and the discriminant loci of f and $\check{f}$ would coincide. These fibrations would then be dual away from the discriminant locus.

There are examples of special Lagrangian fibrations on non-compact toric varieties X with discriminant locus looking very similar to what we have described in the topological case. In particular, if X is an n-dimensional Ricci-flat Kähler manifold with a T^{n-1}-action preserving the metric and holomorphic n-form, then X will have a very nice special Lagrangian fibration with codimension two discriminant locus. (See [21] and [16]). However, Dominic Joyce (see [48] and other papers cited therein) began studying some three-dimensional S^1-invariant examples, and discovered quite different behaviour. There is an argument in [19] that if a special Lagrangian fibration is C^∞, then the discriminant locus will be (Hausdorff) codimension two. However, Joyce discovered examples which were not differentiable, but only piecewise differentiable, and furthermore, had a codimension one discriminant locus:

EXAMPLE 5.1. Define $F : \mathbb{C}^3 \to \mathbb{R} \times \mathbb{C}$ by $F(z_1, z_2, z_3) = (a, c)$ with $2a = |z_1|^2 - |z_2|^2$ and

$$c = \begin{cases} z_3 & a = z_1 = z_2 = 0 \\ z_3 - \bar{z}_1 \bar{z}_2 / |z_1| & a \geq 0, z_1 \neq 0 \\ z_3 - \bar{z}_1 \bar{z}_2 / |z_2| & a < 0. \end{cases}$$

It is easy to see that if $a \neq 0$, then $F^{-1}(a, c)$ is homeomorphic to $\mathbb{R}^2 \times S^1$, while if $a = 0$, then $F^{-1}(a, c)$ is a cone over T^2: essentially, one copy of S^1 in $\mathbb{R}^2 \times S^1$ collapses to a point. In addition, all fibres of this map are special Lagrangian, and it is obviously only piecewise smooth. The discriminant locus is the entire plane given by $a = 0$.

This example forces a reevaluation of the strong form of the SYZ conjecture. In further work Joyce found evidence for a more likely picture for general special Lagrangian fibrations in three dimensions. The discriminant locus, instead of being a codimension two graph, will be a codimension one blob. Typically the union of the singular points of singular fibres will be a Riemann surface, and it will map to an amoeba-shaped set in B, i.e., the discriminant locus looks like the picture on the right rather than the left in Figure 4, and will be a fattening of the old picture of a codimension two discriminant.

Joyce made some additional arguments to suggest that this fattened discriminant locus must look fundamentally different in a neighbourhood of the two basic types of vertices we saw in §3, with the two types of vertices expected to appear pretty much as depicted in Figure 4. Thus the strong form of duality mentioned above, where we expect the discriminant loci of the special Lagrangian fibrations

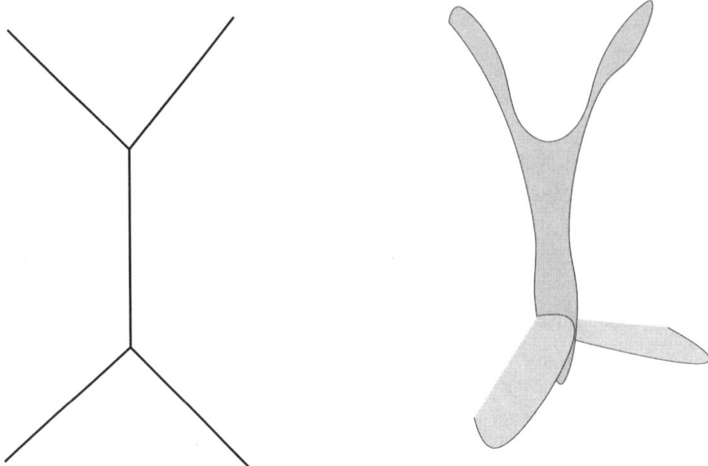

FIGURE 4

on a mirror pair to be the same, cannot hold. If this is the case, one needs to replace this strong form of duality with a weaker form.

It seems likely that the best way to rephrase the SYZ conjecture is in a limiting form. Mirror symmetry as we currently understand it has to do with degenerations of Calabi-Yau manifolds. Given a flat family $f : \mathcal{X} \to D$ over a disk D, with the fibre $\mathcal{X}_0$ over 0 singular and all other fibres n-dimensional Calabi-Yau manifolds, we say the family is *maximally unipotent* if the monodromy transformation $T : H^n(\mathcal{X}_t, \mathbb{Q}) \to H^n(\mathcal{X}_t, \mathbb{Q})$ ($t \in D$ non-zero) satisfies $(T - I)^{n+1} = 0$ but $(T - I)^n \neq 0$. It is a standard expectation of mirror symmetry that mirrors should be associated to maximally unipotent degenerations of Calabi-Yau manifolds. In particular, given two different maximally unipotent degenerations in a single complex moduli space for some Calabi-Yau manifold, one might obtain different mirror manifolds. Such degenerations are usually called "large complex structure limits" in the physics literature, although sometimes this phrase is used to impose some additional conditions on the degeneration, see [63].

We recall the definition of Gromov-Hausdorff convergence, a notion of convergence of a sequence of metric spaces.

DEFINITION 5.2. Let (X, d_X), (Y, d_Y) be two compact metric spaces. Suppose there exists maps $f : X \to Y$ and $g : Y \to X$ (not necessarily continuous) such that for all $x_1, x_2 \in X$,

$$|d_X(x_1, x_2) - d_Y(f(x_1), f(x_2))| < \epsilon$$

and for all $x \in X$,

$$d_X(x, g \circ f(x)) < \epsilon,$$

and the two symmetric properties for Y hold. Then we say the Gromov–Hausdorff distance between X and Y is at most ϵ. The Gromov–Hausdorff distance $d_{GH}(X, Y)$ is the infimum of all such ϵ.

It follows from results of Gromov (see for example [68], pg. 281, Cor. 1.11) that the space of compact Ricci-flat manifolds with diameter $\leq C$ is precompact with respect to Gromov-Hausdorff distance, i.e., any sequence of such manifolds has a subsequence converging with respect to the Gromov-Hausdorff distance to a metric space. This metric space could be quite bad; this is quite outside the realm of algebraic geometry! Nevertheless, this raises the following natural question. Given a maximally unipotent degeneration of Calabi-Yau manifolds $\mathcal{X} \to D$, take a sequence $t_i \in D$ converging to 0, and consider a sequence $(\mathcal{X}_{t_i}, g_{t_i})$, where g_{t_i} is a choice of Ricci-flat metric chosen so that $Diam(g_{t_i})$ remains bounded. What is the Gromov-Hausdorff limit of $(\mathcal{X}_{t_i}, g_{t_i})$, or the limit of some convergent subsequence?

EXAMPLE 5.3. Consider a degenerating family of elliptic curves parameterized by t, given by $\mathbb{C}/(\mathbb{Z}+\mathbb{Z}\tau)$ where 1 and $\tau = \frac{1}{2\pi i}\log t$ are periods of the elliptic curves. If we take t approaching 0 along the positive real axis, then we can just view this as a family of elliptic curves $\mathcal{X}_\alpha$ with period 1 and $i\alpha$ with $\alpha \to \infty$. If we take the standard Euclidean metric g on $\mathcal{X}_\alpha$, then the diameter of $\mathcal{X}_\alpha$ is unbounded. To obtain a bounded diameter, we replace g by g/α^2; equivalently, we can keep g fixed on $\mathbb{C}$ but change the periods of the elliptic curve to $1/\alpha, i$. It then becomes clear that the Gromov-Hausdorff limit of such a sequence of elliptic curves is a circle $\mathbb{R}/\mathbb{Z}$.

This simple example motivates the first conjecture about maximally unipotent degenerations, conjectured independently by myself and Wilson on the one hand [**38**] and Kontsevich and Soibelman [**53**] on the other.

CONJECTURE 5.4. *Let $\mathcal{X} \to D$ be a maximally unipotent degeneration of simply-connected Calabi-Yau manifolds with full $SU(n)$ holonomy, $t_i \in D$ with $t_i \to 0$, and let g_i be a Ricci-flat metric on $\mathcal{X}_{t_i}$ normalized to have fixed diameter C. Then a convergent subsequence of $(\mathcal{X}_{t_i}, g_i)$ converges to a metric space (X_∞, d_∞), where X_∞ is homeomorphic to S^n. Furthermore, d_∞ is induced by a Riemannian metric on $X_\infty \setminus \Gamma$, where $\Gamma \subseteq X_\infty$ is a set of codimension two.*

Here the topology of the limit depends on the nature of the non-singular fibres $\mathcal{X}_t$; for example, if instead $\mathcal{X}_t$ was hyperkähler, then we would expect the limit to be a projective space. Also, even in the case of full $SU(n)$ holonomy, if $\mathcal{X}_t$ is not simply connected, we would expect limits such as $\mathbb{Q}$-homology spheres to arise.

Conjecture 5.4 is directly inspired by the SYZ conjecture. Suppose we had special Lagrangian fibrations $f_i : \mathcal{X}_{t_i} \to B_i$. Then as the maximally unipotent degeneration is approached, one can see that the volume of the fibres of these fibrations goes to zero. This would suggest these fibres collapse, hopefully leaving the base as the limit.

This conjecture was proved by myself and Wilson in 2000 for K3 surfaces in [**38**]. The proof relied on a number of pleasant facts about K3 surfaces. First, they are hyperkähler manifolds, and a special Lagrangian torus fibration becomes an elliptic fibration after a hyperkähler rotation of the complex structure. Since it is easy to construct elliptic fibrations on K3 surfaces, and indeed such a fibration arises from the data of the maximally unipotent degeneration, it is easy to obtain a special Lagrangian fibration. Once this is done, one needs to carry out a detailed analysis of the behaviour of Ricci-flat metrics in the limit. This is done by creating good approximations to Ricci-flat metric, using the existence of explicit local models for these metrics near singular fibres of special Lagrangian fibrations in complex dimension two.

Most of the techniques used are not available in higher dimension. However, much more recently, weaker collapsing results in the hyperkähler case were obtained in work with V. Tosatti and Y. Zhang in [**36**], assuming the existence of abelian variety fibrations analogous to the elliptic fibrations in the K3 case. Rather than getting an explicit approximate Ricci-flat metric, we make use of a priori estimates of Tosatti in [**76**].

In the general Calabi-Yau case, the only progress towards the conjecture has been work of Zhang in [**78**] showing existence of special Lagrangian fibrations in regions of Calabi-Yau manifolds with bounded injectivity radius and sectional curvature and deduces local collapsing from the existence of special Lagrangian fibrations.

The motivation for Conjecture 5.4 from SYZ also provides a limiting form of the conjecture. There are any number of problems with trying to prove the existence of special Lagrangian fibrations on Calabi-Yau manifolds. Even the existence of a single special Lagrangian torus near a maximally unipotent degeneration is unknown, but we expect it should be easier to find them as we approach the maximally unipotent point. Furthermore, even if we find a special Lagrangian torus, we know that it moves in an n-dimensional family, but we don't know its deformations fill out the entire manifold. In addition, there is no guarantee that even if it does, we obtain a foliation of the manifold: nearby special Lagrangian submanifolds may intersect. (For an example, see [**60**].) So instead, we will just look at the moduli space of special Lagrangian tori.

Given a maximally unipotent degeneration of Calabi-Yau manifolds of dimension n, it is known that the image of $(T - I)^n : H_n(\mathcal{X}_t, \mathbb{Q}) \to H_n(\mathcal{X}_t, \mathbb{Q})$ is a one-dimensional subspace W_0. Suppose, given a sequence t_i with $t_i \to 0$ as $i \to \infty$, that for t_i sufficiently close to zero, there is a special Lagrangian T^n which generates W_0. This is where we expect to find fibres of a special Lagrangian fibration associated to a maximally unipotent degeneration. Let $B_{0,i}$ be the moduli space of deformations of this torus; every point of $B_{0,i}$ corresponds to a smooth special Lagrangian torus in $\mathcal{X}_{t_i}$. This manifold then comes equipped with the McLean metric and affine structures defined in §2. One can then compactify $B_{0,i} \subseteq B_i$, (probably by taking the closure of $B_{0,i}$ in the space of special Lagrangian currents; the details aren't important here). This gives a series of metric spaces (B_i, d_i) with the metric d_i induced by the McLean metric. If the McLean metric is normalized to keep the diameter of B_i constant independent of i, then we can hope that (B_i, d_i) converges to a compact metric space (B_∞, d_∞). Here then is the limiting form of SYZ:

CONJECTURE 5.5. *If $(\mathcal{X}_{t_i}, g_i)$ converges to (X_∞, g_∞) and (B_i, d_i) is non-empty for large i and converges to (B_∞, d_∞), then B_∞ and X_∞ are isometric up to scaling. Furthermore, there is a subspace $B_{\infty,0} \subseteq B_\infty$ with $\Gamma := B_\infty \setminus B_{\infty,0}$ of Hausdorff codimension 2 in B_∞ such that $B_{\infty,0}$ is a Monge-Ampère manifold, with the Monge-Ampère metric inducing d_∞ on $B_{\infty,0}$.*

Essentially what this is saying is that as we approach the maximally unipotent degeneration, we expect to have a special Lagrangian fibration on larger and larger subsets of $\mathcal{X}_{t_i}$. Furthermore, in the limit, the codimension one discriminant locus suggested by Joyce converges to a codimension two discriminant locus, and (the not necessarily Monge-Ampère, see [**60**]) Hessian metrics on $B_{0,i}$ converge to a Monge-Ampère metric.

The main point I want to get at here is that it is likely the SYZ conjecture is only "approximately" correct, and one needs to look at the limit to have a hope of proving anything. On the other hand, the above conjecture seems likely to be accessible by currently understood techniques. I remain hopeful that this conjecture will be proved, though much additional work will be necessary.

How do we do mirror symmetry using this modified version of the SYZ conjecture? Essentially, we would follow these steps:

(1) We begin with a maximally unipotent degeneration of Calabi-Yau manifolds $\mathcal{X} \to D$, along with a choice of polarization. This gives us a Kähler class $[\omega_t] \in H^2(\mathcal{X}_t, \mathbb{R})$ for each $t \in D \setminus 0$, represented by ω_t the Kähler form of a Ricci-flat metric g_t.

(2) Identify the Gromov-Hausdorff limit of a sequence $(\mathcal{X}_{t_i}, r_i g_{t_i})$ where $t_i \to 0$ and r_i is a scale factor which keeps the diameter of $\mathcal{X}_{t_i}$ constant. The limit will be, if the above conjectures work, an affine manifold with singularities B along with a Monge-Ampère metric.

(3) Perform a Legendre transform to obtain a new affine manifold with singularities $\check{B}$, though with the same metric.

(4) Try to construct a compactification of $X_\epsilon(\check{B}_0)$ for small $\epsilon > 0$ to obtain a complex manifold $X_\epsilon(\check{B})$. This will be the mirror manifold.

As we shall see, we do not expect that we will need the full strength of steps (2) and (3) to carry out mirror symmetry; some way of identifying the base B will be sufficient. Nevertheless, (2) is interesting from the point of view of understanding the differential geomtry of Ricci-flat Kähler manifolds.

Step (4), on the other hand, is crucial, and we need to elaborate on this last step a bit more. The problem is that while we expect that it should be possible in general to construct symplectic compactifications of the symplectic manifold $\check{X}(B_0)$ (and hence get the mirror as a symplectic manifold, see [**8**] for the three-dimensional case), we don't expect to be able to compactify $X_\epsilon(\check{B}_0)$ as a complex manifold. Instead, the expectation is that a small deformation of $X_\epsilon(\check{B}_0)$ is necessary before it can be compactified. Furthermore, this small deformation is critically important in mirror symmetry: *it is this small deformation which provides the B-model instanton corrections.*

Because this last item is so important, let's give it a name:

QUESTION 5.6 (The reconstruction problem, Version I). *Given a tropical affine manifold with singularities B, construct a complex manifold $X_\epsilon(B)$ which is a compactification of a small deformation of $X_\epsilon(B_0)$.*

We will return to this question later in the paper. However, I do not wish to dwell further on the differential-geometric versions of the SYZ conjecture here. Instead I will move on to describing how the above discussion motivated the algebro-geometric program developed by myself and Siebert for understanding mirror symmetry, and then describe recent work and ideas coming out of this program.

6. Gromov-Hausdorff limits, algebraic degenerations, and mirror symmetry

We now have two notions of limit: the familiar algebro-geometric notion of a degenerating family $\mathcal{X} \to D$ over a disk on the one hand, and the Gromov-Hausdorff limit on the other. In 2000 Kontsevich and Soibelman had an important insight (see [**53**]) into the connection between these two. In this section I will give a rough idea of how and why this works.

Very roughly speaking, the Gromov-Hausdorff limit $(\mathcal{X}_{t_i}, g_{t_i})$ as $t_i \to 0$, or equivalently, the base of the putative SYZ fibration, should coincide, topologically, with the *dual intersection complex* of the singular fibre $\mathcal{X}_0$. More precisely, in a relatively simple situation, suppose $f : \mathcal{X} \to D$ is relatively minimal (in the sense of Mori) and normal crossings, with $\mathcal{X}_0$ having irreducible components $X_1, \ldots, X_m$. The dual intersection complex of $\mathcal{X}_0$ is the simplicial complex with vertices $v_1, \ldots, v_m$, and which contains a simplex $\langle v_{i_0}, \ldots, v_{i_p} \rangle$ if $X_{i_0} \cap \cdots \cap X_{i_p} \neq \emptyset$. The idea that the dual intersection complex should play a role in describing the base of the SYZ fibration was perhaps first suggested by Leung and Vafa in [**56**].

Let us explain roughly why this should be, first by looking at a standard family of degenerating elliptic curves with periods 1 and $\frac{n}{2\pi i} \log t$ for n a positive integer. Such a family over the punctured disk is extended to a family over the disk by adding a Kodaira type I_n (a cycle of n rational curves) fibre over the origin.

Taking a sequence $t_i \to 0$ with t_i real and positive gives a sequence of elliptic curves of the form $X_{\epsilon_i}(B)$ where $B = \mathbb{R}/n\mathbb{Z}$ and $\epsilon_i = -\frac{2\pi}{\ln t_i}$. In addition, the metric on $X_{\epsilon_i}(B)$, properly scaled, comes from the constant Hessian metric on B. So we wish to explain how B is related to the geometry near the singular fibre. To this end, let $X_1, \ldots, X_n$ be the irreducible components of $\mathcal{X}_0$; these are all $\mathbb{P}^1$'s. Let $P_1, \ldots, P_n$ be the singular points of $\mathcal{X}_0$.

We'll consider two sorts of open sets in $\mathcal{X}$. For the first type, choose a coordinate z on X_i, with P_i given by $z = 0$ and P_{i+1} given by $z = \infty$. Let $U_i \subseteq X_i$ be the open set $\{z \,|\, \delta \leq |z| \leq 1/\delta\}$ for some small fixed δ. Then one can find a neighbourhood $\widetilde{U}_i$ of U_i in $\mathcal{X}$ such that $\widetilde{U}_i$ is biholomorphic to $U_i \times D_\rho$ for $\rho > 0$ sufficiently small, D_ρ a disk of radius ρ in $\mathbb{C}$, and $f|_{\widetilde{U}_i}$ is the projection onto D_ρ.

On the other hand, each P_i has a neighbourhood $\widetilde{V}_i$ in $\mathcal{X}$ biholomorphic to a polydisk $\{(z_1, z_2) \in \mathbb{C}^2 \,|\, |z_1| \leq \delta', |z_2| \leq \delta'\}$ on which f takes the form $z_1 z_2$.

If δ and δ' are chosen correctly, then for t sufficiently close to zero,

$$\{\widetilde{V}_i \cap \mathcal{X}_t \,|\, 1 \leq i \leq n\} \cup \{\widetilde{U}_i \cap \mathcal{X}_t \,|\, 1 \leq i \leq n\}$$

form an open cover of $\mathcal{X}_t$. Now each of the sets in this open cover can be written as $X_\epsilon(U)$ for some U a one-dimensional (non-compact) affine manifold and $\epsilon = -2\pi/\ln|t|$. If U is an open interval $(a, b) \subseteq \mathbb{R}$, then $X_\epsilon(U)$ is biholomorphic to the annulus

$$\{z \in \mathbb{C} \,|\, e^{-2\pi b/\epsilon} \leq |z| \leq e^{-2\pi a/\epsilon}\}$$

as $q = e^{2\pi i(x+iy)/\epsilon}$ is a holomorphic coordinate on $X_\epsilon((a, b))$. Thus

$$\widetilde{U}_i \cap \mathcal{X}_t \cong X_\epsilon\left(\left(\frac{\epsilon \ln \delta}{2\pi}, -\frac{\epsilon \ln \delta}{2\pi}\right)\right)$$

with $\epsilon = -2\pi/\ln|t|$. As $t \to 0$, the interval $(\epsilon \ln \delta/2\pi, -\epsilon \ln \delta/2\pi)$ shrinks to a point. So $\widetilde{U}_i \cap \mathcal{X}_t$ is a smaller and smaller open subset of $\mathcal{X}_t$ as $t \to 0$ when we view things in this way. This argument suggests that every irreducible component should be associated to a point on B.

Now look at $\widetilde{V}_i \cap \mathcal{X}_t$. This is

$$\{(z_1, z_2) \in \mathbb{C}^2 \,|\, |z_1|, |z_2| < \delta', z_1 z_2 = t\} \;\cong\; \{z \in \mathbb{C} \,|\, |t|/\delta' \leq |z| \leq \delta'\}$$
$$\cong\; X_\epsilon\left(\frac{-\epsilon}{2\pi} \ln \delta', \frac{\epsilon}{2\pi}(\ln \delta' - \ln|t|)\right)$$

with $\epsilon = -2\pi/\ln|t|$. This interval approaches the unit interval $(0,1)$ as $t \to 0$. So the open set $\widetilde{V}_i \cap \mathcal{X}_t$ ends up being a large portion of $\mathcal{X}_t$. We end up with $\mathcal{X}_t$, for small t, being a union of open sets of the form $X_\epsilon((i+\epsilon', i+1-\epsilon'))$ (i.e., $\widetilde{V}_i \cap \mathcal{X}_\epsilon$) and $X_\epsilon((i-\epsilon'', i+\epsilon''))$ (i.e., $\widetilde{U}_i \cap \mathcal{X}_t$) for ϵ', ϵ'' sufficiently small. These should glue, at least approximately, to give $X_\epsilon(B)$. So we see that irreducible components of $\mathcal{X}_0$ seem to coincide with points on B, but intersections of components coincide with lines. In this way we see the dual intersection complex emerge.

Let us make one more observation before beginning with rigorous results in the next section. Suppose more generally we had a *Gorenstein toroidal crossings* degeneration of Calabi-Yau manifolds $f : \mathcal{X} \to D$ (see [**73**]). This means that every point $x \in \mathcal{X}$ has a neighbourhood isomorphic to an open set in an affine Gorenstein (i.e., the canonical class is a Cartier divisor) toric variety, with f given locally by a monomial which vanishes exactly to order 1 on each codimension one toric stratum. This is a generalization of the notion of normal crossings. Very roughly, the above argument suggests that each irreducible component of the central fibre will correspond to a point of the Gromov-Hausdorff limit. The following exercise shows what kind of contribution to B to expect from a point $x \in \mathcal{X}_0$ which is a zero-dimensional stratum in $\mathcal{X}_0$.

EXERCISE 6.1. *Suppose that there is a point $x \in \mathcal{X}_0$ which has a neighbourhood isomorphic to a neighbourhood of a dimension zero torus orbit of an affine Gorenstein toric variety Y_x. Such an affine variety is specified as follows. Set $M = \mathbb{Z}^n$, $M_\mathbb{R} = M \otimes_\mathbb{Z} \mathbb{R}$, $N = \mathrm{Hom}_\mathbb{Z}(M, \mathbb{Z})$, $N_\mathbb{R} = N \otimes_\mathbb{Z} \mathbb{R}$ with $n = \dim \mathcal{X}_t$. Then there is a lattice polytope $\sigma \subseteq M_\mathbb{R}$, $C(\sigma) := \{(rm, r) \,|\, m \in \sigma, r \geq 0\} \subseteq M_\mathbb{R} \oplus \mathbb{R}$, $P := C(\sigma)^\vee \cap (N \oplus \mathbb{Z})$ the monoid determined by the dual of the cone $C(\sigma)$, $Y_x = \mathrm{Spec}\,\mathbb{C}[P]$, and finally f coincides with the monomial $z^{(0,1)}$.*

Now let us take a small neighbourhood of x of the form

$$\widetilde{U}_\delta = \{y \in \mathrm{Spec}\,\mathbb{C}[P] \,|\, |z^p| < \delta \text{ for all } p \in P\}.$$

This is an open set as the condition $|z^p| < \delta$ can be tested on a finite generating set for P, provided that $\delta < 1$. Then show that for a given t, $|t| < 1$ and $\epsilon = -2\pi/\log|t|$, if

$$\sigma_t := \{m \in M_\mathbb{R} \,|\, \langle p, (m, 1) \rangle > \tfrac{\log \delta}{\log |t|} \text{ for all } p \in P\},$$

then

$$f^{-1}(t) \cap \widetilde{U}_\delta \cong X_\epsilon(\sigma_t).$$

Note that

$$\sigma := \{m \in M_\mathbb{R} \,|\, \langle p, (m, 1) \rangle \geq 0 \text{ for all } p \in P\},$$

so σ_t is an open subset of σ, and as $t \to 0$, σ_t converges to the interior of σ. $\square$

This observation hopefully motivates the basic construction of the next section.

7. Toric degenerations, the intersection complex and its dual

I will now introduce the basic objects of the program developed by myself and Siebert to understand mirror symmetry in an algebro-geometric context. This program was announced in [**29**], and has been developed further in a series of papers [**31**], [**32**], [**33**], [**22**], [**34**], [**28**].

The motivation for this program came from two different directions. The first, which was largely my motivation, was the discussion of the limiting form of the

SYZ conjecture of the previous sections. The second arose in work of Schröer and Siebert [**72**], [**73**], which led Siebert to the idea that log structures on degenerations of Calabi-Yau manifolds would allow one to view mirror symmetry as an operation performed on degenerate Calabi-Yau varieties. Siebert observed that at a combinatorial level, mirror symmetry exchanged data pertaining to the log structure and a polarization. This will be explained more clearly in the following section, when I introduce log structures. Together, Siebert and I realised that the combinatorial data he was considering could be encoded naturally in the dual intersection complex of the degeneration, which we saw in the previous section appears to be the base of the SYZ fibration. The combinatorial interchange of data necessary for mirror symmetry then corresponded to a discrete Legendre transform on the dual intersection complex. It became apparent that this approach provided an algebro-geometrization of the SYZ conjecture.

To set this up properly, one has to consider what kind of degenerations to allow. They should be maximally unipotent, of course, but there can be many different birational models of degenerations. Below we define the notion of *toric degeneration*. The class of toric degenerations may seem rather restrictive, but it appears to be the largest class of degenerations closed under mirror symmetry: one can construct the mirror of a toric degeneration as a toric degeneration. It does not appear that there is any other natural family of degenerations with this property. Much of the material in this section comes from [**31**], §4.

Roughly put, a toric degeneration of Calabi-Yau varieties is a degeneration whose central fibre is a union of toric varieties glued along toric strata, and the total space of the degeneration is, off of some well-behaved set Z contained in the central fibre, locally toric with the family locally given by a monomial. The precise technical definition is as follows.

DEFINITION 7.1. Let $f : \mathcal{X} \to D$ be a proper flat family of relative dimension n, where D is a disk and $\mathcal{X}$ is a complex analytic space (not necessarily non-singular). We say f is a *toric degeneration* of Calabi-Yau varieties if

(1) $\mathcal{X}_t$ is an irreducible normal Calabi-Yau variety with only canonical singularities for $t \neq 0$. (The reader may like to assume $\mathcal{X}_t$ is smooth for $t \neq 0$).

(2) If $\nu : \widetilde{\mathcal{X}}_0 \to \mathcal{X}_0$ is the normalization, then $\widetilde{\mathcal{X}}_0$ is a disjoint union of toric varieties, the conductor locus $C \subseteq \widetilde{\mathcal{X}}_0$ is reduced, and the map $C \to \nu(C)$ is unramified and generically two-to-one. (The conductor locus is a naturally defined scheme structure on the set where ν is not an isomorphism.) The square

$$
\begin{array}{ccc}
C & \longrightarrow & \widetilde{\mathcal{X}}_0 \\
\downarrow & & \downarrow{\scriptstyle \nu} \\
\nu(C) & \longrightarrow & \mathcal{X}_0
\end{array}
$$

is cartesian and cocartesian.

(3) $\mathcal{X}_0$ is a reduced Gorenstein space and the conductor locus C restricted to each irreducible component of $\widetilde{\mathcal{X}}_0$ is the union of all toric Weil divisors of that component.

(4) There exists a closed subset $Z \subseteq \mathcal{X}$ of relative codimension ≥ 2 such that Z satisfies the following properties: Z does not contain the image under

ν of any toric stratum of $\widetilde{\mathcal{X}}_0$, and for any point $x \in \mathcal{X} \setminus Z$, there is a neighbourhood $\widetilde{U}_x$ (in the analytic topology) of x, an $n+1$-dimensional affine toric variety Y_x, a regular function f_x on Y_x given by a monomial, and a commutative diagram

$$
\begin{array}{ccc}
\widetilde{U}_x & \xrightarrow{\psi_x} & Y_x \\
\downarrow{\scriptstyle f|_{\widetilde{U}_x}} & & \downarrow{\scriptstyle f_x} \\
D' & \xrightarrow{\varphi_x} & \mathbb{C}
\end{array}
$$

where ψ_x and φ_x are open embeddings and $D' \subseteq D$. Furthermore, f_x vanishes precisely once on each toric divisor of Y_x.

EXAMPLE 7.2. Take $\mathcal{X}$ to be defined by the equation $tf_4 + z_0 z_1 z_2 z_3 = 0$ in $\mathbb{P}^3 \times D$, where D is a disk with coordinate t and f_4 is a general homogeneous quartic polynomial on $\mathbb{P}^3$. It is easy to see that $\mathcal{X}$ is singular at the locus

$$\{t = f_4 = 0\} \cap Sing(\mathcal{X}_0).$$

As $\mathcal{X}_0$ is the coordinate tetrahedron, the singular locus of $\mathcal{X}_0$ consists of the six coordinate lines of $\mathbb{P}^3$, and $\mathcal{X}$ has four singular points along each such line, for a total of 24 singular points. Take $Z = Sing(\mathcal{X})$. Then away from Z, the projection $\mathcal{X} \to D$ is normal crossings, which yields condition (4) of the definition of toric degeneration. It is easy to see all other conditions are satisfied.

Given a toric degeneration $f : \mathcal{X} \to D$, we can build the *dual intersection complex* $(B, \mathscr{P})$ of f, as follows. Here B is an integral affine manifold with singularities, and $\mathscr{P}$ is a *polyhedral decomposition* of B, i.e., a decomposition of B into lattice polytopes. In fact, we will construct B as a union of lattice polytopes. Specifically, let the normalisation of $\mathcal{X}_0$, $\widetilde{\mathcal{X}}_0$, be written as a disjoint union $\coprod X_i$ of toric varieties X_i, $\nu : \widetilde{\mathcal{X}}_0 \to \mathcal{X}_0$ the normalisation. The *strata* of $\mathcal{X}_0$ are the elements of the set

$$Strata(\mathcal{X}_0) = \{\nu(S) \mid S \text{ is a toric stratum of } X_i \text{ for some } i\}.$$

Here by toric stratum we mean the closure of a $(\mathbb{C}^*)^n$ orbit.

Let $\{x\} \in Strata(\mathcal{X}_0)$ be a zero-dimensional stratum. Applying Definition 7.1,(4), to a neighbourhood of x, there is a toric variety Y_x such that in a neighbourhood of x, $f : \mathcal{X} \to D$ is locally isomorphic to $f_x : Y_x \to \mathbb{C}$, where f_x is given by a monomial. Now the condition that f_x vanishes precisely once along each toric divisor of Y_x is the statement that Y_x is Gorenstein, and as such, it arises as in Exercise 6.1. Indeed, let M, N be given in Exercise 6.1, with rank $M = \dim \mathcal{X}_0$. Then there is a lattice polytope $\sigma_x \subseteq M_{\mathbb{R}}$ such that $C(\sigma_x) = \{(rm, r) \mid m \in \sigma, r \geq 0\}$ is the cone defining the toric variety Y_x. As we saw in Exercise 6.1, a small neighbourhood of x in $\mathcal{X}$ should contribute a copy of σ_x to B, which provides the motivation for our construction. We can now describe how to construct B by gluing together the polytopes

$$\{\sigma_x \mid \{x\} \in Strata(\mathcal{X}_0)\}.$$

We will do this in the case that every irreducible component of $\mathcal{X}_0$ is in fact itself normal so that $\nu : X_i \to \nu(X_i)$ is an isomorphism. The reader may be able to imagine the more general construction. With this normality assumption, there is a one-to-one inclusion reversing correspondence between faces of σ_x and elements of $Strata(\mathcal{X}_0)$ containing x. We can then identify faces of σ_x and $\sigma_{x'}$ if they correspond to the same strata of $\mathcal{X}_0$. Some argument is necessary to show that this identification can be done via an integral affine transformation, but again this is not difficult.

Making these identifications, one obtains B. One can then prove

LEMMA 7.3. *If $\mathcal{X}_0$ is complex n-dimensional, then B is a real n-dimensional manifold.*

See [**31**], Proposition 4.10 for a proof.

Now so far B is just a topological manifold, constructed by gluing together lattice polytopes. Let

$$\mathscr{P} = \{\sigma \subseteq B \,|\, \sigma \text{ is a face of } \sigma_x \text{ for some zero-dimensional stratum } x\}.$$

There is a one-to-one inclusion reversing correspondence between strata of $\mathcal{X}_0$ and elements of $\mathscr{P}$.

It only remains to give B an affine structure with singularities. In fact, I shall describe somewhat more structure on B derived from $\mathcal{X}_0$ which in particular gives an affine structure with singularities on B.

First, for $\tau \in \mathscr{P}$, let

$$U_\tau := \bigcup_{\{\sigma \in \mathscr{P} \,|\, \tau \subseteq \sigma\}} \text{Int}(\sigma).$$

A *fan structure along* $\tau \in \mathscr{P}$ is a continuous map $S_\tau : U_\tau \to \mathbb{R}^k$ such that

(1) $S_\tau^{-1}(0) = \text{Int}(\tau)$.
(2) If $e : \tau \to \sigma$ is an inclusion then $S_\tau|_{\text{Int}\,\sigma}$ is an integral affine submersion onto its image.
(3) The collection of cones

$$\{K_e := \mathbb{R}_{\geq 0} S_\tau(\sigma \cap U_\tau) \,|\, e : \tau \to \sigma\}$$

defines a finite fan Σ_τ in $\mathbb{R}^k$.

Two fan structures $S_\tau, S_\tau' : U_\tau \to \mathbb{R}^k$ are considered *equivalent* if they differ only by an integral linear transformation of $\mathbb{R}^k$.

If $S_\tau : U_\tau \to \mathbb{R}^k$ is a fan structure along $\tau \in \mathscr{P}$ and $\sigma \supseteq \tau$ then $U_\sigma \subseteq U_\tau$. The fan structure along σ *induced by* S_τ is the composition

$$U_\sigma \longrightarrow U_\tau \xrightarrow{\ S_\tau\ } \mathbb{R}^k \longrightarrow \mathbb{R}^k / L_\sigma \cong \mathbb{R}^\ell$$

where $L_\sigma \subseteq \mathbb{R}^k$ is the linear span of $S_\tau(\sigma)$.

DEFINITION 7.4. An *integral tropical manifold of dimension* n is a pair $(B, \mathscr{P})$ as above along with a choice of fan structure S_v at each vertex v of $\mathscr{P}$, with the property that if $v, w \in \tau$, then the fan structures along τ induced by S_v and S_w are equivalent.

Such data gives B the structure of an integral affine manifold with singularities. Let $\Gamma \subseteq B$ be the union of those cells of $\text{Bar}(\mathscr{P})$ (the first barycentric subdivision of $\mathscr{P}$) which are not contained in maximal cells of $\mathscr{P}$ nor contain vertices of $\mathscr{P}$. Then $B_0 := B \setminus \Gamma$ can be covered by

$$\{\text{Int}(\sigma) \,|\, \sigma \in \mathscr{P}_{\max}\} \cup \{W_v \,|\, v \in \mathscr{P} \text{ a vertex}\}$$

for certain open neighbourhoods W_v of $v \in B$ contained in U_v. We define an affine structure on B_0 by giving $\text{Int}(\sigma)$ the natural affine structure given by σ being a lattice polytope, while $S_v : U_v \to \mathbb{R}^n$ restricts to an affine chart on W_v.

Finally, the point is that the structure of $\mathcal{X}_0$ gives rise to an integral tropical manifold structure on $(B, \mathscr{P})$. Indeed, each vertex $v \in \mathscr{P}$ corresponds to an irreducible component X_v of $\mathcal{X}_0$ and this irreducible component is a toric variety with fan Σ_v in $\mathbb{R}^n$. Furthermore, there is a one-to-one correspondence between p-dimensional cones of Σ_v and p-dimensional cells of $\mathscr{P}$ containing v as a vertex, as they both correspond to strata of $\mathcal{X}_0$ contained in X_v. There is then a continuous map

$$\psi_v : U_v \to \mathbb{R}^n$$

which takes $U_v \cap \sigma$, for any $\sigma \in \mathscr{P}$ containing v as a vertex, into the corresponding cone of Σ_v *integral affine linearly*. Such a map is uniquely determined by the combinatorial correspondence and the requirement that it be integral affine linear on each cell. These maps define a fan structure at each vertex. Furthermore, these fan structures are compatible in the sense that if $v, w \in \tau$, the two induced fan structures on U_τ are equivalent. This follows because there is a well-defined fan Σ_τ defining the stratum corresponding to τ.

EXAMPLE 7.5. Let $f : \mathcal{X} \to D$ be a degeneration of elliptic curves to an I_n fibre. Then B is the circle $\mathbb{R}/n\mathbb{Z}$, decomposed by $\mathscr{P}$ into n line segments of length one.

EXAMPLE 7.6. Continuing with Example 7.2, the dual intersection complex is the boundary of a tetrahedron, with each face affine isomorphic to a standard two-simplex, and the affine structure near each vertex makes the polyhedral decomposition look locally like the fan for $\mathbb{P}^2$. There is one singularity at the barycenter of each edge, and one can calculate that the monodromy of Λ about each of these singularities is $\begin{pmatrix} 1 & 4 \\ 0 & 1 \end{pmatrix}$ in a suitable basis.

EXAMPLE 7.7. Consider the polytope Δ of Example 3.2. The dual polytope ∇ is the convex hull of the points $(-1,-1,-1,-1), (1,0,0,0), \ldots, (0,0,0,1)$. The corresponding projective toric variety $\mathbb{P}_\nabla$ has a crepant resolution $X_\Sigma \to \mathbb{P}_\nabla$ where Σ is the fan consisting of cones over all elements of the decomposition $\mathscr{P}$ of $\partial\Delta$ as described in Example 3.2. Consider in $\mathbb{P}_\nabla \times \mathbb{A}^1$ the degenerating family $\mathcal{X} \to \mathbb{A}^1$ of Calabi-Yau manifolds given by

$$s_0 + t \sum_{m \in \nabla \cap \mathbb{Z}^4} c_m s_m = 0$$

where s_m is the section of $\mathcal{O}_{\mathbb{P}_\nabla}(1)$ corresponding to $m \in \nabla \cap \mathbb{Z}^4$. Let $\widetilde{\mathcal{X}}$ be the proper transform of $\mathcal{X}$ in $X_\Sigma \times \mathbb{A}^1$. Then the family $\widetilde{\mathcal{X}} \to \mathbb{A}^1$ is a toric degeneration with general fibre the mirror quintic, and its dual intersection complex is the affine manifold B constructed in Example 3.2.

Is the dual intersection complex the right affine manifold with singularities? The following theorem provides evidence for this, and gives the connection between this construction and the SYZ conjecture.

THEOREM 7.8. *Let* $\mathcal{X} \to D$ *be a toric degeneration, with dual intersection complex* $(B, \mathscr{P})$. *Then there is an open set* $U \subseteq B$ *such that* $B \setminus U$ *retracts onto the discriminant locus* Γ *of* B, *and an open subset* $\mathcal{U}_t$ *of* $\mathcal{X}_t$ *which is biholomorphic to a small deformation of a twist of* $X_\epsilon(U)$, *where* $\epsilon = O(-1/\ln|t|)$.

We will not be precise here about what we mean by small deformation; by twist, we mean a twist of the complex structure of $X_\epsilon(U)$ by a B-field. See [**29**] for a much more precise statement; the above statement is meant to give a feel for what is true. The proof, along with much more precise statements, will eventually appear in [**30**].

If $\mathcal{X} \to D$ is a *polarized* toric degeneration, i.e., if there is a relatively ample line bundle $\mathcal{L}$ on $\mathcal{X}$, then we can construct another integral tropical manifold $(\check{B}, \check{\mathscr{P}})$, which we call the *intersection complex*, as follows.

For each irreducible component X_i of $\mathcal{X}_0$, $\mathcal{L}|_{X_i}$ is an ample line bundle on a toric variety. Let $\check{\sigma}_i \subseteq N_\mathbb{R}$ denote the Newton polytope of this line bundle. There is then a one-to-one inclusion preserving correspondence between strata of $\mathcal{X}_0$ contained in X_i and faces of $\check{\sigma}_i$. We can then glue together the $\check{\sigma}_i$'s in the obvious way: if Y is a codimension one stratum of $\mathcal{X}_0$, it is contained in two irreducible components X_i and X_j, and defines faces of $\check{\sigma}_i$ and $\check{\sigma}_j$. These faces are affine isomorphic because they are both the Newton polytope of $\mathcal{L}|_Y$, and we can then identify them in the canonical way. Thus we obtain a topological space $\check{B}$ with a polyhedral decomposition $\check{\mathscr{P}}$.

To define the fan structure at a vertex $v \in \check{\mathscr{P}}$, note that such a vertex corresponds to a zero-dimensional stratum of $\mathcal{X}_0$, giving rise to a maximal cell σ_v of the dual intersection complex. Take the fan structure at v to be defined using the normal fan $\check{\Sigma}_v$ to σ_v. Then there is a one-to-one inclusion preserving correspondence between cones in $\check{\Sigma}_v$ and strata of $\mathcal{X}_0$ containing the stratum corresponding to v. This correspondence allows us to define a fan structure

$$S_v : U_v \to \mathbb{R}^n$$

which takes $U_v \cap \check{\sigma}$, for any $\check{\sigma} \in \check{\mathscr{P}}$ containing v as a vertex, into the corresponding cone of $\check{\Sigma}_v$. One checks easily that this set of fan structures satisfies the definition of integral tropical manifold, and hence defines the intersection complex $(\check{B}, \check{\mathscr{P}})$.

Analogously to Theorem 7.8, we expect

CONJECTURE 7.9. *Let* $\mathcal{X} \to D$ *be a polarized toric degeneration, with intersection complex* $(\check{B}, \check{\mathscr{P}})$. *Let* ω_t *be a Kähler form on* $\mathcal{X}_t$ *representing the first Chern class of the polarization. Then there is an open set* $\check{U} \subseteq \check{B}$ *such that* $\check{B} \setminus \check{U}$ *retracts onto the discriminant locus* Γ *of* $\check{B}$, *such that* $\mathcal{X}_t$ *is a symplectic compactification of* $\check{X}(\check{U})$ *for any* t.

I don't expect this to be particularly difficult: it should be amenable to the techniques of W.-D. Ruan [**71**], but such an approach has not been carried out in general.

The relationship between the intersection complex and the dual intersection complex can be made more precise by introducing multi-valued piecewise linear functions, in analogy with the multi-valued convex functions of Definition 1.3.

DEFINITION 7.10. Let $(B, \mathscr{P})$ be an integral tropical manifold. Then a *multivalued piecewise linear function* φ on B is a collection of continuous functions on an open cover $\{(U_i, \varphi_i)\}$ such that φ_i is affine linear on each cell of $\mathscr{P}$ intersecting U_i, and on $U_i \cap U_j$, $\varphi_i - \varphi_j$ is affine linear. Furthermore, for any $\tau \in \mathscr{P}$, let $S_\tau : U_\tau \to \mathbb{R}^k$ be the induced fan structure. Then there is a piecewise linear function φ_τ on the fan Σ_τ such that on $U_i \cap U_\tau$, $\varphi_i - \varphi_\tau \circ S_\tau$ is affine linear. Here we will always assume that each linear part of φ_i has differential in $\check{\Lambda}$, i.e., φ_i has integral slopes.

The rather technical condition on the local behaviour of each φ_i on U_τ comes from the idea that such a multi-valued piecewise linear function is really just a collection of piecewise linear functions on the fans Σ_τ given by the fan structure of $(B, \mathscr{P})$. These functions need to satisfy some compatibility conditions, and this compatibility is motivated by the following discussion.

Suppose we are given a polarized toric degeneration $\mathcal{X} \to D$. We in fact obtain a multi-valued piecewise linear function φ on the dual intersection complex $(B, \mathscr{P})$ as follows. Restricting to any toric stratum X_τ, $\mathcal{L}|_{X_\tau}$ is determined completely by an integral piecewise linear function φ_τ on Σ_τ, well-defined up to a choice of linear function. Pulling back this piecewise linear function via S_τ to U_τ, we obtain a collection of piecewise linear functions $\{(U_\tau, \varphi_\tau \circ S_\tau) \,|\, \tau \in \mathscr{P}\}$. The fact that $(\mathcal{L}|_{X_\tau})|_{X_\sigma} = \mathcal{L}|_{X_\sigma}$ for $\tau \subseteq \sigma$ implies that on overlaps $\varphi_\sigma \circ S_\sigma$ and $\varphi_\tau \circ S_\tau$ differ by at most an affine linear function. So $\{(U_\tau, \varphi_\tau \circ S_\tau)\}$ defines a multi-valued piecewise linear function. The last condition in the definition of multi-valued piecewise linear function then reflects the need for the function to be locally a pull-back of a function via S_σ in a neighbourhood of σ.

If $\mathcal{L}$ is ample, then the piecewise linear function determined by $\mathcal{L}|_{X_\sigma}$ is strictly convex. So we say a multi-valued piecewise linear function is *strictly convex* if φ_τ is strictly convex for each $\tau \in \mathscr{P}$.

As a consequence, if $\mathcal{X} \to D$ is a polarized toric degeneration, we will write $(B, \mathscr{P}, \varphi)$ for the data of the dual intersection complex and the induced multi-valued function φ. We call this triple the dual intersection complex of the polarized degeneration.

Now suppose we are given abstractly a triple $(B, \mathscr{P}, \varphi)$ with $(B, \mathscr{P})$ an integral tropical manifold and φ a strictly convex multi-valued piecewise linear function on B. Then we construct the *discrete Legendre transform* $(\check{B}, \check{\mathscr{P}}, \check{\varphi})$ of $(B, \mathscr{P}, \varphi)$ as follows.

$\check{B}$ will be constructed by gluing together Newton polytopes. If we view, for v a vertex of $\mathscr{P}$, the fan Σ_v as living in $M_\mathbb{R}$, then the Newton polytope of φ_v is

$$\check{v} := \{x \in N_\mathbb{R} \,|\, \langle x, y \rangle \geq -\varphi_v(y) \quad \forall y \in M_\mathbb{R}\}.$$

There is a one-to-one inclusion reversing correspondence between faces of $\check{v}$ and cells of $\mathscr{P}$ containing v. Furthermore, if σ is the smallest cell of $\mathscr{P}$ containing two vertices v and v', then the corresponding faces of $\check{v}$ and $\check{v}'$ are integral affine isomorphic, as they are both isomorphic to the Newton polytope of φ_σ. Thus we can glue $\check{v}$ and $\check{v}'$ along this common face. After making all these identifications, we obtain a cell complex $(\check{B}, \check{\mathscr{P}})$, which is really just the dual cell complex of $(B, \mathscr{P})$. This is given an integral tropical structure by taking the fan structure at a vertex $\check{\sigma}$, for $\sigma \in \mathscr{P}_{\max}$, to be given by the normal fan to σ.

Finally, the function φ has a discrete Legendre transform $\check{\varphi}$ on $(\check{B}, \check{\mathscr{P}})$. We have no choice but to define $\check{\varphi}$ in a neighbourhood of a vertex $\check{\sigma} \in \check{\mathscr{P}}$ dual to a maximal cell $\sigma \in \mathscr{P}$ to be a piecewise linear function whose Newton polytope is σ, i.e.,

$$\check{\varphi}_{\check{\sigma}}(y) = -\inf\{\langle y, x \rangle \,|\, x \in \sigma \subseteq M_\mathbb{R}\}.$$

This gives $(\check{B}, \check{\mathscr{P}}, \check{\varphi})$, the discrete Legendre transform of $(B, \mathscr{P}, \varphi)$. If B is $\mathbb{R}^n$, then this coincides with the classical notion of discrete Legendre transform. The discrete Legendre transform has several relevant properties:

- The discrete Legendre transform of $(\check{B}, \check{\mathscr{P}}, \check{\varphi})$ is $(B, \mathscr{P}, \varphi)$.

- If we view the underlying topological spaces B and $\check{B}$ as identified by being the underlying space of dual cell complexes, then $\Lambda_{B_0} \cong \check{\Lambda}_{\check{B}_0}$ and $\check{\Lambda}_{B_0} \cong \Lambda_{\check{B}_0}$, where the subscript denotes which affine structure is being used to define Λ or $\check{\Lambda}$.

This hopefully makes it clear that the discrete Legendre transform is a suitable replacement for the duality provided by the Legendre transform of §2.

Note in particular that if $\mathcal{X} \to D$ is a polarized toric degeneration, with dual intersection complex $(B, \mathscr{P}, \varphi)$, then the discrete Legendre transform $(\check{B}, \check{\mathscr{P}}, \check{\varphi})$ satisfies the condition that $(\check{B}, \check{\mathscr{P}})$ is the intersection complex of the polarized degeneration. The function $\check{\varphi}$ is some extra information on $\check{B}$, which from the definition of discrete Legendre transform encodes the cells of $\mathscr{P}$. These cells of the dual intersection complex were defined using the local toric structure of $\mathcal{X} \to D$. So $\check{\varphi}$ can be seen as carrying information about this local toric structure. We will say $(\check{B}, \check{\mathscr{P}}, \check{\varphi})$ is the intersection complex of the polarized toric degeneration $\mathcal{X} \to D$.

So we see that for $(B, \mathscr{P}, \varphi)$, $\mathscr{P}$ carries information about the log structure and φ carries information about the polarization, but for $(\check{B}, \check{\mathscr{P}}, \check{\varphi})$, $\check{\mathscr{P}}$ carries information about the polarization and $\check{\varphi}$ carries information about the log structure. Mirror symmetry interchanges these two pieces of information!

We can now state an *algebro-geometric SYZ procedure*. In analogy with the procedure suggested in §5, we could follow these steps:

(1) We begin with a toric degeneration of Calabi-Yau manifolds $\mathcal{X} \to D$ with an ample polarization.
(2) Construct the dual intersection complex $(B, \mathscr{P}, \varphi)$ from this data, as explained above.
(3) Perform the discrete Legendre transform to obtain $(\check{B}, \check{\mathscr{P}}, \check{\varphi})$.
(4) Try to construct a polarized degeneration of Calabi-Yau manifolds $\check{\mathcal{X}} \to D$ whose dual intersection complex is $(\check{B}, \check{\mathscr{P}}, \check{\varphi})$, or whose intersection complex is $(B, \mathscr{P}, \varphi)$.

EXAMPLE 7.11. The discrete Legendre transform enables us to reproduce Batyrev duality [5]. Let $\Delta \subseteq M_{\mathbb{R}}$ be a reflexive polytope, $\nabla \subseteq N_{\mathbb{R}}$ the polar dual, and assume $0 \in \Delta$ is the unique interior point. We then obtain two toric degenerations given by the equations

$$s_0 + t \sum_{m \in M \cap \Delta} c_m s_m = 0, \qquad s_0 + t \sum_{n \in N \cap \nabla} c_n s_n = 0$$

in $\mathbb{P}_{\Delta} \times \mathbb{A}^1$ and $\mathbb{P}_{\nabla} \times \mathbb{A}^1$ respectively, with s_m (s_n) the section of $\mathcal{O}_{\mathbb{P}_{\Delta}}(1)$ corresponding to m (the section of $\mathcal{O}_{\mathbb{P}_{\nabla}}(1)$ corresponding to n). It is easy to check that the dual intersection complexes of these two degenerations are given as follows. For the first degeneration, $B = \partial \nabla$ with polyhedral decomposition given by the proper faces of ∇. The fan structure at each vertex v is given by projection $U_v \hookrightarrow N_{\mathbb{R}} \to N_{\mathbb{R}}/\mathbb{R}v$. For the second degeneration, one uses Δ instead of ∇. One can then check that if one polarizes the two degenerations using $\mathcal{O}_{\mathbb{P}_{\Delta}}(1)$ and $\mathcal{O}_{\mathbb{P}_{\nabla}}(1)$ respectively, then the corresponding triples $(B, \mathscr{P}, \varphi)$ are Legendre dual. Thus Batyrev duality is a special case of this general approach to a mirror construction.

For a much more general construction which works for the Batyrev-Borisov construction [6] of mirrors of complete intersection Calabi-Yaus in toric varieties, see [22].

The only step missing in this mirror symmetry algorithm is the last:

QUESTION 7.12 (The reconstruction problem, Version II). *Given $(B, \mathscr{P}, \varphi)$, is it possible to construct a polarized toric degeneration $\mathcal{X} \to D$ whose intersection complex is $(B, \mathscr{P}, \varphi)$?*

One could hope to solve this problem via naive deformation theory, by constructing the central fibre $\mathcal{X}_0$ from the data $(B, \mathscr{P}, \varphi)$, and then deforming this to find a smoothing. However, as initially observed in the normal crossings case by Kawamata and Namikawa in [**51**], one needs to put some additional structure on $\mathcal{X}_0$ before it has good deformation theory. This structure is a *log structure*, and introducing log structures allows us to study many aspects of mirror symmetry directly on the degenerate fibre itself. So let us turn to a review of the theory of logarithmic structures.

8. Log structures

We review the notion of log structures of Fontaine-Illusie and Kato ([**45**], [**50**]). These play a key role in trying to understand mirror symmetry via degenerations.

DEFINITION 8.1. A log structure on a scheme (or analytic space) X is a (unital) homomorphism

$$\alpha_X : \mathcal{M}_X \to \mathcal{O}_X$$

of sheaves of (multiplicative and commutative) monoids inducing an isomorphism $\alpha_X^{-1}(\mathcal{O}_X^\times) \to \mathcal{O}_X^\times$. The monoid structure on $\mathcal{O}_X$ is given by multiplication. The triple $(X, \mathcal{M}_X, \alpha_X)$ is then called a *log space*. We often write the whole package as $X^\dagger$.

A morphism of log spaces $F : X^\dagger \to Y^\dagger$ consists of a morphism $\underline{F} : X \to Y$ of underlying spaces together with a homomorphism $F^\# : \underline{F}^{-1}(\mathcal{M}_Y) \to \mathcal{M}_X$ commuting with the structure homomorphisms:

$$\alpha_X \circ F^\# = \underline{F}^* \circ \alpha_Y.$$

The key examples:

EXAMPLES 8.2. (1) Let X be a scheme and $Y \subseteq X$ a closed subset of codimension one. Denote by $j : X \setminus Y \to X$ the inclusion. Then the inclusion

$$\alpha_X : \mathcal{M}_X = j_*(\mathcal{O}_{X\setminus Y}^\times) \cap \mathcal{O}_X \to \mathcal{O}_X$$

of the sheaf of regular functions invertible off of Y is a log structure on X. This is called a *divisorial log structure* on X.

(2) A *prelog structure*, i.e., an arbitrary homomorphism of sheaves of monoids $\varphi : \mathcal{P} \to \mathcal{O}_X$, defines an associated log structure $\mathcal{M}_X$ by

$$\mathcal{M}_X = (\mathcal{P} \oplus \mathcal{O}_X^\times)/\{(p, \varphi(p)^{-1}) \mid p \in \varphi^{-1}(\mathcal{O}_X^\times)\}$$

and $\alpha_X(p, h) = h \cdot \varphi(p)$.

(3) If $f : X \to Y$ is a morphism of schemes and $\alpha_Y : \mathcal{M}_Y \to \mathcal{O}_Y$ is a log structure on Y, then the prelog structure $f^{-1}(\mathcal{M}_Y) \to \mathcal{O}_X$ given as the composition of $\alpha_Y : f^{-1}(\mathcal{M}_Y) \to f^{-1}\mathcal{O}_Y$ and $f^* : f^{-1}\mathcal{O}_Y \to \mathcal{O}_X$ defines an associated log structure on X, the *pull-back log structure*.

(4) In (1) we can pull back the log structure on X to Y using (3). Thus in particular, if $\mathcal{X} \to D$ is a toric degeneration, the inclusion $\mathcal{X}_0 \subseteq \mathcal{X}$ gives a log

structure on $\mathcal{X}$ and an induced log structure on $\mathcal{X}_0$. Similarly the inclusion $0 \in D$ gives a log structure on D and an induced one on 0. Here $\mathcal{M}_0 = \mathbb{C}^\times \oplus \mathbb{N}$, where $\mathbb{N}$ is the (additive) monoid of natural (non-negative) numbers, and

$$\alpha_0(h, n) = \begin{cases} h & n = 0 \\ 0 & n \neq 0. \end{cases}$$

$0^\dagger$ is usually called the *standard log point*.

We then have log morphisms $\mathcal{X}^\dagger \to D^\dagger$ and $\mathcal{X}_0^\dagger \to 0^\dagger$.

(5) If $\sigma \subseteq M_\mathbb{R} = \mathbb{R}^n$ is a strictly convex rational polyhedral cone, $\sigma^\vee \subseteq N_\mathbb{R}$ the dual cone, let $P = \sigma^\vee \cap N$: this is a monoid under addition. The affine toric variety defined by σ can be written as $X = \operatorname{Spec}\mathbb{C}[P]$. We then have a pre-log structure induced by the homomorphism of monoids

$$P \to \mathbb{C}[P]$$

given by $p \mapsto z^p$. There is then an associated log structure on X. This is in fact the same as the log structure induced by $\partial X \subseteq X$, where ∂X is the toric boundary of X, i.e., the union of toric divisors of X.

If $p \in P$, then the monomial z^p defines a map $f : X \to \operatorname{Spec}\mathbb{C}[\mathbb{N}] = \mathbb{A}^1$ which is a log morphism with the log structure on $\operatorname{Spec}\mathbb{C}[\mathbb{N}]$ induced similarly by $\mathbb{N} \to \mathbb{C}[\mathbb{N}]$. The fibre $X_0 = \operatorname{Spec}\mathbb{C}[P]/(z^p)$ is a subscheme of X, there is an induced log structure on X_0, and a map $X_0^\dagger \to 0^\dagger$ as in (4). The log morphism f is an example of a *log smooth* morphism, see Definition 8.3.

Condition (4) of Definition 7.1 in fact implies that locally, away from Z, $\mathcal{X}^\dagger$ and $\mathcal{X}_0^\dagger$ are of the above form. So we should view $\mathcal{X}^\dagger \to D^\dagger$ as log smooth away from Z, and from the log point of view, $\mathcal{X}_0^\dagger$ can be treated much like a non-singular scheme away from Z.

(6) Given a monoid P as in (5) and a morphism $X \to \operatorname{Spec}\mathbb{C}[P]$, we can pull back the log structure defined above on $\operatorname{Spec}\mathbb{C}[P]$ to X. If $X^\dagger$ is a log scheme which étale locally can be described in this way, we say $X^\dagger$ is a *fine saturated log scheme*. The adjective "fine" tells us it is locally described via maps to schemes of the form $\operatorname{Spec}\mathbb{C}[P]$ where P is a finitely generated *integral* monoid, i.e., the canonical homomorphism $P \to P^{\mathrm{gp}}$ is an injection. The adjective "saturated" tells us the monoid P is saturated. This means that P is integral and whenever $p \in P^{\mathrm{gp}}$ satisfies $mp \in P$ for some $m > 0$, $p \in P$. Such monoids arise, e.g., as the intersection of a rational polyhedral cone with a lattice.

Most of the literature on log geometry tends to apply only to fine log structures. In the key example of $\mathcal{X}_0^\dagger \to 0^\dagger$, the log structure is fine saturated away from the set Z. However, it is not in general fine along Z, and this tends to cause many technical problems as new techniques have to be developed to deal properly with the log structure along Z. $\qquad\square$

The notion of log smoothness generalizes the morphisms of Examples 8.2, (5):

DEFINITION 8.3. A morphism $f : X^\dagger \to Y^\dagger$ of fine log schemes is *log smooth* if étale locally on X and Y it fits into a commutative diagram

$$
\begin{array}{ccc}
X & \longrightarrow & \operatorname{Spec} \mathbb{Z}[P] \\
\downarrow & & \downarrow \\
Y & \longrightarrow & \operatorname{Spec} \mathbb{Z}[Q]
\end{array}
$$

with the following properties:

(1) The canonical log structure on $\operatorname{Spec}\mathbb{Z}[P]$ and $\operatorname{Spec}\mathbb{Z}[Q]$ of Examples 8.2, (5), pull-back to the log structures on X and Y respectively.

(2) The induced morphism

$$
X \to Y \times_{\operatorname{Spec}\mathbb{Z}[Q]} \operatorname{Spec}\mathbb{Z}[P]
$$

is a smooth morphism of schemes.

(3) The right-hand vertical arrow is induced by a monoid homomorphism $Q \to P$ with $\ker(Q^{\mathrm{gp}} \to P^{\mathrm{gp}})$ and the torsion part of $\operatorname{coker}(Q^{\mathrm{gp}} \to P^{\mathrm{gp}})$ finite groups of orders invertible on X. Here P^{gp} denotes the Grothendieck group of P.

Log smooth morphisms include, in the simplest case, normal crossings morphisms.

On a log scheme $X^\dagger$ there is always an exact sequence

$$
1 \longrightarrow \mathcal{O}_X^\times \xrightarrow{\alpha^{-1}} \mathcal{M}_X \longrightarrow \overline{\mathcal{M}}_X \longrightarrow 0,
$$

where we write the quotient sheaf of monoids $\overline{\mathcal{M}}_X$ additively. We call $\overline{\mathcal{M}}_X$ the *ghost sheaf* of the log structure. I like to view $\overline{\mathcal{M}}_X$ as specifying the combinatorial information associated to the log structure. For example, if $X^\dagger$ is induced by the Cartier divisor $Y \subseteq X$ with X normal, then the stalk $\overline{\mathcal{M}}_{X,x}$ at $x \in X$ is the monoid of effective Cartier divisors on a neighbourhood of x supported on Y.

It is useful for understanding pull-backs of log structures to note that if $f : Y \to X$ is a morphism with X carrying a log structure, and Y is given the pull-back log structure, then $\overline{\mathcal{M}}_Y = f^{-1}\overline{\mathcal{M}}_X$. In the case that $\mathcal{M}_X$ is induced by an inclusion of $Y \subseteq X$, $\overline{\mathcal{M}}_X$ is supported on Y, so we can equate $\overline{\mathcal{M}}_X$ and $\overline{\mathcal{M}}_Y$, the ghost sheaves for the divisorial log structure on X and its restriction to Y.

EXERCISE 8.4. *Show that in Example 8.2, (5), $\overline{\mathcal{M}}_{X,x} = P$ if $\dim \sigma = \dim M_{\mathbb{R}}$ and x is the unique zero-dimensional torus orbit of X. More generally,*

$$
\overline{\mathcal{M}}_{X,x} = \frac{\tau^\vee \cap N}{\tau^\perp \cap N} = \operatorname{Hom}_{monoid}(\tau \cap M, \mathbb{N}),
$$

when $x \in X$ is in the torus orbit corresponding to a face τ of σ. In particular, τ can be recovered as $\operatorname{Hom}_{monoid}(\overline{\mathcal{M}}_{X,x}, \mathbb{R}_{\geq 0})$, where $\mathbb{R}_{\geq 0}$ is the additive monoid of non-negative real numbers. □

In the sections which follow, the key logarithmic spaces we consider will be those arising from toric degenerations $\mathcal{X} \to D$. As above, the central fibre $\mathcal{X}_0 \subseteq \mathcal{X}$ induces a divisorial log structure on $\mathcal{X}$, and restricting gives a log scheme $\mathcal{X}_0^\dagger$ along with a morphism $\mathcal{X}_0^\dagger \to 0^\dagger$ which is log smooth off of the bad set $Z \subseteq \mathcal{X}_0$.

We can now elaborate on the philosophy we wish to take with the following diagram:

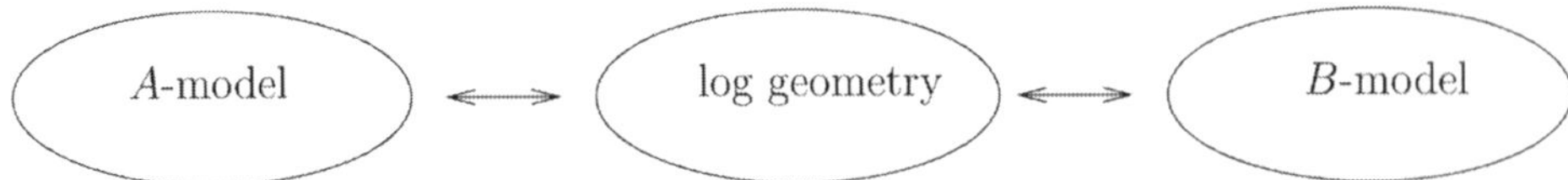

There are two sides to mirror symmetry. The A-model side involves counting curves: we wish to count curves in the general fibre of a toric degeneration $\mathcal{X} \to D$. There are good reasons to believe that this count can in fact be performed on $\mathcal{X}_0^\dagger$, using a theory of logarithmic Gromov-Witten invariants: see §9. The hope is that $\mathcal{X}_0$ is a sufficiently combinatorial object so that such a count can be carried out in a combinatorial manner.

The B-side involves deformations of complex structure. The idea is that to understand deformations of complex structure, we should start with the central fibre $\mathcal{X}_0^\dagger$ and try to construct smoothings, i.e., construct a toric degeneration with this central fibre. The log structure is necessary to find a unique smoothing. If this smoothing can be described sufficiently explicitly, then again one should be able to extract the necessary periods for the B-model calculations purely in terms of combinatorics.

So log geometry will play an important role on both sides of mirror symmetry, but as the above suggests, there should be some combinatorial objects underlying both calculations.

In fact, log geometry is closely related to tropical geometry. We will explore in the following sections how tropical geometry controls both the A- and B-model sides of the above picture, completing the above diagram:

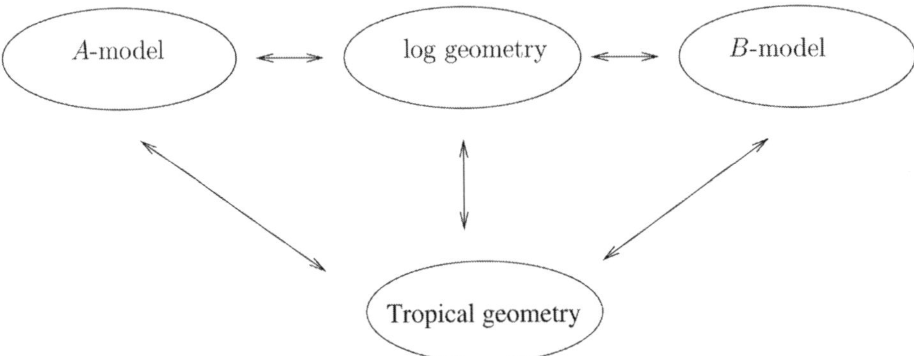

9. The A-model and tropical geometry

The first link between log geometry and tropical geometry comes from an elementary combinatorial construction. Given a log scheme $X^\dagger$, we can construct the *tropicalization* of $X^\dagger$, as follows. For each geometric point $\bar\eta$ of X, we have a monoid $\overline{\mathcal{M}}_{X,\bar\eta}$, and hence a cone $C_{\bar\eta} := \mathrm{Hom}(\overline{\mathcal{M}}_{X,\bar\eta}, \mathbb{R}_{\geq 0})$ (where here Hom denotes monoid homomorphisms and $\mathbb{R}_{\geq 0}$ is given the additive monoid structure). Further, if $\bar{\eta}$ is in the closure of $\bar{\eta}'$, there is a generization map $\overline{\mathcal{M}}_{X,\bar{\eta}} \to \overline{\mathcal{M}}_{X,\bar{\eta}'}$.[1] Dualizing, this gives maps $C_{\bar{\eta}'} \to C_{\bar{\eta}}$. If the log structure on X is fine, then these maps are inclusions of faces of strictly convex rational polyhedral cones. We can then form a cell complex by making identifications given by these inclusions of faces, obtaining a polyhedral cone complex $\mathrm{Trop}(X^{\dagger})$. Actually, in general this may not really make sense as a cell complex because the generization maps may induce many strange self-identifications on faces, but in the situations we want to describe here, this will not cause a problem.

This construction is functorial, so if $f : X^{\dagger} \to Y^{\dagger}$ is a morphism of log schemes, then we obtain $\mathrm{Trop}(f) : \mathrm{Trop}(X^{\dagger}) \to \mathrm{Trop}(Y^{\dagger})$.

For example, consider the case of a toric degeneration $\mathcal{X} \to D$. As we saw in the previous section, this gives a morphism of log schemes $\mathcal{X}_0^{\dagger} \to 0^{\dagger}$. The bad set $Z \subseteq \mathcal{X}_0$ is precisely the locus where the log structure on $\mathcal{X}_0$ is not fine. Thus we can apply the above tropicalization construction to $\mathcal{X}_0^{\dagger} \setminus Z \to 0^{\dagger}$. Now $\mathrm{Trop}(0^{\dagger}) = \mathbb{R}_{\geq 0}$ is a ray. On the other hand, if $x \in \mathcal{X}_0$ is a zero-dimensional stratum, locally a neighbourhood of x looks like the fibre over 0 of $f_x : Y_x \to \mathbb{C}$ where Y_x is a toric variety defined by $C(\sigma_x) \subseteq M_{\mathbb{R}} \oplus \mathbb{R}$ for a lattice polytope $\sigma_x \subseteq M_{\mathbb{R}}$, and the morphism $Y_x \to \mathbb{C}$ is given by the projection $M_{\mathbb{R}} \oplus \mathbb{R} \to \mathbb{R}$. Then $\overline{\mathcal{M}}_{\mathcal{X}_0,x} = C(\sigma_x)^{\vee} \cap (N \oplus \mathbb{Z})$, as follows from Exercise 8.4, and $C_x = C(\sigma_x)$. In particular, the induced map $\mathrm{Trop}(f) : C_x \to \mathrm{Trop}(0^{\dagger})$ has fibre $\mathrm{Trop}(f)^{-1}(1) = \sigma_x$. From this, one checks easily that

$$\mathrm{Trop}(f) : \mathrm{Trop}(\mathcal{X}_0^{\dagger} \setminus Z) \to \mathrm{Trop}(0^{\dagger}) = \mathbb{R}_{\geq 0}$$

has fibre

$$\mathrm{Trop}(f)^{-1}(1) = B,$$

with B coming with the polyhedral decomposition $\mathscr{P}$. So the dual intersection complex comes from a very general construction. In particular, note that B only depends on $\mathcal{X}_0^{\dagger}$, not on $\mathcal{X}$ (although this is obvious without knowing this general construction).

Now let us turn to the A-model, which for the purposes of this discussion means counting curves on Calabi-Yau manifolds. Suppose we have a toric degeneration $\mathcal{X} \to D$. We would like to count curves on the general fibre. Can we do so by counting curves on $\mathcal{X}_0$ instead, where the problem might have a more combinatorial nature?

This question has a long history. The first work on this kind of question was due to Li and Ruan [58] and Ionel and Parker [46],[47]. Essentially they considered a situation where one has a degeneration $\mathcal{X} \to D$ where the special fibre $\mathcal{X}_0 = X_1 \cup X_2$ is a normal crossings union of two smooth irreducible components. They showed that there was a theory of Gromov-Witten invariants of $\mathcal{X}_0$, and that it gave the same answer as Gromov-Witten theory on a general fibre. Further, they gave gluing formulas, which stated that the Gromov-Witten invariants of $\mathcal{X}_0$ could be computed using the Gromov-Witten invariants of the two pairs $(X_i, X_1 \cap X_2)$, $i = 1, 2$. Here the Gromov-Witten theory associated to a pair (X, D) where $D \subseteq X$ is a smooth divisor is the theory of *relative* Gromov-Witten invariants, where one considers curves in X with some imposed orders of tangency at points on the curve with

---

[1]Since we need to work in the étale topology, there can actually be a number of generization maps. For example, if X is a nodal cubic, then there are two generization maps from $\bar{\eta}$ the node to $\bar{\eta}'$ the generic point.

D. This gluing formula has proven to be a very powerful tool in Gromov-Witten theory.

In 2001, Bernd Siebert [74] proposed using log geometry to generalize these results. Meanwhile, Jun Li was working on an algebro-geometric approach to the Li-Ruan and Ionel-Parker theories (which were carried out using symplectic techniques). He gave a satisfactory algebro-geometric definition of relative Gromov-Witten invariants and reproved the gluing formula, using a few techniques from log geometry. However, the theory possesses a technical difficulty. In Gromov-Witten theory, it is standard that one allows the domain curves to develop bubbles. But in relative Gromov-Witten theory, it is also necessary to allow the target space X to develop bubbles. This occurs when an irreducible component of the domain curve falls into the divisor D, so that the order of tangency with D becomes meaningless. So the actual target space for a relative stable map might be X with a chain of $\mathbb{P}^1$-bundles over D glued to $D \subseteq X$. This often makes the analysis more difficult, and was a major stumbling block for extending these techniques to more complicated degenerations.

Several solutions to this problem were completed in 2011. Brett Parker in [66], [67] provided a completely new category, the category of *exploded manifolds*, in which to study Gromov-Witten theory. These manifolds carry information similar to log spaces, but is a somewhat more flexible and "softer" category in which to work. In [67] he provides a definition of Gromov-witten invariants in this setting and gives a gluing formula. Also, Siebert and I [34] completed a theory of logarithmic Gromov-Witten invariants, as did Abramovich and Chen [11],[1], working with Siebert's original suggestion. I will summarize the basic ideas here.

DEFINITION 9.1. A *log curve* over a fine saturated log scheme $W^\dagger$ is a fine saturated log scheme $C^\dagger$ with a morphism $C^\dagger \to W^\dagger$ which is flat of relative dimension one, log smooth, and with all geometric fibres reduced.

Here log smoothness implies that the geometric fibres of $C \to W$ are nodal curves, which is pleasant as this is precisely the sort of curve which is allowed as the domain of a stable map. The log structure can also be viewed as incorporating marked points. For example, given a smooth curve C over $W = \mathrm{Spec}\,\mathbb{C}$, one can take a finite number of points $x_1, \ldots, x_k \in C$ and give C the divisorial log structure associated to the subset $\{x_1, \ldots, x_k\} \subseteq C$. Then $C^\dagger$ is log smooth over W with the trivial log structure $\mathcal{M}_W = \mathcal{O}_W^\times$.

DEFINITION 9.2. Let $X^\dagger \to S^\dagger$ be a morphism of fine saturated log schemes. A *log curve in $X^\dagger$ with base $W^\dagger$* is a log curve $C^\dagger/W^\dagger$ together with a morphism $f : C^\dagger \to X^\dagger$ fitting into a commutative diagram of log schemes

$$
\begin{array}{ccc}
C^\dagger & \xrightarrow{\;f\;} & X^\dagger \\
\downarrow & & \downarrow \\
W^\dagger & \longrightarrow & S^\dagger
\end{array}
$$

A log curve in $X^\dagger$ is a *stable log map* if for every geometric point $\bar{w} \to W$, the restriction of f to the underlying marked curve $C_{\bar{w}} \to \bar{w}$ is an ordinary stable map. We write the data as $(C^\dagger/W^\dagger, f)$.

This definition can be further decorated in the usual way by labelling marked points.

The main work of [**34**] is to construct a well-behaved moduli space of stable log maps. There is a technical issue which arises whenever one tries to construct a moduli space of log objects; this was explored by Martin Olsson in his thesis [**65**]. The problem is as follows. Suppose we are given a stable log map with domain $\pi : (C, \mathcal{M}_C) \to (W, \mathcal{M}_W)$. Then $\pi' : (C, \mathcal{M}_C \oplus \mathbb{N}^r) \to (W, \mathcal{M}_W \oplus \mathbb{N}^r)$ also gives the domain of a stable log map. Here the structure map α_C (or α_W) takes the value 0 on the non-zero elements of the constant sheaf $\mathbb{N}^r$, and the map π' acting on monoids just takes $\mathbb{N}^r$ isomorphically to $\mathbb{N}^r$. The new map $f^{\#}$ is the composition of the old $f^{\#} : f^{-1}\mathcal{M}_X \to \mathcal{M}_C$ and the inclusion $\mathcal{M}_C \to \mathcal{M}_C \oplus \mathbb{N}^r$. As a result, a single stable log map gives rise to a countable number of other maps, so the stack of stable log maps has no chance of being finite type, and hence cannot be proper.

The solution is to identify log structures on W which are universal in a suitable sense. In the above example, all the log curves in question arise as a cartesian diagram of log schemes:

$$
\begin{array}{ccc}
(C, \mathcal{M}_C \oplus \mathbb{N}^r) & \longrightarrow & (C, \mathcal{M}_C) \\
\downarrow & & \downarrow \\
(W, \mathcal{M}_W \oplus \mathbb{N}^r) & \longrightarrow & (W, \mathcal{M}_W)
\end{array}
$$

Thus all these extraneous log curves can be viewed as obtained by pull-back from the initial choice of log curve via a logarithmic base-change.

To solve this problem, we introduce a property of stable log maps called *basic*. I do not wish to give the definition here, as it is very involved, but the important properties of basic stable log maps are universality and boundedness, as expressed in the following two theorems, a summation of the main results of [**34**]:

THEOREM 9.3. *Given a stable log map $(C^{\dagger}/W^{\dagger}, f)$, there is a basic stable log map $(C_b^{\dagger}/W_b^{\dagger}, f_b)$ fitting into a commutative diagram*

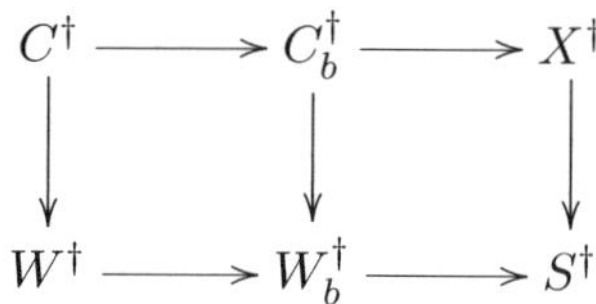

$$
\begin{array}{ccccc}
C^{\dagger} & \longrightarrow & C_b^{\dagger} & \longrightarrow & X^{\dagger} \\
\downarrow & & \downarrow & & \downarrow \\
W^{\dagger} & \longrightarrow & W_b^{\dagger} & \longrightarrow & S^{\dagger}
\end{array}
$$

where the left-hand square is cartesian in the category of fine saturated log schemes and the maps $W \to W_b$ and $C \to C_b$ of underlying schemes are isomorphisms. Furthermore, $(C_b^{\dagger}/W_b^{\dagger}, f_b)$ and the maps in the above diagram are determined by $(C^{\dagger}/W^{\dagger}, f)$ uniquely up to unique isomorphism.

THEOREM 9.4. *Let $\mathscr{M}(X^{\dagger}/S^{\dagger})$ denote the stack of basic stable log maps in $X^{\dagger}$ over $S^{\dagger}$. Then:*

(1) *$\mathscr{M}(X^{\dagger}/S^{\dagger})$ is a Deligne-Mumford stack.*

(2) *Let β denote a choice of genus g, number of marked points k, homology class in $H_2(X, \mathbb{Z})$, along with a collection of tangency data for the marked points (this notion can be made precise). Let $\mathscr{M}(X^{\dagger}/S^{\dagger}, \beta)$ denote the substack of $\mathscr{M}(X^{\dagger}/S^{\dagger})$ of basic stable log maps of curves of genus g and k marked points, representing the given homology class, and satisfying the given tangency conditions. Then modulo some technical hypotheses on $X^{\dagger}$, $\mathscr{M}(X^{\dagger}/S^{\dagger}, \beta)$ is proper over S if X is proper over S.*

(3) *Assuming further that $X^\dagger \to S^\dagger$ is log smooth, $\mathscr{M}(X^\dagger/S^\dagger, \beta)$ carries a virtual fundamental class, allowing for the definition of logarithmic Gromov-Witten invariants.*

Similar results were also obtained by Abramovich and Chen in [1],[11].

This is a promising start to the problem of understanding the A-model by working entirely on the central fibre of a toric degeneration. There are, however, still two major gaps in the theory which need to be filled.

First, one needs an analogue of the gluing formula. This should allow us to break down a calculation of curves on the central fibre of a degeneration into simpler pieces. This is expected to be quite subtle, however, and is still work in progress. I will say a bit more shortly about what one expects such a formula to look like.

Second, as observed earlier, the central fibre of a toric degeneration $\mathcal{X}_0^\dagger \to 0^\dagger$ is only fine saturated off of the set Z. As a result none of the above theorems about stable log maps apply. It is quite likely that even the definition of stable log map is not the correct one in this case. So the theory still needs to be extended. This is also work in progress of Michael Kasa.

Let us return to the tropicalization functor. Suppose we have a degeneration $q : \mathcal{X} \to D$, which we assume to be log smooth (say a normal crossings degeneration), so that we don't have to worry about the singular set Z where the log structure on $\mathcal{X}$ is not fine. As usual, this gives $\mathcal{X}_0^\dagger \to 0^\dagger$. Suppose we have a basic stable log map over a point, i.e., a diagram

$$
\begin{array}{ccc}
C^\dagger & \xrightarrow{\quad f \quad} & \mathcal{X}_0^\dagger \\
{\scriptstyle \pi}\downarrow & & \downarrow{\scriptstyle q} \\
W^\dagger = (\operatorname{Spec}\mathbb{C}, Q \oplus \mathbb{C}^\times) & \xrightarrow{\quad g \quad} & 0^\dagger
\end{array}
$$

Here Q is a monoid given by $Q = \sigma_Q \cap \mathbb{Z}^n$ for a strictly convex rational polyhedral cone σ_Q, and the log structure on W is given by $\alpha : Q \oplus \mathbb{C}^\times \to \mathbb{C}$ defined by

$$
\alpha(p, s) = \begin{cases} s & p = 0 \\ 0 & p \neq 0 \end{cases}
$$

Here, the monoid Q is determined by the fact the curve is basic. We then tropicalize this, so get a diagram

$$
\begin{array}{ccc}
\operatorname{Trop}(C^\dagger) & \xrightarrow{\operatorname{Trop}(f)} & \operatorname{Trop}(\mathcal{X}_0^\dagger) \\
{\scriptstyle \operatorname{Trop}(\pi)}\downarrow & & \downarrow{\scriptstyle \operatorname{Trop}(q)} \\
\operatorname{Trop}(W^\dagger) = \sigma_Q^\vee & \xrightarrow{\operatorname{Trop}(g)} & \operatorname{Trop}(0^\dagger) = \mathbb{R}_{\geq 0}
\end{array}
$$

The fibres of $\operatorname{Trop}(\pi)$ are in general one-dimensional graphs, while $\operatorname{Trop}(q)^{-1}(1)$ is the dual intersection complex B of $\mathcal{X}_0^\dagger$. (In general, this is only a polyhedral complex and does not carry an affine structure in codimension one, unlike the case of a toric degeneration.) Thus $\operatorname{Trop}(g)^{-1}(1) \subseteq \sigma_Q^\vee$ can be viewed as a space parameterizing maps from graphs (fibres of $\operatorname{Trop}(\pi)$) into B. These will be tropical curves. In fact, where B does carry an affine structure, these curves satisfy the tropical balancing condition.

The fundamental property that the monoid Q associated with the basic log structure must satisfy is that $\mathrm{Trop}(g)^{-1}(1)$ must parameterize all tropical curves in B of the same "combinatorial type". This makes precise the correspondence between tropical curves and log curves.

We can also describe the expected shape of a gluing formula, in keeping with the formula developed by Brett Parker in his setting [67]. One considers tropical curves in B as above. These in general move in families, but there will be, for any given set of data β, a finite number of tropical curves representing β which cannot be deformed without changing the domain graph. We call such tropical curves *rigid*. The actual moduli space $\mathscr{M}(\mathcal{X}_0^\dagger/0^\dagger, \beta)$ can then be viewed to have a "decomposition into virtual irreducible components" indexed by these rigid curves. Furthermore, the "virtual irreducible component" associated to any rigid curve can be further related to moduli spaces of curves associated to each vertex of the tropical curve. This should ultimately allow an expression for the Gromov-Witten invariants of $\mathcal{X}_0^\dagger/0^\dagger$, and hence the Gromov-Witten invariants of a smoothing of $\mathcal{X}_0^\dagger$, in terms of much simpler invariants. This is an ongoing joint project with Abramovich, Chen and Siebert.

10. The B-model and tropical geometry

Let us turn to the B-model, and understand how tropical geometry may be visible in the variation of complex structures which is necessary for B-model computations.

This problem is closely related to the reconstruction problem, as stated in Question 7.12. If given $(B, \mathscr{P}, \varphi)$, one can find an explicit description of a toric degeneration $\mathcal{X} \to D$, with dual intersection complex $(B, \mathscr{P}, \varphi)$, then one could use this explicit description to calculate periods and extract B-model predictions for the mirror.

Before describing the solution to this problem, let me give a bit of history of the reconstruction problem. The version as stated in Question 5.6 was first studied by Fukaya in [13]. There he considered directly the question of perturbing the complex structure on $X_\epsilon(B_0)$ by looking at the Kodaira-Spencer equation governing deformations of complex structure. Arguing informally in the case that $\dim B = 2$, he suggested that the perturbations should be concentrated along trees made of gradient flow lines, with the lines emanating initially from singular points of B. This gave the first hint that a nice solution to the reconstruction problem might actually see something related to curves. However, Fukaya's work contained no definite theorems, and the analysis looked likely to be very difficult.

In 2004, Siebert and I were considering how to solve the reconstruction problem using our program. Given $(B, \mathscr{P})$, we had shown in [31] how to construct log schemes $X_0(B, \mathscr{P}, s)^\dagger$ along with a morphism to $0^\dagger$ which had all the properties one would want for a central fibre of a toric degeneration $\mathcal{X}_0^\dagger \to 0^\dagger$. Our original hope was that a generalization of the Bogomolov-Tian-Todorov unobstructedness theorem would allow us to show such log schemes smoothed. In particular, Kawamata and Namikawa [51] had had success with this point of view in the normal crossings case. While this approach works easily in dimension 2, we couldn't make it work in higher dimension. Furthermore, this approach fails to give the explicit description of the smoothing which would be needed to describe the B-model. As a consequence, we turned towards a more explicit approach, which involved gluing together explicit local models.

While we were working on this approach, Kontsevich and Soibelman in [**54**] got around the difficult analysis of Fukaya's approach by replacing complex manifolds with rigid analytic manifolds. They were able to show that given a tropical affine surface B with 24 singularities of focus-focus type (the simplest type of singularity which occurs in affine surfaces, to be described shortly) one could construct a rigid analytic K3 surface $X^{an}(B)$. This was done by gluing together standard pieces via automorphisms attached to lines on B. These lines were given as gradient flow lines, giving a similar, but much more precise, picture to the one given by Fukaya.

Combining our approach of gluing local models with one of the central ideas of Kontsevich and Soibelman's work [**54**], we were then able to complete a construction in all dimensions, giving a satisfactory solution to the reconstruction problem within algebraic geometry. This was carried out in [**33**].

Before surveying this approach, let me make a philosophical remark. Note that when we were discussing the A-model, we observed that tropical curves on B should correspond to holomorphic curves on $X(B)$. If we want to see these same tropical curves playing a role on the B-model of the mirror, then we should think of the B-model not on a complex manifold of the form $X(B)$, but rather on $\check{X}(B)$. This is a slightly confusing reversal of roles. Normally counting curves is done in the symplectic category, here expected to mean on the symplectic manifold $\check{X}(B)$, while anything having to do with complex structures should be done on $X(B)$. This reversal can be explained as follows. If we were to study pseudo-holomorphic curves on $\check{X}(B)$, we would need to put an almost complex structure on $\check{X}(B)$. One way to do this is to choose a metric on B; this induces an almost complex structure on $\check{X}(B)$ constant on fibres of the torus fibration, generalizing the construction of a complex structure from a Hessian metric described in §2. Then in a suitable adiabatic limit where the almost complex structure is rescaled, pseudo-holomorphic curves are expected to tend towards trees of gradient flow lines. If the chosen metric was in fact Hessian, these gradient flow lines would in fact be straight lines with respect to the Legendre dual affine structure, so these trees of gradient flow lines can be viewed as a generalization of tropical curves. However, tropical geometry is linear and much easier to control. We take the attitude that we should work on the side in which tropical geometry appears. Indeed, this turns out to be very helpful.

Given this, we can then present a somewhat revised version of the mirror symmetry program:

(1) We begin with a toric degeneration of Calabi-Yau manifolds $\mathcal{X} \to D$ with an ample polarization.
(2) Construct the dual intersection complex $(B, \mathscr{P}, \varphi)$ from this data.
(3) Construct a new toric degeneration $\check{\mathcal{X}} \to D$ whose *intersection complex* is $(B, \mathscr{P}, \varphi)$. This degeneration should be controlled by tropical data.
(4) Understand genus 0 holomorphic curves (or whatever other aspect of the A-model one is interested in) on the general fibre of $\mathcal{X} \to D$ in terms of tropical geometry of B.
(5) Understand the variation of Hodge structures for $\check{\mathcal{X}} \to D$ in terms of tropical geometry of B.
(6) Use the fact that the A- and B-models of $\mathcal{X}$ and $\check{\mathcal{X}}$ respectively are controlled by the same tropical geometry on B to prove mirror symmetry.

Here we outline the completion of step (3) as carried out in [33].

The first step is as follows. Given $(B, \mathscr{P}, \varphi)$, we wish to construct the central fibre $\mathcal{X}_0^\dagger$ of the degeneration. This in fact was carried out in §5 of [31], assuming certain genericity assumptions on the singular locus of B. As a scheme, it is fairly obvious what $\mathcal{X}_0$ should be. For each maximal cell σ, one has an associated projective toric variety $\mathbb{P}_\sigma$ with Newton polytope σ. Any face $\tau \subseteq \sigma$ specifies a toric strata $\mathbb{P}_\tau \subseteq \mathbb{P}_\sigma$, and given $\tau = \sigma_1 \cap \sigma_2$ for σ_1, σ_2 maximal, we can glue together the toric strata $\mathbb{P}_\tau \subseteq \mathbb{P}_{\sigma_1}, \mathbb{P}_{\sigma_2}$ in a torus equivariant manner. There is of course a whole family of possible gluings, parameterized by what we call *closed gluing data* in [31]. Given closed gluing data s, we obtain a scheme $\check{X}_0(B, \mathscr{P}, s)$.

Now $\check{X}_0(B, \mathscr{P}, s)$ cannot be a central fibre of a toric degeneration unless it carries a log structure of the correct sort. There are many reasons this may not happen. If s is poorly chosen, there may be zero-dimensional strata of $\check{X}_0(B, \mathscr{P}, s)$ which do not have neighbourhoods locally étale isomorphic to the toric boundary of an affine toric variety; this is a minimal prerequisite. As a result, we have to restrict attention to closed gluing data induced by what we call *open gluing data*. Explicitly, each vertex v of $\mathscr{P}$ defines local models $V(v) \subseteq U(v)$ as follows. The piecewise linear function φ is defined locally up to affine linear functions. Choose a representative φ_v for φ in a neighbourhood of v which takes the value 0 at v. By extending the function linearly on each cell, we can view this as a piecewise linear function on the fan Σ_v, viewed as a fan in some $\mathbb{R}^n$. We can then set

$$P_v := \{(m, r) \in \mathbb{Z}^n \times \mathbb{Z} \,|\, r \geq \varphi_v(m)\}.$$

Noting that $(0, 1) \in P_v$, we set

$$U(v) := \operatorname{Spec} \mathbb{C}[P_v],$$
$$V(v) := \operatorname{Spec} \mathbb{C}[P_v]/(z^{(0,1)}).$$

Note that $z^{(0,1)}$ vanishes to order one on every toric divisor of $U(v)$, so in fact $V(v)$ is the toric boundary of $U(v)$. It turns out, as we show in [31], that a necessary condition for $\check{X}_0(B, \mathscr{P}, s)$ to be the central fibre of a toric degeneration is that it is obtained by dividing out $\coprod_{v \in \mathscr{P}} V(v)$ by an equivalence relation. In other words, we are gluing together the $V(v)$'s along Zariski open subsets to obtain a scheme.[2] Again, there is some choice of gluing, but now the gluing data are given by equivariant identifications of open subsets of the various $V(v)$'s. We call this *open gluing data*.

The advantage of using open gluing data is that each $V(v)$ carries a log structure induced by the divisorial log structure $V(v) \subseteq U(v)$. These log structures are not identified under the open gluing maps, but the ghost sheaves of the log structures are isomorphic. So the ghost sheaves $\overline{\mathcal{M}}_{V(v)}$ glue to give a ghost sheaf of monoids $\overline{\mathcal{M}}_{\check{X}_0(B, \mathscr{P}, s)}$. Thus we see how φ influences the log structure.

One then tries to construct a log structure with this ghost sheaf. This is done in [31] by building suitable extensions of the ghost sheaf with $\mathcal{O}_{\check{X}_0(B, \mathscr{P}, s)}^\times$, and this extension depends on some moduli (which may in general be empty). The good situation is that one can find a closed subset $Z \subseteq \check{X}(B, \mathscr{P}, s)$ of complex codimension at least two and a log structure on $\check{X}_0(B, \mathscr{P}, s)$ along with a morphism $\check{X}_0(B, \mathscr{P}, s)^\dagger \to 0^\dagger$ which is log smooth away from Z. Furthermore, the ghost sheaf

---

[2] [31] allowed the case that the cells of $\mathscr{P}$ self-intersect. As a consequence, the equivalence relation is merely étale and one obtains an algebraic space.

on $\check{X}_0(B, \mathscr{P}, s) \setminus Z$ should be the given ghost sheaf of monoids $\overline{\mathcal{M}}_{\check{X}_0(B,\mathscr{P},s)}$ restricted to $\check{X}_0(B, \mathscr{P}, s) \setminus Z$. We call such a log scheme with morphism to $0^\dagger$ a *log Calabi-Yau space*.

The technical heart of [**31**] is an explicit classification of log Calabi-Yau spaces with given intersection complex $(B, \mathscr{P}, \varphi)$, modulo some assumptions on the singularities of B called *simplicity*. The definition of simplicity is rather involved, so we will not give it here, but it essentially says that not too much topology of $\check{X}(B)$ (or $X(B)$) can be hiding over the singular locus of B.

A main result of [**31**], (Theorem 5.4) is then

THEOREM 10.1. *Given $(B, \mathscr{P}, \varphi)$ simple, the set of log Calabi-Yau spaces with intersection complex $(B, \mathscr{P}, \varphi)$ modulo isomorphism preserving B (i.e., does not interchange irreducible components) is $H^1(B, i_* \Lambda \otimes \mathbb{C}^\times)$. An isomorphism is said to preserve B if it induces the identity on the intersection complex.*

So the moduli space is an algebraic torus (or a disjoint union of algebraic tori) of dimension equal to $\dim_\mathbb{C} H^1(B, i_* \check{\Lambda} \otimes \mathbb{C})$. In [**32**], we in fact show the dimension of this torus is the dimension of $H^1(\mathcal{X}_t, \mathcal{T}_{\mathcal{X}_t}) \cong H^{n-1,1}(\mathcal{X}_t)$ for a smooth fibre $\mathcal{X}_t$ of a smoothing $\mathcal{X} \to D$ of $X_0(B, \mathscr{P}, s)^\dagger$. This is the expected dimension, as this latter vector space is the tangent space to the moduli space of $\mathcal{X}_t$.

Now assume given a log Calabi-Yau space $X_0 := X_0(B, \mathscr{P}, s)^\dagger \to 0^\dagger$. Our goal is to use the log structure to provide "initial conditions" to produce k-th order deformations $X_k \to \operatorname{Spec} \mathbb{C}[t]/(t^{k+1})$, order by order. To do so, we will glue together standard thickenings of "pieces" of X_0, modifying standard gluings by a complicated system of data we call a *structure*.

First, the "pieces" of X_0 we consider are toric open affine subsets of strata of X_0. Recall that strata of X_0 are indexed by cells $\tau \in \mathscr{P}$, corresponding to a projective toric variety $\mathbb{P}_\tau$. Recall also that if $\omega \subseteq \tau$, the normal cone to τ along ω is a cone in the fan defining $\mathbb{P}_\tau$ and hence defines an open affine subset of $\mathbb{P}_\tau$. We call this open affine subset $V_{\omega,\tau} \subseteq \mathbb{P}_\tau$; note

$$V_{\omega,\tau} = \mathbb{P}_\tau \setminus \bigcup_{\substack{\rho \subseteq \tau \\ \omega \not\subseteq \rho}} \mathbb{P}_\rho.$$

For example, if ω is a vertex of τ, then $V_{\omega,\tau}$ is the standard toric open affine subset of $\mathbb{P}_\tau$ containing the zero-dimensional stratum of $\mathbb{P}_\tau$ corresponding to ω.

Second, what are the thickenings of the sets $V_{\omega,\tau}$? These can be described explicitly as follows. Choose a point x in the interior of ω not contained in the singular locus Γ of B. We obtain a fan Σ_x in the tangent space $\Lambda_x \otimes_\mathbb{Z} \mathbb{R}$ of not necessarily strictly convex cones consisting of the tangent cones at x of each cell σ containing ω. We can choose a representative φ_x for φ in a small neighbourhood of x which is zero along ω, and this can then be extended linearly on each cone of Σ_x to view φ_x as a piecewise linear function $\varphi_x : \Lambda_x \otimes \mathbb{R} \to \mathbb{R}$. This in turn defines a monoid

$$P_x := \{(m, r) \in \Lambda_x \times \mathbb{Z} \mid r \geq \varphi_x(m)\},$$

completely analogous to the definition of P_v.

For each maximal cell σ containing τ, let $n_\sigma \in \check{\Lambda}_x$ denote the slope of φ_x restricted to the tangent cone of σ. We then define a monomial ideal in the ring $\mathbb{C}[P_x]$ given by

$$I_{\omega,\tau}^{>k} = \langle z^{(m,r)} \mid (m,r) \in P_x, \, r - \langle n_\sigma, m \rangle > k \text{ for some } \sigma \in \mathscr{P}_{\max} \text{ with } \sigma \supseteq \tau \rangle.$$

Then the desired standard thickening of $V_{\omega,\tau}$ is

$$V_{\omega,\tau}^k := \operatorname{Spec} \mathbb{C}[P_x]/I_{\omega,\tau}^{>k}.$$

One checks easily that if $k = 0$, this recovers $V_{\omega,\tau}$, and if $k > 0$, then the reduced space of $V_{\omega,\tau}^k$ is $V_{\omega,\tau}$. Thus this is indeed a thickening of $V_{\omega,\tau}$.

There is one point we have to be quite careful about. This definition would appear to depend on the point x, and identifications of different tangent spaces Λ_x, $\Lambda_{x'}$ via parallel transport depend on the path because of the presence of the singular locus. We deal with this issue not by choosing a specific point x, but choosing a specific maximal reference cell σ containing τ. We then can identify any Λ_x with Λ_σ, the well-defined tangent space to σ, via parallel transport from x directly into σ. We will notate this additional choice of reference cell by writing $V_{\omega,\tau,\sigma}^k$. A different choice of reference cell σ' gives a space $V_{\omega,\tau,\sigma'}^k$ abstractly, but not canonically, isomorphic to $V_{\omega,\tau,\sigma}^k$. This will prove important below. We also use the notation for the coordinate rings

$$R_{\omega,\tau,\sigma}^k := \mathbb{C}[P_x]/I_{\omega,\tau}^{>k},$$

again keeping in mind this choice of reference cell.

There are also natural gluings between these various thickened schemes. One notes that given $\tau_1 \subseteq \tau_2 \subseteq \tau_3$ there are natural surjections

$$R_{\tau_1,\tau_3,\sigma}^k \to R_{\tau_1,\tau_2,\sigma}^k$$

giving a closed embedding $V_{\tau_1,\tau_2,\sigma}^k \to V_{\tau_1,\tau_3,\sigma}^k$, and natural inclusions

$$R_{\tau_1,\tau_3,\sigma}^k \to R_{\tau_2,\tau_3,\sigma}^k,$$

giving open embeddings $V_{\tau_2,\tau_3,\sigma}^k \to V_{\tau_1,\tau_3,\sigma}^k$.

If B has no singularities, then the reference cell σ is not important, and we drop this from the notation in this case. In particular, it is easy to check that if we take, say, τ_1 to be a fixed vertex v, and we take the limit of the directed system $\{V_{v,\tau}^k \mid v \in \tau\}$ of schemes, we obtain a k-th order thickening $V^k(v)$ of $V(v)$ given by $V^k(v) = U(v) \times_{\mathbb{A}^1} \operatorname{Spec} \mathbb{C}[t]/(t^{k+1})$, with $U(v) \to \mathbb{A}^1$ the morphism given by $z^{(0,1)}$. This is precisely the kind of vanilla smoothing the log structure leads us to expect. Note we can write this direct limit of schemes as

$$\operatorname{Spec} \varprojlim_\tau R_{v,\tau}^k.$$

The basic idea then will be to modify the various maps above by some additional data.

To understand why we need these modifications, let us consider the single most important example, that of an isolated singularity of focus-focus type in a two-dimensional B.

We suppose $\mathscr{P}$ contains two maximal cells σ_1, σ_2, with $\sigma_1 \cap \sigma_2 = \tau$, as depicted in Figure 5. Note that the intersection of the two coordinate charts is $(\sigma_1 \cup \sigma_2) \setminus \tau$, and the transition map is then the identity on $\sigma_1 \setminus \tau$ and is given by the linear transformation $\begin{pmatrix} 1 & 0 \\ 1 & 1 \end{pmatrix}$ on $\sigma_2 \setminus \tau$. Together, these two charts define an integral affine structure on $(\sigma_1 \cup \sigma_2) \setminus \Gamma$, where $\Gamma = \{p\}$ is the common point of the two cuts.

We then take φ to be single-valued, identically 0 on σ_1 and taking the value 1 at the right-hand vertex.

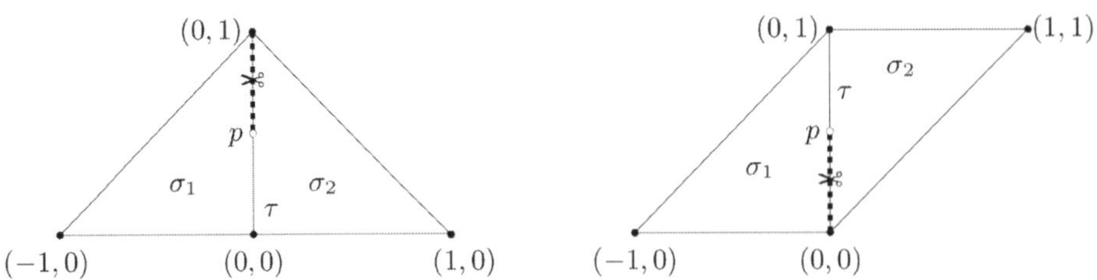

FIGURE 5. The fundamental example. The diagram shows the affine embeddings of two charts, obtained by cutting the union of two triangles as indicated in two different ways. Each triangle is a standard simplex.

One now finds

$$R^k_{\tau,\sigma_1,\sigma_1} = \mathbb{C}[x,y,w^{\pm 1}]/(y^{k+1})$$
$$R^k_{\tau,\sigma_2,\sigma_2} = \mathbb{C}[x,y,w^{\pm 1}]/(x^{k+1})$$
$$R^k_{\tau,\tau,\sigma_i} = \mathbb{C}[x,y,w^{\pm 1}]/(x^{k+1},y^{k+1}).$$

Here, if we use the chart on the left, i.e., choose a point s below p and work in $P_s \subseteq \Lambda_s \oplus \mathbb{Z}$, the variables x, y and w are identified with elements of $\mathbb{C}[P_s]$ as

$$x = z^{(-1,0,0)}, \quad y = z^{(1,0,1)}, \quad w = z^{(0,1,0)}.$$

We have the natural surjections $R^k_{\tau,\sigma_i,\sigma_i} \to R^k_{\tau,\tau,\sigma_i}$, and we identify R^k_{τ,τ,σ_1} with R^k_{τ,τ,σ_2} by identifying Λ_{σ_1} and Λ_{σ_2} by parallel transport through s. Since we have written everything in the left-hand chart, where Λ_{σ_1} and Λ_{σ_2} are identified via parallel transport through s, this identification is the trivial one. We can thus glue together the coordinate rings of the thickenings as

$$R^k_{\tau,\sigma_1,\sigma_1} \times_{R^k_{\tau,\tau,\sigma_i}} R^k_{\tau,\sigma_2,\sigma_2}.$$

This fibred product of rings is easily seen to be isomorphic to the ring

$$\mathbb{C}[X,Y,W^{\pm 1},t]/(t - XY, t^{k+1}),$$

where $X = (x,x)$, $Y = (y,y)$, $W = (w,w)$, and $t = (xy,xy)$ as elements of the Cartesian product of rings.

On the other hand, suppose we instead identified R^k_{τ,τ,σ_1} and R^k_{τ,τ,σ_2} by parallel transport through a point s' lying above p. To do this, we can work in the right-hand chart. Again, x, y and w are defined using the tangent vectors $(-1,0), (1,0)$ and $(0,1)$ in σ_1, and these are transported to the same tangent vectors in σ_2 in the second chart. However, to compare this with our original description of R^k_{τ,τ,σ_2}, we need to think of these as tangent vectors in σ_2 in the original chart, i.e., the left-hand chart. There, these tangent vectors are $(-1,1)$, $(1,-1)$ and $(0,1)$ respectively. Thus we obtain an isomorphism $R^k_{\tau,\tau,\sigma_1} \to R^k_{\tau,\tau,\sigma_2}$ given by

(10.1) $$x \mapsto xw, \quad y \mapsto yw^{-1}, \quad w \mapsto w.$$

Using this identification, we obtain a composed map $R^k_{\tau,\sigma_1,\sigma_1} \to R^k_{\tau,\tau,\sigma_1} \to R^k_{\tau,\tau,\sigma_2}$, leading to a fibred product

$$R^k_{\tau,\sigma_1,\sigma_1} \times_{R^k_{\tau,\tau,\sigma_2}} R^k_{\tau,\sigma_2,\sigma_2} \cong \mathbb{C}[X,Y,W^{\pm 1},t]/(XY - tW, t^{k+1}),$$

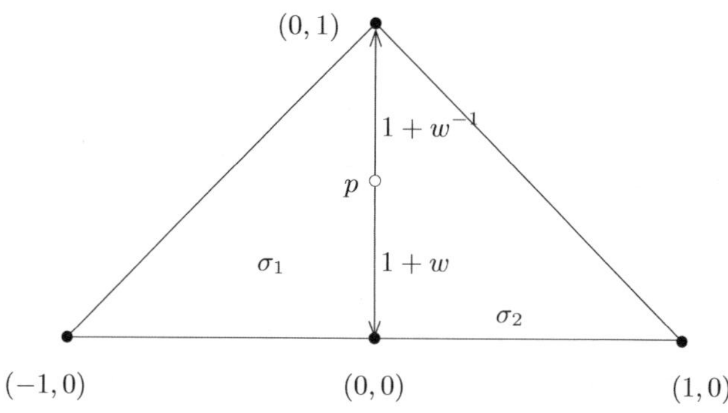

FIGURE 6

where now

$$X = (x, xw), \quad Y = (yw, y), \quad W = (w, w), \quad t = (xy, xy).$$

Note that while this new ring is abstractly isomorphic to the previous ring, there is no isomorphism as $\mathbb{C}[t]/(t^{k+1})$-algebras.

So the gluing is not well-defined, and this is caused by the singularities of B. The correct smoothing in this case will depend on the choice of log structure, but in any event we expect it should be a family of the form $\operatorname{Spec} \mathbb{C}[X, Y, W^{\pm 1}, t]/(XY - f(W)t)$ for some function $f(W)$ which vanishes along the W-axis precisely at the points where the given log structure on $X_0(B, \mathscr{P}, s)$ is not fine. Clearly f is then determined by the log structure up to invertible functions. Let us take for the sake of this example the function $f(W) = 1 + W$, noting that $f(W) = 1 + W^{-1}$ would do just as well. We can now modify the gluings using Figure 6.

In this figure, we have drawn two rays contained in τ emanating from the singular point, and labelled these two arrows with the functions $1 + w^{-1}$ and $1 + w$ respectively. These rays tell us that if we try to identify $R^k_{\tau, \tau, \sigma_1}$ with $R^k_{\tau, \tau, \sigma_2}$ using parallel transport between the two maximal cells, we need to modify the identification via an automorphism given by the crossing of one of these rays. Here, we will get different automorphisms depending on whether we cross above or below the singularity p. If we cross below, the ray tells us to use an automorphism of $R^k_{\tau, \tau, \sigma_1}$ given by

$$(10.2) \qquad x \mapsto x(1 + w), \quad y \mapsto y(1 + w)^{-1}, \quad w \mapsto w,$$

while if we cross above the singularity, we use the automorphism

$$(10.3) \qquad x \mapsto x(1 + w^{-1}), \quad y \mapsto y(1 + w^{-1})^{-1}, \quad w \mapsto w.$$

Actually, note that $1 + w$ or $1 + w^{-1}$ is not invertible in $R^k_{\tau, \tau, \sigma_i}$, so we need to modify this ring by localizing it at $1 + w$ (or equivalently $1 + w^{-1}$). Let's see how this affects the fibred products $R^k_{\tau, \sigma_1, \sigma_1} \times_{R^k_{\tau, \tau, \sigma_2}} R^k_{\tau, \sigma_2, \sigma_2}$.

If we use parallel transport below the singular point, then the map $R^k_{\tau, \sigma_1, \sigma_1} \to R^k_{\tau, \tau, \sigma_2}$ is just given by (10.2), while $R^k_{\tau, \sigma_2, \sigma_2} \to R^k_{\tau, \tau, \sigma_2}$ remains the canonical one. One then finds

$$R^k_{\tau, \sigma_1, \sigma_1} \times_{R^k_{\tau, \tau, \sigma_2}} R^k_{\tau, \sigma_2, \sigma_2} \cong \mathbb{C}[X, Y, W^{\pm}, t]/(XY - (1 + W)t, t^{k+1}),$$

with

$$X = (x, x(1 + w)), \quad Y = (y(1 + w), y), \quad W = (w, w), \quad t = (xy, xy).$$

On the other hand, if we use parallel transport above the singular point, we need to compose the automorphism (10.3) with the isomorphism (10.1), giving a map $R^k_{\tau,\sigma_1,\sigma_1} \to R^k_{\tau,\tau,\sigma_2}$ given by

$$x \mapsto xw(1 + w^{-1}) = x(1 + w), \quad y \mapsto yw^{-1}(1 + w^{-1})^{-1} = y(1 + w)^{-1}, \quad w \mapsto w.$$

Thus this map is exactly the same as (10.2), and hence we get the same fibred product. The glued thickenings are independent of choices. The introduction of the extra automorphisms removes the problems caused by monodromy.

This is a very local situation. The next problem which arises is that more globally, we need to propagate the automorphisms attached to the rays. Indeed, imagine now that the picture we are looking at is contained in a more complex situation, as on the left-hand side of Figure 7. Here we have two singularities, and rays emanate in each direction from the singularity. Let us follow the rule that any identification of rings which involves parallel transport through a ray must be modified by the appropriate automorphism as described above. Then looking at the vertex v_1, say, we need to glue together five irreducible components, but only one of these gluings is modified. These gluings would not be compatible. To correct for this, one can extend the ray indefinitely, and "parallel transport" the automorphism along the ray. There is a precise sense in which this can be done. This is shown on the right-hand picture in Figure 7, with the dotted lines showing the extension of the rays. Now if crossing a ray in one direction produces the inverse of the automorphism given by crossing the ray the other direction, one finds that gluing at the vertices v_1 and v_2 have now become compatible.

A new problem arises, however, at the intersection point of the two rays. Again, when we try to identify various rings using parallel transport and automorphisms induced by crossing rays, we don't want the choice to depend on the particular path we take. Because in general the two automorphisms attached to the rays don't commute, we again have trouble at the point of intersection.

This is in fact where our thinking stood in early 2004, shortly before the release of Kontsevich and Soibelman's paper [54]. The solution to this problem, really the key part of Kontsevich and Soibelman's argument, is to add new rays emanating from the point of intersection of the old rays, as depicted in Figure 8. These rays are added in such a way as to guarantee that the composition of automorphisms given by a loop around the intersection point is in fact the identity, and thus the identifications will be independent of the choice of path.

The description here is somewhat vague, but demonstrates the basic idea. We've seen how we obtain our degeneration by gluing together basic pieces. Other than these different basic pieces, in two dimensions, the main distinction between our approach and the one taken by Kontsevich and Soibelman in [54] is that we work in the affine structure dual to the one [54] works with. They propogate automorphisms along gradient flow lines, but we are able to propogate automorphisms along straight lines with respect to the affine structure. This saves a great deal of trouble in higher dimensions, where gradient flow lines will be much more difficult to control. That makes it possible for us to obtain results in all dimensions.

We of course have not made it particularly clear how we really encode automorphisms and how they propagate, but we will make at least the first point clearer in

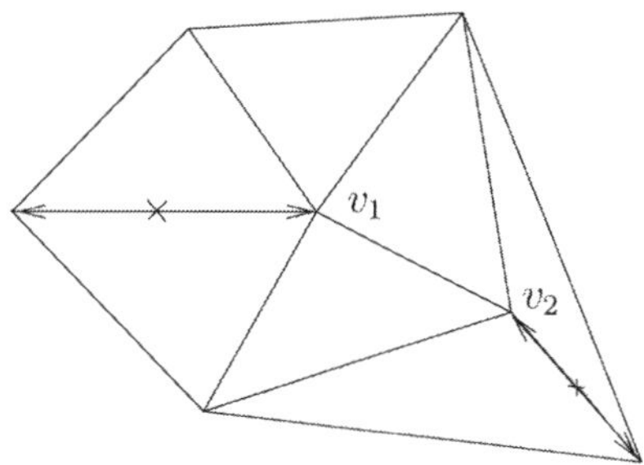
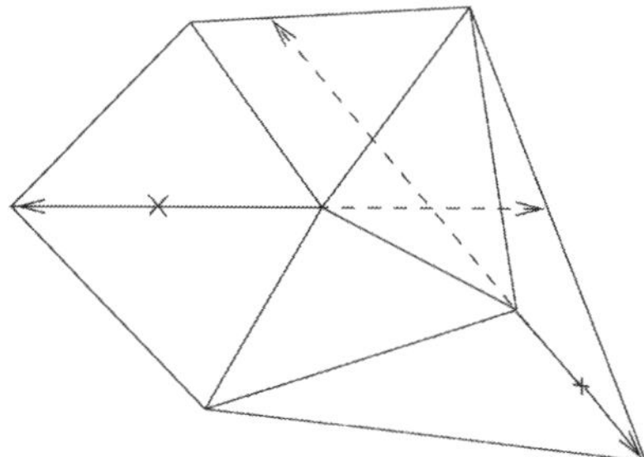

FIGURE 7

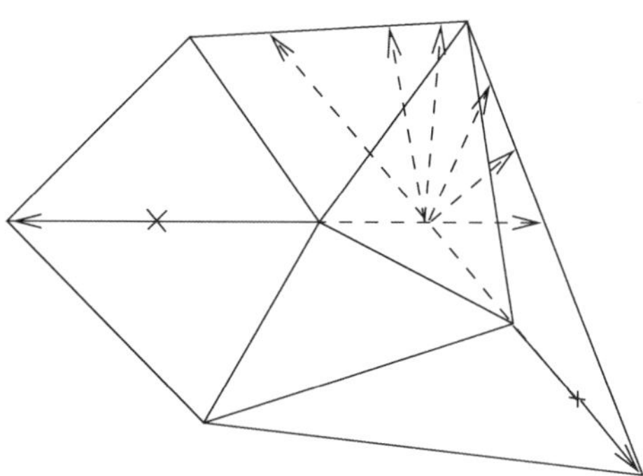

FIGURE 8

the next section. For the second point, the main thing is that they propagate along straight lines; this in fact is crucial for guaranteeing that the automorphisms don't start to involve monomials with poles on irreducible components of X_0. So here we see something which looks tropical already, with the union of rays looking like a tropical tree. Again, we will make this more precise in the next section.

In higher dimensions, the argument becomes much more subtle. Instead of rays carrying automorphisms, codimension one walls carry automorphisms, and one needs to be very careful about how these walls propagate. Furthermore, there are great technical difficulties concerning convergence of the algorithm near the discriminant locus. This was handled in [54] in two dimensions via an argument showing new rays added can be guaranteed to avoid a neighbourhood of each singularity, but this is done by choosing the metric carefully. In higher dimensions, this is not true, and instead we used algebraic methods to prove convergence. All these difficulties were overcome in [33].

In [25] I wrote down a complete version of the proof in two dimensions; this has the advantage of avoiding most of the really technical issues. Hopefully, [25] provides a gentler entry point into the ideas outlined here than the main paper [33].

We now turn to a more precise description of the automorphisms involved, and give evidence that the description of the explicit deformations (which we view as B-model information) really encodes A-model information on the mirror.

11. The tropical vertex

To simplify the discussion, we will work in this section only with the simplest rings which occur in the previous section, of the form $R^k_{\sigma,\sigma,\sigma}$ where σ is a maximal (two-dimensional) cell. This ring is isomorphic to $\mathbb{C}[x^{\pm 1}, y^{\pm 1}, t]/(t^{k+1})$. Let us work formally instead, setting

$$R = \mathbb{C}[x^{\pm 1}, y^{\pm 1}][\![t]\!].$$

This is the ring of formal power series in t with coefficients Laurent polynomials in x and y. Let $f \in R$ be of the form

$$f = 1 + tx^a y^b \cdot g(x^a y^b, t), \quad g(z,t) \in \mathbb{C}[z][\![t]\!].$$

Then this defines an automorphism $\theta_{(a,b),f}$ of R as a $\mathbb{C}[\![t]\!]$-algebra given by

$$\theta_{(a,b),f}(x) = x \cdot f^b, \quad \theta_{(a,b),f}(y) = y \cdot f^{-a}.$$

Note that $\theta^{-1}_{(a,b),f} = \theta_{(a,b),f^{-1}}$. These automorphisms have the further property that they preserve the holomorphic symplectic form $\frac{dx}{x} \wedge \frac{dy}{y}$.

We define the *tropical vertex group* $\mathbb{V}$ to be the completion with respect to the maximal ideal $(t) \subseteq \mathbb{C}[\![t]\!]$ of the subgroup of $\mathbb{C}[\![t]\!]$-algebra automorphisms of R generated by all such automorphisms. Note that infinite products are defined in $\mathbb{V}$ only if only finitely many factors are non-trivial modulo t^k for every $k > 0$. This is a slight modification of a group originally introduced by Kontsevich and Soibelman in [**54**].

We now describe a local version of the rays described in the previous section. For convenience, set $M = \mathbb{Z}^2$, $M_{\mathbb{R}} = M \otimes_{\mathbb{Z}} \mathbb{R}$, and identify $\mathbb{C}[x^{\pm 1}, y^{\pm 1}]$ with $\mathbb{C}[M]$.

DEFINITION 11.1. A *ray* or *line* in $M_{\mathbb{R}}$ is a pair $(\mathfrak{d}, f_{\mathfrak{d}})$ for some $\mathfrak{d} = \mathbb{R}_{\leq 0}m$ if $\mathfrak{d}$ is a ray and $\mathfrak{d} = \mathbb{R}m$ if $\mathfrak{d}$ is a line, where $m \in M \setminus \{0\}$. Furthermore,

$$f_{\mathfrak{d}} = 1 + tz^m \cdot g(z^m, t) \in R, \quad g(z,t) \in \mathbb{C}[z][\![t]\!].$$

A *scattering diagram* $\mathfrak{D}$ is a collection of rays and lines $\{(\mathfrak{d}, f_{\mathfrak{d}})\}$ with the property that for any $k > 0$, $f_{\mathfrak{d}} \equiv 1 \mod t^k$ for all but a finite number of elements of $\mathfrak{D}$.

Given a scattering diagram $\mathfrak{D}$, let

$$\operatorname{Supp}\mathfrak{D} = \bigcup_{(\mathfrak{d}, f_{\mathfrak{d}}) \in \mathfrak{D}} \mathfrak{d}.$$

If we are given a path $\gamma : [0, 1] \to M_{\mathbb{R}} \setminus \{0\}$ with $\gamma(0), \gamma(1) \notin \operatorname{Supp}(\mathfrak{D})$ and γ being transversal to each ray it crosses, then we can define the *path-ordered product* $\theta_{\gamma,\mathfrak{D}} \in \mathbb{V}$ which is a composition of automorphisms associated to each ray that γ crosses. We define the path-ordered product for each power $k > 0$, and take the limit. For any given $k > 0$, we can find numbers

$$0 < t_1 \leq t_2 \leq \cdots \leq t_s < 1$$

and elements $\mathfrak{d}_i \in \mathfrak{D}$ with $f_{\mathfrak{d}_i} \not\equiv 1 \mod t^k$ such that $\gamma(t_i) \in \mathfrak{d}_i$, $\mathfrak{d}_i \neq \mathfrak{d}_j$ if $t_i = t_j$, and s taken as large as possible. For each i define $\theta_{\mathfrak{d}_i}$ to be the automorphism defined as follows. Let $n \in N = \operatorname{Hom}(M, \mathbb{Z})$ be the unique primitive element which vanishes on $\mathfrak{d}_i$ and is negative on $\gamma'(t_i)$. Then define $\theta_{\mathfrak{d}_i}$ to be the automorphism

$$\theta_{\mathfrak{d}_i}(z^m) = z^m f_{\mathfrak{d}_i}^{\langle n,m \rangle}.$$

Note this is of the form $\theta_{(a,b),f_{\mathfrak{d}}}$ for suitable choice of (a,b). We then define

$$\theta^k_{\gamma,\mathfrak{D}} = \theta_{\mathfrak{d}_s} \circ \cdots \circ \theta_{\mathfrak{d}_1}.$$

Note that if γ crosses two rays at the same time, the order doesn't matter as one checks easily that two automorphisms commute if they are associated with the same underlying $\mathfrak{d} \subseteq M_{\mathbb{R}}$. Finally, we define

$$\theta_{\gamma,\mathfrak{D}} = \lim_{k \to \infty} \theta^k_{\gamma,\mathfrak{D}}.$$

We can then express the essential lemma of [**54**] in this context:

PROPOSITION 11.2. *Let $\mathfrak{D}$ be a scattering diagram. Then there is a scattering diagram $\mathsf{S}(\mathfrak{D})$ such that $\mathsf{S}(\mathfrak{D}) \setminus \mathfrak{D}$ consists just of rays and $\theta_{\gamma,\mathsf{S}(\mathfrak{D})}$ is the identity for any loop γ around the origin.*

The proof is very simple and algorithmic; I give a quick outline. One constructs a sequence of scattering diagrams $\mathfrak{D}_1 = \mathfrak{D}, \mathfrak{D}_2, \ldots$ with the property that $\theta_{\gamma,\mathfrak{D}_k} \equiv \mathrm{id}$ mod t^k. This is clearly true for $\mathfrak{D}_1$, so we proceed inductively, assuming we have constructed $\mathfrak{D}_k$. Then one shows (by looking at the Lie algebra of $\mathbb{V}$) that

$$\theta_{\gamma,\mathfrak{D}_k}(x) = x \sum_{i=1}^{n} b_i c_i t^k x^{a_i} y^{b_i}$$

$$\theta_{\gamma,\mathfrak{D}_k}(y) = -y \sum_{i=1}^{n} a_i c_i t^k x^{a_i} y^{b_i}$$

for integers a_i, b_i (with a_i, b_i not both zero) and $c_i \in \mathbb{C}$. Then one obtains $\mathfrak{D}_{k+1}$ by adding rays

$$(\mathbb{R}_{\leq 0}(a_i, b_i), 1 \pm c_i t^k x^{a_i} y^{b_i}), \quad 1 \leq i \leq n$$

with the sign chosen so that when γ crosses this ray, it produces the automorphism

$$x \mapsto x(1 - b_i c_i t^k x^{a_i} y^{b_i}) \quad \mathrm{mod} \ t^{k+1}, \quad y \mapsto y(1 + a_i c_i t^k x^{a_i} y^{b_i}) \quad \mathrm{mod} \ t^{k+1}.$$

Since this automorphism will commute with all other automorphisms in $\mathfrak{D}_k$ modulo t^{k+1}, inserting these rays will precisely cancel out the contributions to $\theta_{\gamma,\mathfrak{D}_k}$ to order k, and thus $\theta_{\gamma,\mathfrak{D}_{k+1}} \equiv \mathrm{id} \mod t^{k+1}$.

It is very easy to program this algorithm and explore these scattering diagrams. They appear to have a very rich and fascinating structure. The following simple examples show their complexity.

EXAMPLE 11.3. Consider the case that

$$\mathfrak{D} = \{(\mathbb{R}(1,0), (1 + tx^{-1})^{\ell}), (\mathbb{R}(0,1), (1 + ty^{-1})^{\ell})\}$$

for ℓ some positive integer. For $\ell = 1$, it is easy to check that

$$\mathsf{S}(\mathfrak{D}) \setminus \mathfrak{D} = \{(\mathbb{R}_{\geq 0}(1,1), 1 + t^2 x^{-1} y^{-1})\}.$$

Figure 9 shows explicitly what the automorphisms are as one traverses the depicted loop; the reader can easily check that the composition of the five automorphisms is the identity.

If $\ell = 2$, then one finds

$$\mathsf{S}(\mathfrak{D}) \setminus \mathfrak{D} = \quad \{(\mathbb{R}(n+1,n), (1 + t^{2n+1} x^{-(n+1)} y^{-n})^2)| n \in \mathbb{Z}, n \geq 1\}$$
$$\cup \quad \{(\mathbb{R}(n,n+1), (1 + t^{2n+1} x^{-n} y^{-(n+1)})^2)| n \in \mathbb{Z}, n \geq 1\}$$
$$\cup \quad \{(\mathbb{R}(1,1), (1 - t^2 x^{-1} y^{-1})^{-4})\}.$$

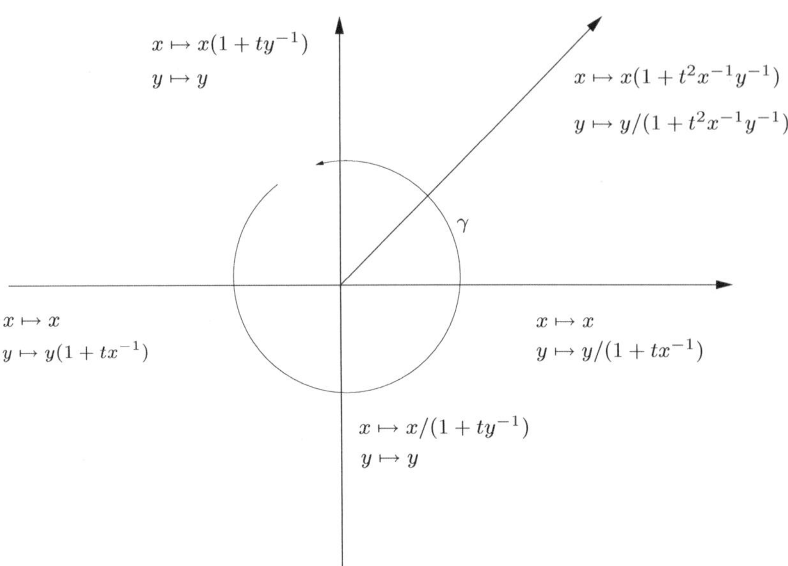

FIGURE 9. $S(\mathfrak{D})$ for $\ell = 1$. Here the automorphisms are given explicitly, and the identity $\theta_{\gamma, S(\mathfrak{D})}$ is just the composition of the given automorphisms.

This was first found experimentally by myself and Siebert via a computer program, and the first verification of this was given in [14]. It also follows immediately from the results of [28] which will be explained in what follows.

If $\ell = 3$, the situation becomes even more complicated. First, as noticed by Kontsevich, $S(\mathfrak{D})$ has a certain periodicity. Namely,

$$(\mathbb{R}_{\geq 0}(m_1, m_2), f(x^{-m_1} y^{-m_2})) \in S(\mathfrak{D})$$

if and only if

$$(\mathbb{R}_{\geq 0}(3m_1 - m_2, m_1), f(x^{-(3m_1 - m_2)} y^{-m_1})) \in S(\mathfrak{D}),$$

provided that m_1, m_2 and $3m_1 - m_2$ are all positive. In addition, there are rays with support $\mathbb{R}_{\geq 0}(3, 1)$ and $\mathbb{R}_{\geq 0}(1, 3)$, hence by the periodicity, there are also rays with support

$$\mathbb{R}_{\geq 0}(8, 3), \ \mathbb{R}_{\geq 0}(21, 8), \ \ldots \quad \text{and} \quad \mathbb{R}_{\geq 0}(3, 8), \ \mathbb{R}_{\geq 0}(8, 21), \ \ldots$$

which converge to the rays of slope $(3 \pm \sqrt{5})/2$, corresponding to the two distinct eigenspaces of the linear transformation $\begin{pmatrix} 3 & -1 \\ 1 & 0 \end{pmatrix}$. Each of these rays is of the form

$$\left(\mathbb{R}_{\geq 0}(m_1, m_2), (1 + t^{m_1 + m_2} x^{-m_1} y^{-m_2})^3 \right).$$

These are the only rays appearing outside of the cone generated by the rays of slope $(3 \pm \sqrt{5})/2$. On the other hand, inside this cone, every rational slope occurs, and the attached functions are very complicated. For example, the function attached to the line of slope 1 is

$$\left(\sum_{n=0}^{\infty} \frac{1}{3n + 1} \binom{4n}{n} t^{2n} x^{-n} y^{-n} \right)^9.$$

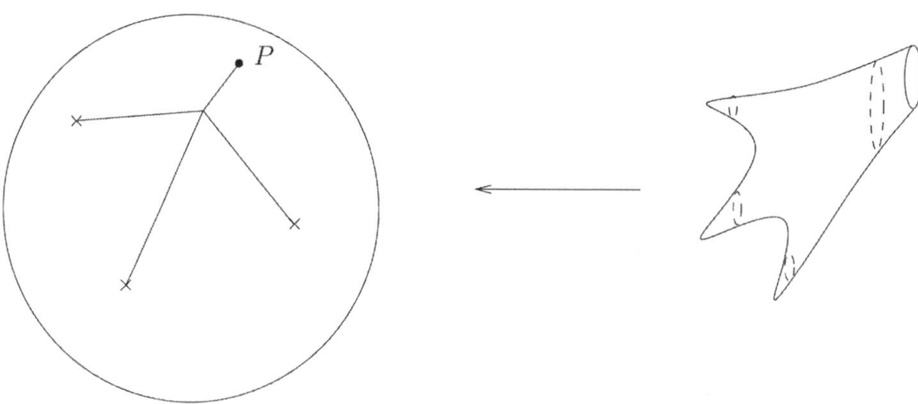

FIGURE 10. A tropical disk on an affine K3 surface. Here the ×'s indicate singular points, while the disk "ends" at the point P.

Again, Siebert and I found this form via computer experiment, but it was verified by Reineke in [69]. Recently, Kontsevich has shown the functions attached to all these rays are algebraic. For example, if g denotes the 9-th root of the above function, it satisfies the equation

$$t^2 x^{-1} y^{-1} g^4 - g + 1 = 0.$$

This series of examples also makes contact with a number of other interesting objects. On the one hand, Reineke in [69] gave an interpretation of the attached functions in terms of Euler characteristics of moduli spaces of representations of the Kronecker ℓ-quiver, the quiver with two vertices and ℓ arrows between them. On the other hand, these diagrams are also closely related to the cluster algebras defined by these quivers. This connection will be studied in more detail in work with Hacking, Keel, and Kontsevich [27].

We will now explain the enumerative interpretation for the functions which arise in $\mathsf{S}(\mathfrak{D})$. To motivate this, let us return to the tropical interpretation of §4. Begin, say, with a tropical manifold B which corresponds to a K3 surface, as depicted in Figure 10, along with what we will call a *tropical disk*. This is almost a tropical curve, but it just ends at the point P without any balancing condition at P; meanwhile, it has other legs terminating at the singularities of B. This is legal behaviour as explained at the end of §4. Following the description at the end of §4, one can imagine disks over each leg terminating at a singular point. Where these legs meet, one would like to glue these disks together and continue along a cylinder over the segment adjacent to P. Terminating at P, we roughly obtain a disk in $X(B)$ with boundary contained in the torus fibre over P, as depicted. It is natural to ask how many ways the initial disks (possibly taking multiple covers of these disks) can be glued together to give a new disk.

Now compare this picture with what we have seen on the mirror side. Our explicit degeneration really gives, as generic fibre, something like $\check{X}(B)$. However, it is controlled by similar tropical information: rays emanate from the singularities in the monodromy invariant direction, just as in the case of the tropical curves. They collide, and the Kontsevich-Soibelman result in Proposition 11.2 gives new rays. So one may hope that this process precisely reflects holomorphic disks in $X(B)$ with boundary on fibres of $X(B) \to B$.

It is also worth mentioning work of Auroux [4], which makes more precise the notion that the complex structure on one side should be determined by holomorphic disks on the other. This also provides a posteriori support for the idea that there must be an enumerative interpretation for the process of generating new rays.

It is usually difficult to work with holomorphic disks. It is often easier to translate problems involving holomorphic disks into problems involving genuine Gromov-Witten invariants. We can do so for the problems being discussed here. Here then is the enumerative interpretation, in the simplest situation, as explained in [28].

Suppose we are given distinct non-zero primitive vectors $m_1, \ldots, m_p \in M$ and positive integers $\ell_1, \ldots, \ell_p$. Consider the scattering diagram

$$\mathfrak{D} = \{(\mathbb{R}m_i, (1 + tz^{-m_i})^{\ell_i}) \mid 1 \leq i \leq p\}.$$

Let $(\mathfrak{d}, f_\mathfrak{d}) \in S(\mathfrak{D}) \setminus \mathfrak{D}$. We can always assume that this is the only ray in $S(\mathfrak{D}) \setminus \mathfrak{D}$ with a given underlying ray $\mathfrak{d}$. This is because if there are rays $(\mathfrak{d}_1, f_{\mathfrak{d}_1}), (\mathfrak{d}_2, f_{\mathfrak{d}_2}), \ldots$ in $S(\mathfrak{D}) \setminus \mathfrak{D}$ with $\mathfrak{d}_1 = \mathfrak{d}_2 = \cdots$, we can replace this collection of rays with a single ray $(\mathfrak{d}_1, \prod f_{\mathfrak{d}_i})$ without affecting $\theta_{\gamma, S(\mathfrak{D})}$. With this assumption, $f_\mathfrak{d}$ is uniquely determined by $\mathfrak{D}$. We wish to interpret $f_\mathfrak{d}$ enumeratively.

To do this, consider a complete fan Σ in $M_\mathbb{R}$ whose one-dimensional rays are

$$\mathbb{R}_{\leq 0}m_1, \ldots, \mathbb{R}_{\leq 0}m_p, \mathfrak{d}.$$

Assume for the sake of simplicity in this discussion that $\mathfrak{d}$ does not coincide with the other p rays. Let X be the toric variety defined by Σ, with toric divisors $D_1, \ldots, D_p, D_{\text{out}}$ corresponding to the above rays. Next, choose ℓ_i general points on the divisor D_i, say labelled $P_{i1}, \ldots, P_{i\ell_i}$. Let $\nu : \widetilde{X} \to X$ be the blow-up of these $\sum_i \ell_i$ points, with exceptional divisor E_{ij} over P_{ij}. Let $\widetilde{D}_i, \widetilde{D}_{\text{out}}$ denote the proper transforms of D_i, D_{out}.

In what follows, we will use the notation $\mathbf{P}_i = (p_{i1}, \cdots, p_{i\ell_i})$ for a partition of length ℓ_i of some non-negative integer $|\mathbf{P}_i| = p_{i1} + \cdots + p_{i\ell_i}$, allowing some of the p_{ij}'s to be zero. Fix a class $\beta \in H^2(X, \mathbb{Z})$ with the property that $a_i := \beta \cdot D_i$ are non-negative and $k := \beta \cdot D_{\text{out}}$ is positive. It is an easy exercise in toric geometry that this implies a relationship

$$\sum_{i=1}^{p} a_i m_i = km_{\text{out}},$$

where m_{out} is a primitive generator of $\mathfrak{d}$. If one chooses a collection of partitions $\mathbf{P} = (\mathbf{P}_1, \ldots, \mathbf{P}_p)$ where $\mathbf{P}_i$ is a partition of a_i, let

$$\beta_\mathbf{P} := \nu^*\beta - \sum_{i=1}^{p}\sum_{j=1}^{\ell_i} p_{ij}E_{ij}.$$

This can be thought of as the class of a curve on X which passes through the point P_{ij} precisely p_{ij} times.

We would now like to associate a number to this cohomology class. This will be a Gromov-Witten count of one-pointed rational curves in $\widetilde{X}$ which (1) represent the class $\beta_\mathbf{P}$; (2) are tangent to $\widetilde{D}_{\text{out}}$ at the marked point with order k; and (3) are otherwise disjoint from any of the divisors $\widetilde{D}_i$. This is a relative Gromov-Witten invariant. However, the classical theory of relative Gromov-Witten invariants works relative to a smooth divisor, and of course the union of the boundary divisors here is singular. One can instead encode the above conditions using log Gromov-Witten theory. At the time [28] was written, log Gromov-Witten theory was not yet available, and as a consequence, we used a technical work-around to reduce to the classical theory. I give this description here since it does not require knowing log Gromov-Witten theory.

One defines $\widetilde{X}^o := \widetilde{X} \setminus \bigcup_{i=1}^p \widetilde{D}_i$. One then considers the moduli space $\mathfrak{M}(\widetilde{X}^o/\widetilde{D}^o_{\text{out}}, \beta_{\mathbf{P}})$ of relative stable maps of genus zero with target space $\widetilde{X}^o$, relative to the divisor $\widetilde{D}^o_{\text{out}} = \widetilde{D}_{\text{out}} \cap \widetilde{X}^o$. These curves have one marked point with order of tangency k with $\widetilde{D}^o_{\text{out}}$. The only problem is that the target space is non-proper, but one shows this doesn't cause any problems because nevertheless the moduli space is proper. One finds it is virtual dimension zero, and since it carries a virtual fundamental class, we can define

$$N_{\mathbf{P}} := \int_{[\mathfrak{M}(\widetilde{X}^o/\widetilde{D}^o, \beta_{\mathbf{P}})]^{vir}} 1 \in \mathbb{Q}.$$

We can then state the enumerative result ([28]):

THEOREM 11.4. *We have*

$$\log f_{\mathfrak{d}} = \sum_{\beta} \sum_{\mathbf{P}} k(\beta) N_{\mathbf{P}} t^{\sum_i |\mathbf{P}_i|} z^{-k(\beta) m_{\text{out}}},$$

where the sum is over all $\beta \in H^2(X, \mathbb{Z})$ with $\beta \cdot D_i \geq 0$, $k(\beta) := \beta \cdot D_{\text{out}} > 0$, and partitions $\mathbf{P}$ with $|\mathbf{P}_i| = \beta \cdot D_i$.

EXAMPLE 11.5. Returning to Example 11.3, consider the function $f_{\mathfrak{d}}$ attached to the ray of slope 1 for the cases $\ell = 1, 2$ and 3. In each case, the surface X is $\mathbb{P}^2$, with coordinate axes D_1, D_2 and D_{out}. Then $\widetilde{X}$ is obtained by blowing up ℓ points on each of D_1 and D_2.

Considering first the case of $\ell = 1$, we note that for $\beta = dH$, the class of a degree d curve in $\mathbb{P}^2$, the only relevant choice of $\mathbf{P}$ is $\mathbf{P}_1 = d$, $\mathbf{P}_2 = d$, and thus we have

$$\beta_{\mathbf{P}} = d\nu^* H - dE_{11} - dE_{21}.$$

This represents the class of a curve of degree d passing through the two blown-up points d times each. It is easy to see that the only choice for such a curve is a d-fold cover of a line passing through the two points. Furthermore, this cover must be totally ramified over D_{out} to guarantee the required order of tangency with D_{out}. This requires a virtual count, and the relevant localization calculations are carried out in [28], giving a value of $N_{\mathbf{P}} = (-1)^{d+1}/d^2$. Thus we get

$$\log f_{\mathfrak{d}} = \sum_{d=1}^{\infty} d \left(\frac{(-1)^{d+1}}{d^2} \right) t^{2d} x^{-d} y^{-d}.$$

Exponentiating one finds $f_{\mathfrak{d}} = 1 + t^2 x^{-1} y^{-1}$, agreeing with Example 11.3. So here we are just counting the one line through two points in $\mathbb{P}^2$ along with certain multiple covers of this line.

Going to $\ell = 2$, and $\beta = dH$, one finds four choices for the partition in the case $d = 1$, $\mathbf{P} = (1 + 0, 1 + 0), (1 + 0, 0 + 1), (0 + 1, 1 + 0)$, and $(0 + 1, 0 + 1)$. Each corresponds to a choice of one point on each of D_1, D_2, and one has one line through each of these pairs of points. Thus $N_{\mathbf{P}} = 1$ for each choice of such $\mathbf{P}$. As in the case $\ell = 1$, each of these lines also contributes to higher degree via multiple covers, with, say, $\mathbf{P} = (d+0, d+0)$ contributing $N_{\mathbf{P}} = (-1)^{d+1}/d^2$. For $d = 2$, one sees there are no curves for $\mathbf{P} = (2+0, 1+1)$, say, as this would require a conic with a node on D_1 and tangent to D_{out}; such does not exist. But with $\mathbf{P} = (1+1, 1+1)$, we look at conics passing through all four points and tangent to D_{out}. It is very easy to see there are two such conics.

One can then check that the only other curves contributing are multiple covers of one of the four lines or two conics. The multiple cover contribution for conics is actually different than for lines, because the order of tangency with D_{out} is different. It turns out the correct contribution for a d-fold cover of a conic is $1/d^2$. Hence we find

$$\log f_{\mathfrak{d}} = 4 \sum_{d=1}^{\infty} d \left(\frac{(-1)^{d+1}}{d^2} \right) t^{2d} x^{-d} y^{-d} + 2 \sum_{d=1}^{\infty} 2d \left(\frac{1}{d^2} \right) t^{4d} x^{-2d} y^{-2d}$$

and exponentiating we get

$$f_{\mathfrak{d}} = \frac{(1 + t^2 xy)^4}{(1 - t^4 x^{-2} y^{-2})^4} = (1 - t^2 x^{-1} y^{-1})^{-4}.$$

In the case that $\ell = 3$, one expects $3 \times 3 = 9$ lines, as there is one line passing through each pair of choices of one point on D_1 and one point on D_2. For conics, one has double covers of these lines, for a contribution of $-9/4$, and $2 \times 3 \times 3 = 18$ conics. Here one needs to choose two points on D_1 and two points on D_2, and then there are two conics passing through these four points tangent to D_{out}.

For cubics, there is the contribution of triple covers of lines, for a total of $9/9$, and a number of contributions from plane cubics. It turns out that for $\mathbf{P} = (1+1+1, 1+1+1)$, $N_{\mathbf{P}} = 18$. Note this gives a count of nodal plane cubics passing through 6 fixed points and for which D_{out} is a tri-tangent. On the other hand, for $\mathbf{P} = (1+2+0, 1+1+1)$, $N_{\mathbf{P}} = 3$. Note that there are a total of 12 partitions of this shape. This latter count represents nodal cubics with the node at one of the chosen points, passing also through four other chosen points, with D_{out} being tritangent. One concludes that

$$\log f_{\mathfrak{d}} = 9t^2 x^{-1} y^{-1} + 2(-9/4 + 18) t^4 x^{-2} y^{-2} + 3(9/9 + 54) t^6 x^{-3} y^{-3} + \cdots .$$

A direct comparision with the value given in Example 11.3 gives agreement.

We end this section with brief additional motivation for Theorem 11.4 and a word about the proof.

Suppose we have a piece of an integral affine manifold as depicted in Figure 11. Here we imagine a situation with two singular points in a surface, with local monodromy around the singularities contained in the horizontal and vertical line segments being $\begin{pmatrix} 1 & \ell_1 \\ 0 & 1 \end{pmatrix}$ and $\begin{pmatrix} 1 & \ell_2 \\ 0 & 1 \end{pmatrix}$ in suitably chosen bases (different for each segment). This is a slightly more general situation than was considered in §10, where we only discussed singularities with monodromy of the form $\begin{pmatrix} 1 & 1 \\ 0 & 1 \end{pmatrix}$. Nevertheless, the techniques of that section still apply, but the functions attached to the initial rays emanating from the singularities towards the central vertex v can be taken to be of the form $(1 + x^{-1})^{\ell_1}$ and $(1 + y^{-1})^{\ell_2}$. This is roughly the shape of the examples discussed above. Applying the scattering procedure would then produce a smoothing of $\check{X}_0(B, \mathscr{P}, s)$. However, on the mirror side, we interpret B as a dual intersection complex, which means it should arise from a degeneration $\mathcal{X} \to D$ where the central fibre $\mathcal{X}_0$ has an irreducible component Y_v isomorphic to $\mathbb{P}^2$ (corresponding to the vertex v). Furthermore, the total space $\mathcal{X}$ should have $\ell_1 + \ell_2$ ordinary double points lying on the toric boundary of Y_v. If one blows up the Weil divisor Y_v inside of $\mathcal{X}$, one obtains a small resolution $\widetilde{\mathcal{X}} \to \mathcal{X}$ of these ordinary double points, and in particular, the proper transform $\widetilde{Y}_v$ of Y_v is the blow-up of Y_v at the points $Y_v \cap \mathrm{Sing}(\mathcal{X})$. This operation blows up ℓ_1 points on one coordinate axis of Y_v and ℓ_2 on the other. This is exactly the same surface considered in Theorem 11.4.

Now consider the kind of curves on $\widetilde{Y}_v$ counted by Theorem 11.4. These are curves in $\widetilde{Y}_v$ which only intersect the third coordinate axis at one point. These can be viewed as curves in $\widetilde{\mathcal{X}}_0$, but not ones which deform to holomorphic curves in a general fibre of the family $\widetilde{\mathcal{X}} \to D$. Rather, roughly, we expect such curves to deform to holomorphic disks, with the point of intersection with the singular locus of $\widetilde{\mathcal{X}}_0$ (i.e., the point of intersection with the third axis of $\widetilde{Y}_v$) expanding into an S^1, giving the boundary of the holomorphic disk. Approximately, we expect this boundary to lie in a fibre of an SYZ fibration on a general fibre of the family $\widetilde{\mathcal{X}} \to D$. The homology class of this boundary inside the fibre is determined by the order of tangency of the curve with the third axis.

This correspondence between the relative curves considered in Theorem 11.4 is only a moral one; there is no proof yet that we are really counting such holomorphic disks. However, this argument served as the primary motivation for Theorem 11.4.

Finally, as far as the proof is concerned, there are several steps. First, we show that scattering diagrams can be deformed to look like a union of tropical curves, and use a variant of Mikhalkin's fundamental curve-counting results [62] as developed by Nishinou and Siebert [64] to show that scattering diagrams perform certain curve counts on toric surfaces. This is then related to the Gromov-Witten counts of the blown-up surfaces using Jun Li's gluing formula [57].

12. Other recent results and the future

I will close with a brief discussion of applications and future developments of the methods discussed here.

Recently a variant of the smoothing mechanism described here was used by myself, Hacking and Keel [26] to give a very general construction of mirrors of pairs (Y, D) where Y is a rational surface and D is an effective anti-canonical divisor forming a cycle of rational curves. We make use of [28] to write down what we call the *canonical scattering diagram,* which can be described entirely in terms of the Gromov-Witten theory of the pair (Y, D) (and more specifically, counts of curves intersecting D at only one point). This scattering diagram determines the mirror family. However, there is an additional crucial tool used to partially compactify the family constructed. This is necessary because unlike the affine manifolds considered in this paper, the natural one to associate to the pair (Y, D) has a singularity at a vertex of the polyhedral decomposition. There is no local model for a smoothing at this vertex, and as a consequence, one constructs families which are "missing" a point. To add this point back, one needs to be sure there are enough functions on the family constructed, and it turns out homological mirror symmetry suggests a natural way to construct such functions. This can be done tropically, creating what we call *theta functions.* The same construction applied to the case of degenerating

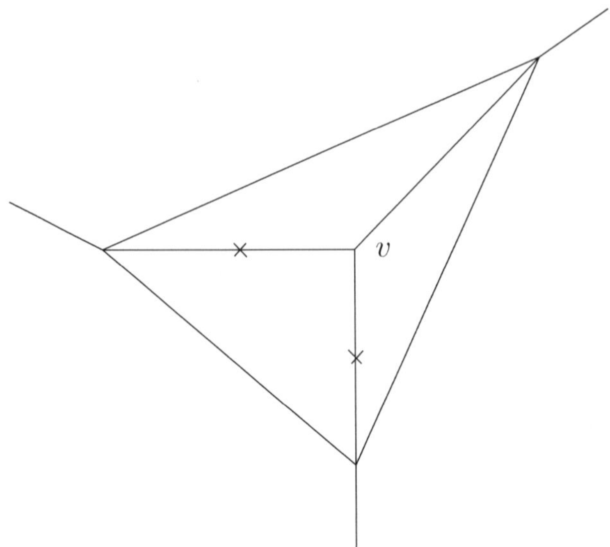

FIGURE 11

abelian varieties indeed produces ordinary theta functions, and we anticipate the functions we construct in these other contexts will be similarly useful. See [**35**] for a survey of these ideas.

The construction of [**26**] then also solves a problem which pre-dates mirror symmetry. In particular, we prove a conjecture of Looijenga concerning smoothability of cusp singularities.

Theta functions can be viewed as canonical bases for rings of functions on an affine variety or spaces of sections of line bundles on projective varieties. As such, they make contact with cluster algebra theory, providing a framework for constructing canonical bases of cluster algebras.

We also expect that the techniques for surface pairs (Y, D) will generalize. Indeed, a mirror partner to any maximally unipotent normal crossings degeneration of K3 surfaces can be constructed along similar lines, in work in progress with Hacking, Keel and Siebert. The expectation is that with an additional helping of log Gromov-Witten theory, one should be able write down a general construction in all dimensions for mirror partners to maximally unipotent degenerations of Calabi-Yau manifolds.

There still remains the question of extracting enumerative information from periods which provided the original excitement in mirror symmetry. Here we showed how enumerative geometry can be reflected in the mirror, but in a rather local way. We expect that it should be possible to carry out the computation of period integrals to extract genus zero Gromov-Witten invariants of the mirror, but some technical issues remain in this direction. Nevertheless, the program of understanding mirror symmetry via degenerations, inspired by the SYZ conjecture, seems to provide a powerful framework of thinking about mirror symmetry inside the realm of algebraic and tropical geometry.

References

[1] D. Abromovich, Q. Chen: *Stable logarithmic maps to Deligne–Faltings pairs II,* preprint, 2011.

[2] S. Amari: *Differential-geometric methods in statistics,* Lecture Notes in Statistics, **28**, Springer-Verlag, 1985.

[3] P. Aspinwall, T. Bridgeland, A. Craw, M. Douglas, M. Gross, A. Kapustin, G. Moore, G. Segal, B. Szendroi, P. Wilson, Dirichlet branes and mirror symmetry. Clay Mathematics Monographs, 4. American Mathematical Society, Providence, RI; Clay Mathematics Institute, Cambridge, MA, 2009. x+681 pp.

[4] D. Auroux: *Mirror symmetry and T-duality in the complement of an anticanonical divisor,* J. Gökova Geom. Topol. GGT **1** (2007), 51–91.

[5] V. Batyrev: *Dual polyhedra and mirror symmetry for Calabi-Yau hypersurfaces in toric varieties.* J. Algebraic Geom. **3** (1994), 493–535.

[6] V. Batyrev, and L. Borisov: *On Calabi-Yau complete intersections in toric varieties,* in *Higher-dimensional complex varieties (Trento, 1994),* 39–65, de Gruyter, Berlin, 1996.

[7] O. Ben-Bassat, *Mirror symmetry and generalized complex manifolds,* preprint, 2004, math.AG/0405303.

[8] R. Castaño-Bernard and D. Matessi, *Lagrangian 3-torus fibration,* J. Differential Geom., **81** (2009), 483–573.

[9] P. Candelas, M. Lynker, R. Schimmrigk, *Calabi-Yau manifolds in weighted* $\mathbf{P}_4$, Nuclear Phys. B **341** (1990), 383–402.

[10] P. Candelas, X. de la Ossa, P. Green, and L. Parkes, *A pair of Calabi-Yau manifolds as an exactly soluble superconformal theory,* Nuclear Phys. B **359** (1991), 21–74.

[11] Q. Chen: *Stable logarithmic maps to Deligne–Faltings pairs I,* preprint, 2010.

[12] S.-Y. Cheng and S.-T. Yau, *The real Monge-Ampère equation and affine flat structures,* in *Proceedings of the 1980 Beijing Symposium on Differential Geometry and Differential Equations,* Vol. 1, 2, 3 (Beijing, 1980), 339–370, Science Press, Beijing, 1982.

[13] K. Fukaya, *Multivalued Morse theory, asymptotic analysis and mirror symmetry,* in *Graphs and patterns in mathematics and theoretical physics,* 205–278, Proc. Sympos. Pure Math., **73**, Amer. Math. Soc., Providence, RI, 2005.

[14] D. Gaiotto, G. Moore, A. Neitzke: *Four-dimensional wall-crossing via three-dimensional field theory,* Comm. Math. Phys. **299** (2010), 163–224.

[15] A. Givental, *Equivariant Gromov-Witten invariants,* Internat. Math. Res. Notices **13**, (1996), 613–663.

[16] E. Goldstein: *A construction of new families of minimal Lagrangian submanifolds via torus actions,* J. Differential Geom. **58** (2001), 233–261.

[17] B. Greene and M. Plesser, *Duality in Calabi-Yau moduli space,* Nuclear Phys. B **338** (1990), 15–37.

[18] M. Gross: *Special Lagrangian Fibrations I: Topology,* in: Integrable Systems and Algebraic Geometry, (M.-H. Saito, Y. Shimizu and K. Ueno eds.), World Scientific 1998, 156–193.

[19] M. Gross: *Special Lagrangian Fibrations II: Geometry,* in: Surveys in Differential Geometry, Somerville: MA, International Press 1999, 341–403.

[20] M. Gross: *Topological Mirror Symmetry,* Invent. Math. **144** (2001), 75–137.

[21] M. Gross: *Examples of special Lagrangian fibrations,* in *Symplectic geometry and mirror symmetry (Seoul, 2000),* 81–109, World Sci. Publishing, River Edge, NJ, 2001.

[22] M. Gross: *Toric Degenerations and Batyrev-Borisov Duality,* Math. Ann. **333**, (2005) 645–688.

[23] M. Gross: *The Strominger-Yau-Zaslow conjecture: from torus fibrations to degenerations,* Algebraic Geometry, Seattle 2005. Part 1, 149–192, Proc. Sympos. Pure Math., **80**, Part 1, Amer. Math. Soc., Providence, RI, 2009.

[24] M. Gross: *Mirror symmetry for* $\mathbb{P}^2$ *and tropical geometry,* Adv. Math., **224** (2010), 169–245.

[25] M. Gross: *Tropical geometry and mirror symmetry,* CBMS Regional Conference Series in Mathematics, **114**. American Mathematical Society, Providencem RI, 2011. xvi+317 pp.

[26] M. Gross, P. Hacking, S. Keel: *Mirror symmetry for log Calabi-Yau surfaces I,* preprint, 2011.

[27] M. Gross, P. Hacking, S. Keel, M. Kontsevich: *Mirror symmetry and cluster algebras,* in preparation.

[28] M. Gross, R. Pandharipande, B. Siebert: *The tropical vertex*, Duke Math. J. **153** (2010), 297–362.

[29] M. Gross, and B. Siebert: *Affine manifolds, log structures, and mirror symmetry*, Turkish J. Math. **27** (2003), 33–60.

[30] M. Gross, and B. Siebert: *Torus fibrations and toric degenerations*, in preparation.

[31] M. Gross, and B. Siebert: *Mirror symmetry via logarithmic degeneration data I*, J. Diff. Geom., **72**, (2006) 169–338.

[32] M. Gross, and B. Siebert: *Mirror symmetry via logarithmic degeneration data II*, J. Algebraic Geom., **19**, (2010) 679–780

[33] M. Gross and B. Siebert: *From affine geoemtry to complex geometry*, Annals of Mathematics, **174**, (2011), 1301–1428.

[34] M. Gross and B. Siebert: *Logarithmic Gromov-Witten invariants*, preprint, 2011, to appear in J. of the AMS.

[35] M. Gross and B. Siebert: *Theta functions and mirror symmetry*, preprint, 2011.

[36] M. Gross, V. Tosatti, Y. Zhang: *Collapsing of abelian fibred Calabi-Yau manifolds*, preprint, 2011. To appear in Duke Math. J.

[37] M. Gross, and P.M.H. Wilson: *Mirror symmetry via 3-tori for a class of Calabi-Yau three-folds*, Math. Ann. **309** (1997), 505–531.

[38] M. Gross, and P.M.H. Wilson: *Large complex structure limits of K3 surfaces*, J. Differential Geom. **55** (2000), 475–546.

[39] M. Gualtieri, *Generalized complex geometry*, Oxford University DPhil thesis, `math.DG/0401221`.

[40] C. Haase, and I. Zharkov: *Integral affine structures on spheres and torus fibrations of Calabi-Yau toric hypersurfaces I*, preprint 2002, `math.AG/0205321`.

[41] C. Haase, and I. Zharkov: *Integral affine structures on spheres III: complete intersections*, preprint, `math.AG/0504181`.

[42] N. Hitchin: *The Moduli Space of Special Lagrangian Submanifolds*, Ann. Scuola Norm. Sup. Pisa Cl. Sci. (4) **25** (1997), 503–515.

[43] N. Hitchin: *Generalized Calabi-Yau manifolds*, Q. J. Math. **54** (2003), 281–308.

[44] D. Huybrechts: *Generalized Calabi-Yau structures, K3 surfaces, and B-fields*, Internat. J. Math. **16** (2005), 13–36.

[45] L. Illusie: *Logarithmic spaces (according to K. Kato)*, in Barsotti Symposium in Algebraic Geometry (Abano Terme 1991), 183–203, Perspect. Math. 15, Academic Press 1994.

[46] E. Ionel, T. Parker, *Relative Gromov-Witten invariants*, Ann. of Math. (2) **157** (2003), 45–96.

[47] E. Ionel, T. Parker, *The symplectic sum formula for Gromov-Witten invariants*, Ann. of Math. (2) **159** (2004), 935–1025.

[48] D. Joyce, *Singularities of special Lagrangian fibrations and the SYZ conjecture*, Comm. Anal. Geom. **11** (2003), 859–907.

[49] F. Kato: *Log smooth deformation theory*, Tohoku Math. J. **48** (1996), 317–354.

[50] K. Kato: *Logarithmic structures of Fontaine-Illusie*, in: Algebraic analysis, geometry, and number theory (J.-I. Igusa et. al. eds.), 191–224, Johns Hopkins Univ. Press, Baltimore, 1989.

[51] Y. Kawamata, Y. Namikawa: *Logarithmic deformations of normal crossing varieties and smooothing of degenerate Calabi-Yau varieties*, Invent. Math. **118** (1994), 395–409.

[52] M. Kontsevich: *Homological algebra of mirror symmetry*, Proceedings of the International Congress of Mathematicians, Vol. 1, 2 (Zürich, 1994), 120–139, Birkhäuser, Basel, 1995.

[53] M. Kontsevich, and Y. Soibelman: *Homological mirror symmetry and torus fibrations*, in: Symplectic geometry and mirror symmetry (Seoul, 2000), 203–263, World Sci. Publishing, River Edge, NJ, 2001.

[54] M. Kontsevich, and Y. Soibelman: *Affine structures and non-archimedean analytic spaces*, The unity of mathematics, 321–385, Progr. Math., 244, Birkhäuser Boston, Boston, MA, 2006.

[55] N.C. Leung: *Mirror symmetry without corrections*, Comm. Anal. Geom. **13** (2005), 287–331.

[56] N.C. Leung, C. Vafa: *Branes and toric geometry*, Adv. Theor. Math. Phys. **2** (1998), 91–118.

[57] J. Li: *A degeneration formula of GW-invariants*, J. Differential Geom. **60** (2002), 199–293.

[58] A.-M. Li, Y. Ruan: *Symplectic surgery and Gromov-Witten invariants of Calabi-Yau 3-folds*, Invent. Math., **145** (2001), 151–218.

[59] B. Lian, K. Liu, S-T. Yau, *Mirror principle. I*, Asian J. Math. **1** (1997), 729–763.

[60] D. Matessi, *Some families of special Lagrangian tori*, Math. Ann. **325** (2003), 211–228.

[61] R. McLean, *Deformations of calibrated submanifolds*, Comm. Anal. Geom. **6** (1998), 705–747.

[62] G. Mikhalkin, *Enumerative tropical algebraic geometry in* $\mathbb{R}^2$, J. Amer. Math. Soc. **18** (2005), 313–377.

[63] D. Morrison, *Compactifications of moduli spaces inspired by mirror symmetry*, in *Journées de Géométrie Algébrique d'Orsay (Orsay, 1992)*, Astérisque **218** (1993), 243–271.

[64] T. Nishinou, B. Siebert: *Toric degenerations of toric varieties and tropical curves*, Duke Math. J., **135** (2006), 1–51.

[65] M. Olsson: *Log algebraic stacks and moduli of log schemes*, Ph.D. thesis, Berkeley.

[66] B. Parker: *Exploded manifolds*, Adv. in Math., **229** (2012), 3256–3319.

[67] B. Parker: *Gromov-Witten invariants of exploded manifolds*, preprint, 2011.

[68] P. Petersen: *Riemannian geometry*, Graduate Texts in Mathematics, 171. Springer-Verlag, New York, 1998.

[69] M. Reineke: *Poisson automorphisms and quiver moduli*, J. Inst. Math. Jussieu **9**, (2010), 653–667.

[70] W.-D. Ruan: *Lagrangian torus fibration and mirror symmetry of Calabi-Yau hypersurface in toric variety*, preprint 2000, math.DG/0007028.

[71] W.-D. Ruan: *Lagrangian torus fibration of quintic Calabi-Yau hypersurfaces. II. Technical results on gradient flow construction*, J. Symplectic Geom. **1** (2002), no. 3, 435–521.

[72] S. Schröer, B. Siebert: *Irreducible degenerations of primary Kodaira surfaces*, in Complex geometry (Göttingen, 2000), 193–222, Springer, Berlin, 2002.

[73] S. Schröer, B. Siebert: *Toroidal crossings and logarithmic structures*, Adv. Math. **202** (2006), 189–231.

[74] B. Siebert: *Gromov-Witten invariants in relative and singular cases*. Lecture given at the workshop "Algebraic aspects of mirror symmetry," Univ. Kaiserslautern, Germany, June 2001.

[75] A. Strominger, S.-T. Yau, and E. Zaslow, *Mirror Symmetry is T-duality*, Nucl. Phys. **B479**, (1996) 243–259.

[76] V. Tosatti, *Adiabatic limits of Ricci-flat Kähler metrics*, J. Differential Geom., **84** (2010), 427–453.

[77] S.-T. Yau: *On the Ricci curvature of a compact Kähler manifold and the complex Monge-Ampère equation. I*, Comm. Pure Appl. Math., **31** (1978), 339–411.

[78] Y. Zhang: *Collapsing of Calabi-Yau manifolds and special Lagrangian submanifolds*, preprint, 2009.

UCSD MATHEMATICS, 9500 GILMAN DRIVE, LA JOLLA, CA 92093-0112, USA
E-mail address: mgross@math.ucsd.edu

Perfectoid spaces: A survey

Peter Scholze

ABSTRACT. This paper, written in relation to the Current Developments in Mathematics 2012 Conference, discusses the recent papers on perfectoid spaces. Apart from giving an introduction to their content, it includes some open questions, as well as complements to the results of the previous papers.

CONTENTS

1. Introduction

The original aim of the theory of perfectoid spaces was to prove Deligne's weight-monodromy conjecture over p-adic fields by reduction to the case of local fields of equal characteristic p, where the result is known. In order to do so, the theory of perfectoid spaces establishes a general framework relating geometric questions over local fields of mixed characteristic with geometric questions over local fields of equal characteristic. One application of this theory is a general form of Faltings's almost purity theorem. For the moment, the weight-monodromy conjecture is not proved in full generality using these methods; however, we can reduce the conjecture to a statement on approximating a 'fractal' by algebraic varieties.

However, the theory of perfectoid spaces has proved to be useful in other situations, and certainly many more will be found. On one hand, perfectoid spaces embody Faltings's almost purity theorem; as this is the crucial technical ingredient to Faltings's approach to p-adic Hodge theory, it is not surprising that one can prove

---

2010 *Mathematics Subject Classification.* 14G22, 14G20, 14C30, 14L05.

new results in p-adic Hodge theory. For example, it becomes possible to analyze general (proper smooth) rigid-analytic varieties, instead of just algebraic varieties. That p-adic Hodge theory for rigid-analytic varieties should be possible was already conjectured by Tate, [**33**], when he first conjectured the existence of a Hodge-Tate decomposition, and of course it aligns well with the situation over $\mathbb{C}$.

In these first papers, the perfectoid spaces had more of an auxiliary role. However, it turns out that many natural constructions, which so far could not be given any geometric meaning, are perfectoid spaces in a natural way: For example, Shimura varieties with infinite level at p, and Rapoport-Zink spaces (local analogues of Shimura varieties) with infinite level. For Rapoport-Zink spaces, there has been a duality conjecture (proved by Faltings) which says that certain 'dual' pairs of Rapoport-Zink spaces are isomorphic at infinite level. Until now, the formulation of such an isomorphism has been very ad hoc; however, it can now be formulated as an isomorphism of perfectoid spaces. This has all expected consequences, such as comparisons of étale cohomology.

In the case of Shimura varieties, interesting applications (related to torsion in the cohomology of locally symmetric varieties and Emerton's completed cohomology groups) arise, which are work in progress ([**28**]); some of this is sketched at the end of this survey.

This paper was written in relation to the talks of the author at the Current Developments in Mathematics conference 2012 at Harvard. The author wants to thank the organizers heartily for the invitation, and the opportunity to write this survey. The original intention was to write a survey paper about perfectoid spaces and the weight-monodromy conjecture. However, as this is exactly the content of [**30**], the author decided instead to give some introduction to the general content of the three papers [**30**], [**29**] and [**31**], and mention some open questions, as well as some complements on the results of these papers. These are new results, but they are immediate applications of the theory built up there. Thus, some parts of this paper do not have the nature of a survey, and assume familiarity with the content of these papers.

Acknowledgments. This work was done while the author was a Clay Research Fellow.

2. Perfectoid Spaces

2.1. Introduction. This introduction is essentially identical to a post of the author on MathOverflow, [**27**].

DEFINITION 2.1. A perfectoid field K is a complete non-archimedean field K of residue characteristic p, equipped with a non-discrete valuation of rank 1, such that the Frobenius map $\Phi : \mathcal{O}_K/p \to \mathcal{O}_K/p$ is surjective, where $\mathcal{O}_K \subset K$ is the subring of elements of norm ≤ 1.

Some authors, e.g. Gabber-Ramero in their book on almost ring theory, [**16**], call such fields deeply ramified (although they do not require that they are complete).

EXAMPLE 2.2. Standard examples of perfectoid fields are given by the completion of $\mathbb{Q}_p(p^{1/p^\infty})$, $\mathbb{Q}_p(\mu_{p^\infty})$, $\overline{\mathbb{Q}}_p$, or $\mathbb{F}((t))(t^{1/p^\infty})$.

Given a perfectoid field K, one can form a second perfectoid field $K^\flat$, always of characteristic p, given as the fraction field of

$$\mathcal{O}_{K^\flat} = \varprojlim_{\Phi} \mathcal{O}_K/p \ ,$$

where the transition maps are given by Frobenius. Concretely, if K is the completion of $\mathbb{Q}_p(p^{1/p^\infty})$, then $K^\flat$ is given by the completion of $\mathbb{F}_p((t))(t^{1/p^\infty})$, where t is the element

$$(p, p^{1/p}, p^{1/p^2}, \ldots) \in \mathcal{O}_{K^\flat} = \varprojlim \mathcal{O}_K/p \ .$$

In particular, we have a canonical identification

$$\mathcal{O}_{K^\flat}/t = \mathbb{F}_p[t^{1/p^\infty}]/t \cong \mathbb{Z}_p[p^{1/p^\infty}]/p = \mathcal{O}_K/p \ .$$

In this situation, one has the following theorem, due to Fontaine-Wintenberger, [**15**], in most examples.

THEOREM 2.3. *There is a canonical isomorphism of absolute Galois group* $\mathrm{Gal}(\bar{K}/K) \cong \mathrm{Gal}(\bar{K}^\flat/K^\flat)$.

At this point, it may be instructive to explain this theorem in the example where K is the completion of $\mathbb{Q}_p(p^{1/p^\infty})$; in all examples to follow, we make this choice of K. It says that there is a natural equivalence of categories between the category of finite extensions L of K and the category of finite extensions M of $K^\flat$. Let us give an example: Say M is the extension of $K^\flat$ given by adjoining a root of $X^2 - 7tX + t^5$. Basically, the idea is that one replaces t by p, so that one would like to define L as the field given by adjoining a root of $X^2 - 7pX + p^5$. However, this is obviously not well-defined: If $p = 3$, then $X^2 - 7tX + t^5 = X^2 - tX + t^5$, but $X^2 - 7pX + p^5 \neq X^2 - pX + p^5$, and one will not expect in general that the fields given by adjoining roots of these different polynomials are the same.

However, there is the following way out: M can be defined as the splitting field of $X^2 - 7t^{1/p^n}X + t^{5/p^n}$ for all $n \geq 0$ (using that $K^\flat$ is perfect), and if we choose n very large, then one can see that the fields L_n given as the splitting field of $X^2 - 7p^{1/p^n}X + p^{5/p^n}$ will stabilize as $n \to \infty$; this is the desired field L. Basically, the point is that the discriminant of the polynomials considered becomes very small, and the difference between any two different choices one might make when replacing t by p becomes comparably small.

This argument can be made precise by using Faltings's almost mathematics, as developed systematically by Gabber-Ramero, [**16**]. Consider $K \supset \mathcal{O}_K \supset \mathfrak{m}$, where $\mathfrak{m}$ is the maximal ideal; in the example, it is the one generated by all p^{1/p^n}, and it satisfies $\mathfrak{m}^2 = \mathfrak{m}$, because the valuation on K is non-discrete. We have a sequence of localization functors:

$$\mathcal{O}_K - \mathrm{mod} \to \mathcal{O}_K - \mathrm{mod}/\mathfrak{m} - \mathrm{torsion} \to \mathcal{O}_K - \mathrm{mod}/p - \mathrm{power\ torsion} \ .$$

The last category is equivalent to K-mod, and the composition of the two functors is like taking the generic fibre of an object with an integral structure.

In this sense, the category in the middle can be seen as a slightly generic fibre, sitting strictly between an integral structure and an object over the generic fibre. Moreover, an object like $\mathcal{O}_K/p$ is nonzero in this middle category, so one can talk about torsion objects, neglecting only very small objects. The official name for this middle category is $\mathcal{O}_K^a$-mod: almost $\mathcal{O}_K$-modules.

This category is an abelian tensor category, and hence one can define in the usual way the notion of an $\mathcal{O}_K^a$-algebra ($=$ almost $\mathcal{O}_K$-algebra), etc. . With some work, one also has notions of almost finitely presented modules and (almost) étale maps. In the following, we will often use the notion of an almost finitely presented étale map, which is the almost analogue of a finite étale map: We will use the term almost finite étale map in the following.

THEOREM 2.4 (Tate([**33**]), Gabber-Ramero([**16**])). *If L/K is a finite extension, then $\mathcal{O}_L/\mathcal{O}_K$ is almost finite étale. Similarly, if $M/K^\flat$ is finite, then $\mathcal{O}_M/\mathcal{O}_{K^\flat}$ is almost finite étale.*

As an example, assume $p \neq 2$ and $L = K(p^{1/2})$. For convenience, we look at the situation at a finite level, so let $K_n = \mathbb{Q}_p(p^{1/p^n})$ and $L_n = K_n(p^{1/2})$. Then $\mathcal{O}_{L_n} = \mathcal{O}_{K_n}[X]/(X^2 - p^{1/p^n})$. To check whether this is étale, look at $f(X) = X^2 - p^{1/p^n}$ and look at the ideal generated by f and its derivative f'. This contains p^{1/p^n}, so in some sense $\mathcal{O}_{L_n}$ is etale over $\mathcal{O}_{K_n}$ up to p^{1/p^n}-torsion. Now take the limit as $n \to \infty$ to see that $\mathcal{O}_L$ is almost etale over $\mathcal{O}_K$.

In fact, in the case of equal characteristic, i.e. $K^\flat$, the theorem is easy. Indeed, there will be some big N such that $\Omega^1_{\mathcal{O}_M/\mathcal{O}_{K^\flat}}$ is killed by t^N (where $t \in K^\flat$ is some element with $|t| < 1$). But $K^\flat$ is perfect, and thus also M; it follows that the whole situation is invariant under Frobenius. Thus, $\Omega^1_{\mathcal{O}_M/\mathcal{O}_{K^\flat}}$ is killed by $t^{N/p}$, and thus by t^{N/p^2}, ..., i.e. t^{N/p^k} for all k. This is exactly saying that $\Omega^1_{\mathcal{O}_M/\mathcal{O}_{K^\flat}}$ is almost zero. From here, one can deduce that $\mathcal{O}_M$ is almost finite étale over $\mathcal{O}_{K^\flat}$.

Now we can prove Theorem 2.3 above:

$$\{\text{finite étale covers of } K\} \cong \{\text{almost finite étale covers of } \mathcal{O}_K\}$$
$$\cong \{\text{almost finite étale covers of } \mathcal{O}_K/p\}$$
$$= \{\text{almost finite étale covers of } \mathcal{O}_{K^\flat}/t\}$$
$$\cong \{\text{almost finite étale covers of } \mathcal{O}_{K^\flat}\}$$
$$\cong \{\text{finite étale covers of } K^\flat\}$$

Here, we use that almost finite étale covers lift uniquely over nilpotents. Also, one can always find some element $t \in K^\flat$ as in the example such that $\mathcal{O}_K/p = \mathcal{O}_{K^\flat}/t$.

Now we want to generalize the theory to the relative situation. Here, the basic claim is the following.

CLAIM 2.5. *The affine line $\mathbb{A}^1_{K^\flat}$ 'equals' $\varprojlim \mathbb{A}^1_K$, where the transition maps are the p-th power map.*

As a first step towards understanding this, we check this on points. Here it says that $K^\flat = \varprojlim K$. In particular, there should be map $K^\flat \to K$ by projection to the last coordinate, which is usually denoted $x \mapsto x^\sharp$ and again this can be explained in an example:

Say $x = t^{-1} + 5 + t^3$. Basically, we want to replace t by p, but this is not well-defined. But we have just learned that this problem becomes less serious as we take p-power roots. So we look at $t^{-1/p^n} + 5 + t^{3/p^n}$, replace t by p, get $p^{-1/p^n} + 5 + p^{3/p^n}$, and then we take the p^n-th power again, so that the expression has the chance of being independent of n. Now, it is in fact not difficult to see that

$$\lim_{n \to \infty} (p^{-1/p^n} + 5 + p^{3/p^n})^{p^n} \in K$$

exists, and this defines $x^\sharp \in K$. Now the map $K^\flat \to \varprojlim K$ is given by

$$x \mapsto (x^\sharp, (x^{1/p})^\sharp, (x^{1/p^2})^\sharp, \dots) \, .$$

In order to prove that this is a bijection, just note that

$$\mathcal{O}_{K^\flat} = \varprojlim \mathcal{O}_{K^\flat}/t^{p^n} = \varprojlim_{\Phi} \mathcal{O}_{K^\flat}/t = \varprojlim_{\Phi} \mathcal{O}_K/p \leftarrow \varprojlim_{x \mapsto x^p} \mathcal{O}_K \, .$$

Here, the last map is the obvious projection, and in fact is a bijection, which amounts to the same verification as that the limit above exists. Afterwards, just invert t to get the desired identification.

One sees immediately that the explicit description of the isomorphism involves p-adic limits, so a formalization will necessarily need to use some framework of rigid geometry. In the paper [19] of Kedlaya-Liu, where they are doing closely related things, they choose to work with Berkovich spaces. The author favors the language of Huber's adic spaces, as this language is capable of expressing more (e.g., Berkovich only considers rank-1-valuations, whereas Huber considers also the valuations of higher rank). In the language of adic spaces, the spaces are actually locally ringed topological spaces (equipped with valuations) (and affinoids are open, in contrast to Berkovich's theory, making it easier to glue). There is an analytification functor $X \mapsto X^{\mathrm{ad}}$ from schemes of finite type over K to adic spaces over K (similar to the functor associating to a scheme of finite type over C a complex-analytic space). Then we have the following theorem:

THEOREM 2.6. *There is a homeomorphism of underlying topological spaces* $|(\mathbb{A}^1_{K^\flat})^{\mathrm{ad}}| \cong \varprojlim |(\mathbb{A}^1_K)^{\mathrm{ad}}|$.

At this point, the following question naturally arises: Both sides of this homeomorphism are locally ringed topological spaces: So is it possible to compare the structure sheaves? There is the obvious problem that on the left-hand side, we have characteristic p-rings, whereas on the right-hand side, we have characteristic 0-rings. How can one pass from one to the other side?

DEFINITION 2.7. A perfectoid K-algebra is a complete Banach K-algebra R such that the set of power-bounded elements $R^\circ \subset R$ is bounded and the Frobenius map $\Phi : R^\circ/p \to R^\circ/p$ is surjective.

Similarly, one defines perfectoid $K^\flat$-algebras S. The last condition is then equivalent to requiring that S perfect, whence the name. Examples are K, any finite extension L of K, and $K\langle T^{1/p^\infty}\rangle$, which is $\mathcal{O}_K\langle T^{1/p^\infty}\rangle \otimes_{\mathcal{O}_K} K$, where $\mathcal{O}_K\langle T^{1/p^\infty}\rangle$ is the p-adic completion of $\mathcal{O}_K[T^{1/p^\infty}]$.

Recall that in classical rigid geometry, one considers rings like $K\langle T\rangle$, which is interpreted as the ring of convergent power series on the closed annulus $|T| \leq 1$. Now in the example of the $\mathbb{A}^1$ above, we take p-power roots of the coordinate, so after completion the rings on the inverse limit are in fact perfectoid.

In characteristic p, one can pass from usual affinoid algebras to perfectoid algebras by taking the completed perfection; the difference between the two is small, at least as regards topological information on associated spaces: Frobenius is a homeomorphism on topological spaces, and even on étale topoi. This is also the reason that we did not take the inverse limit $\varprojlim \mathbb{A}^1_{K^\flat}$ above: It does not change the topological space. In order to compare structure sheaves, one should however take this inverse limit.

Now we can state the tilting equivalence.

THEOREM 2.8. *The category of perfectoid K-algebras is canonically equivalent to the category of perfectoid $K^\flat$-algebras.*

The functor is given by $R \mapsto R^\flat = (\varprojlim R^\circ/p)[t^{-1}]$. Again, one also has $R^\flat = \varprojlim R$ as multiplicative monoids, where the transition maps are the p-th power map, giving also the map $R^\flat \to R$, $f \mapsto f^\sharp$.

There are two different proofs for this. One is to write down the inverse functor, given by $S \mapsto W(S^\circ) \otimes_{W(\mathcal{O}_{K^\flat})} K$, using the map

$$\Theta : W(\mathcal{O}_{K^\flat}) \to K$$

known from p-adic Hodge theory. The other proof is similar to what we did above for finite étale K-algebras: Perfectoid K-algebras are equivalent to almost $\mathcal{O}_K$-algebras A s.t. A is flat, p-adically complete and Frobenius induces an isomorphism $A/p^{1/p} \cong A/p$; these are in turn equivalent to almost $\mathcal{O}_K/p$-algebras $\overline{A}$ s.t. $\overline{A}$ is flat and Frobenius induces an isomorphism $\overline{A}/p^{1/p} \cong \overline{A}$. From here, one can go to $\mathcal{O}_{K^\flat}/t$, and reverse the steps to get to $K^\flat$.

The first identification between perfectoid K-algebras and certain almost $\mathcal{O}_K$-algebras is not difficult. The second identification between certain almost $\mathcal{O}_K$-algebras and certain almost $\mathcal{O}_K/p$-algebras relies on the fact (already in the book by Gabber-Ramero) that the cotangent complex $\mathbb{L}_{\overline{A}/(\mathcal{O}_K/p)}$ vanishes, and hence one gets unique deformations of objects and morphisms. At least on differentials Ω^1, this is easy to see: Every element x has the form y^p because Frobenius is surjective; but then $dx = dy^p = py^{p-1}dy = 0$ because $p = 0$ in $\overline{A}$.

Finally, we summarize briefly the main theorems on the basic nature of perfectoid spaces. An affinoid perfectoid space is associated to a perfectoid affinoid K-algebra, which is a pair (R, R^+) consisting of a perfectoid K-algebra R and an open and integrally closed subring $R^+ \subset R^\circ$. In most cases, one will just take $R^+ = R^\circ$. Then also the categories of perfectoid affinoid K-algebras and perfectoid affinoid $K^\flat$-algebras are equivalent. Huber associates to such pairs (R, R^+) a topological space $X = \mathrm{Spa}(R, R^+)$ consisting of continuous valuations on R that are ≤ 1 on R^+, with the topology generated by the rational subsets $\{x \in X \mid \forall i : |f_i(x)| \leq |g(x)|\}$, where $f_1, \ldots, f_n, g \in R$ generate the unit ideal. Moreover, he defines a structure presheaf $\mathcal{O}_X$ on X, and the subpresheaf $\mathcal{O}_X^+$, consisting of functions which have absolute value ≤ 1 everywhere.

THEOREM 2.9. *Let (R, R^+) be a perfectoid affinoid K-algebra, with tilt $(R^\flat, R^{\flat+})$. Let $X = \mathrm{Spa}(R, R^+)$, with $\mathcal{O}_X$, $\mathcal{O}_X^+$, and $X^\flat = \mathrm{Spa}(R^\flat, R^{\flat+})$, with $\mathcal{O}_{X^\flat}$, $\mathcal{O}_{X^\flat}^+$.*

(i) There is a canonical homeomorphism $X \cong X^\flat$, given by mapping x to $x^\flat$ defined via $|f(x^\flat)| = |f^\sharp(x)|$. Rational subsets are identified under this homeomorphism.

(ii) For any rational subset $U \subset X$, the pair $(\mathcal{O}_X(U), \mathcal{O}_X^+(U))$ is perfectoid affinoid with tilt $(\mathcal{O}_{X^\flat}(U), \mathcal{O}_{X^\flat}^+(U))$.

(iii) The presheaves $\mathcal{O}_X$, $\mathcal{O}_X^+$ and $\mathcal{O}_{X^\flat}$, $\mathcal{O}_{X^\flat}^+$ are sheaves.

(iv) For all $i > 0$, the cohomology group $H^i(X, \mathcal{O}_X) = 0$. Moreover, the cohomology group $H^i(X, \mathcal{O}_X^+)$ is almost zero, i.e. $\mathfrak{m}$-torsion. The same holds true for $\mathcal{O}_{X^\flat}$ and $\mathcal{O}_{X^\flat}^+$.

This allows one to define general perfectoid spaces by gluing affinoid perfectoid spaces. Further, one can define étale morphisms of perfectoid spaces, and then étale topoi. This leads to an improvement on Faltings's almost purity theorem:

THEOREM 2.10. *Let R be a perfectoid K-algebra, and let S/R be finite etale. Then S is perfectoid and S° is almost finite étale over R°.*

In particular, no sort of semistable reduction hypothesis is required anymore. Also, the proof is much easier. In characteristic p, the result is easy, and is a consequence of repeated application of Frobenius (roughly, S° is finite étale over R° up to t^N-torsion for some N; but then it is finite étale up to t^{N/p^k}-torsion for all k, i.e. almost finite étale). Moreover, following the ideas for the case of perfectoid fields, one has a fully faithful functor from finite étale $R^\flat$-algebras to finite étale R-algebras. It is enough to prove that this functor is an equivalence of categories, as for finite étale R-algebras in the image, one can deduce the result from the case in characteristic p, which is already known. In order to prove that the functor is an equivalence of categories, one makes a localization argument on $X = \mathrm{Spa}(R, R^+)$ (for any $R^+ \subset R^\circ$), to reduce to the case of perfectoid fields.

Tilting also identifies the étale topoi of a perfectoid space and its tilt, and as an application, one gets the following theorem.

THEOREM 2.11. *There is an equivalence of étale topoi of adic spaces*

$$(\mathbb{P}^n_{K^\flat})^{\mathrm{ad}}_{\mathrm{\acute{e}t}} \cong \varprojlim (\mathbb{P}^n_K)^{\mathrm{ad}}_{\mathrm{\acute{e}t}} \ .$$

Here the transition maps are the p-th power map on homogeneous coordinates.

In a sense, the theorem gives rise to a projection map

$$\pi : \mathbb{P}^n_{K^\flat} \to \mathbb{P}^n_K$$

defined on étale topoi of adic spaces, which is given on coordinates by

$$\pi(x_0 : \ldots : x_n) = (x_0^\sharp : \ldots : x_n^\sharp) \ .$$

It is also defined on the topological spaces underlying the adic spaces.

This discussion has the following consequence.

THEOREM 2.12. *Let k be a finite extension of $\mathbb{Q}_p$, and let $X \subset \mathbb{P}^n_k$ be a smooth complete intersection. Then the weight-monodromy conjecture holds true for X.*

In [**30**], this result is proved for complete intersections in more general toric varieties in place of $\mathbb{P}^n$.

Let us first recall the statement of the weight-monodromy conjecture. Let q be the cardinality of the residue field of k, and fix a geometric Frobenius $\Phi \in \mathrm{Gal}(\bar{k}/k)$, where $\bar{k}$ is a fixed geometric closure of k. The following are known about the étale cohomology group $V = H^i(X_{\bar{k}}, \bar{\mathbb{Q}}_\ell)$, where $\ell \neq p$ is some prime, and $i \geq 0$.

(i) There is a decomposition

$$V = \bigoplus_{j=0}^{2i} V_j \ ,$$

where all eigenvalues of Φ on V_j are Weil numbers of weight j, i.e. elements $\alpha \in \bar{\mathbb{Q}}$ such that $|\iota(\alpha)| = q^{j/2}$, under all embeddings $\iota : \bar{\mathbb{Q}} \to \mathbb{C}$.

(ii) There is a nilpotent operator $N : V \to V$ given by the logarithm of the action of the ℓ-adic inertia subgroup, inducing maps

$$N : V_j \to V_{j-2} \ .$$

Part (i) is a consequence of the Rapoport-Zink spectral sequence, [23], in the case of semistable reduction, and de Jong's alterations, [7], in general, to reduce to this case. Part (ii) is a consequence of Grothendieck's quasi-unipotence theorem, which says that after a finite extension of k, the action of the inertia subgroup becomes unipotent. The decomposition in part (i) depends on the choice of Φ, but the (weight) filtration given by $\mathrm{Fil}^m = \bigoplus_{j \leq m} V_j$ does not.

We note that if k has good reduction, then by base-change, the action of $\mathrm{Gal}(\bar{k}/k)$ is unramified, so $N = 0$, and by the Weil conjectures, $V = V_i$, i.e. all eigenvalues of the Frobenius are Weil numbers of weight i. In general, several different weights can appear, but this is supposed to be made good for by a nontrivial monodromy operator, connecting the different weights:

CONJECTURE 2.13 (Weight-monodromy conjecture, [8]). *For all $j \geq 0$, the map*

$$N^j : V_{i+j} \to V_{i-j}$$

is an isomorphism.

This is reminiscent of the Lefschetz decomposition. In fact, over $\mathbb{C}$, there is an analogue of this conjecture, by looking at a family of projective smooth complex varieties over the punctured unit disc. In that case, there is a limiting Hodge structure, which is not in general pure, and a monodromy operator (coming from a loop around the puncture). In that case, the result is known by work of Schmid, [25], cf. also a paper of Steenbrink, [32] (who introduced what is now called the Rapoport-Zink spectral sequence).

Deligne, [9], proved the analogue of Conjecture 2.13 for a finite extension of $\mathbb{F}_p((t))$ in place of k. Over p-adic fields, Conjecture 2.13 is known for $i = 1$ (this reduces to abelian varieties, where one uses Néron models), and for $i = 2$ (this reduces to surfaces, where it is proved in [23]). Apart from that, little is known.

Let us end this introduction by giving a short sketch of the proof of the weight-monodromy conjecture for a smooth hypersurface $X \subset \mathbb{P}^n$, which is already a new result; the proof in the general case is identical. Let K be the completion of $k(\pi^{1/p^\infty})$, where π is a uniformizer of k; this is a perfectoid field, and its tilt $K^\flat$ is the completion of $\mathbb{F}_q((t))(t^{1/p^\infty})$.[1] Both the weights and the monodromy operator are defined by the action of $\mathrm{Gal}(\bar{K}/K) \subset \mathrm{Gal}(\bar{k}/k)$. We have the projection

$$\pi : \mathbb{P}^n_{K^\flat} \to \mathbb{P}^n_K \ ,$$

and we can look at the preimage $\pi^{-1}(X)$. One has a $\mathrm{Gal}(\bar{K}/K) \cong \mathrm{Gal}(\bar{K}^\flat/K^\flat)$-equivariant injective map $H^i(X) \to H^i(\pi^{-1}(X))$ on ℓ-adic cohomology, and if $\pi^{-1}(X)$ were an algebraic variety, then one could deduce the result from Deligne's theorem in equal characteristic. However, the map π is highly transcendental, and $\pi^{-1}(X)$ will not be given by equations. In general, it will look like some sort of fractal, have infinite-dimensional cohomology, and will have infinite degree in the sense that it will meet a line in infinitely many points. As an easy example, let

$$X = \{x_0 + x_1 + x_2 = 0\} \subset (\mathbb{P}^2_K)^{\mathrm{ad}} \ .$$

Then the homeomorphism

$$|(\mathbb{P}^2_{K^\flat})^{\mathrm{ad}}| \cong \varprojlim_{\varphi} |(\mathbb{P}^2_K)^{\mathrm{ad}}|$$

---

[1] If $k = \mathbb{Q}_p$, then we are just back to our standard example.

means that $\pi^{-1}(X)$ is topologically the inverse limit of the subvarieties

$$X_n = \{x_0^{p^n} + x_1^{p^n} + x_2^{p^n} = 0\} \subset (\mathbb{P}_K^2)^{\mathrm{ad}} .$$

However, we have the following crucial approximation lemma.

LEMMA 2.14. *Let $\tilde{X} \subset (\mathbb{P}_K^n)^{\mathrm{ad}}$ be a small open neighborhood of the hypersurface X. Then there is a hypersurface $Y \subset \pi^{-1}(\tilde{X})$.*

The proof of this lemma is by an explicit approximation algorithm for the homogeneous polynomial defining X, and is the reason that we have to restrict to complete intersections. Using a result of Huber, one finds some $\tilde{X}$ such that $H^i(X) = H^i(\tilde{X})$, and hence gets a $\mathrm{Gal}(\bar{K}/K) \cong \mathrm{Gal}(\bar{K}^\flat/K^\flat)$-equivariant map $H^i(X) = H^i(\tilde{X}) \to H^i(Y)$. As before, one checks (using $\cup$-products and a direct analysis in top degree) that it is injective and concludes.

2.2. Some open questions. Of course, the most urgent question would be to see whether the proof of the weight-monodromy conjecture can be extended to more general situations. The only problem is in proving an analogue of the last lemma; let me state this as a conjecture.

CONJECTURE 2.15. *Let $X \subset \mathbb{P}_K^n$ be a smooth closed subscheme. Let $\tilde{X} \subset (\mathbb{P}_K^n)^{\mathrm{ad}}$ be a small open neighborhood of X. Then there exists a closed subscheme $Y \subset \mathbb{P}_{K^\flat}^n$ such that $Y^{\mathrm{ad}} \subset \pi^{-1}(\tilde{X})$, and $\dim Y = \dim X$.*

If this conjecture is true, the weight-monodromy conjecture follows, using the same argument. We recall that $\pi^{-1}(X^{\mathrm{ad}}) \subset (\mathbb{P}_{K^\flat}^n)^{\mathrm{ad}}$ can be regarded as a fractal. The question is whether this fractal can be approximated (globally) by an algebraic variety. We note that it can be approximated locally by an algebraic variety, as X is smooth (and thus locally a complete intersection, so the argument in the case of global complete intersections works locally).

On the other hand, there are a couple of foundational questions on perfectoid spaces, where a positive answer would simplify many arguments. The key problem is whether the property of being perfectoid is local.

CONJECTURE 2.16. *Let K be some perfectoid field, (A, A^+) a complete affinoid K-algebra. Assume that there is a cover of $X = \mathrm{Spa}(A, A^+)$ by rational subsets $U_i \subset X$ such that $\mathcal{O}_X(U_i)$ is a perfectoid K-algebra. Then A is a perfectoid K-algebra.*

COROLLARY 2.17. *Assume Conjecture 2.16. Let X be a perfectoid space over K, and assume that $U \subset X$ is affinoid, i.e. there is a complete affinoid K-algebra (A, A^+) and a map*

$$(A, A^+) \to (H^0(U, \mathcal{O}_X), H^0(U, \mathcal{O}_X^+))$$

such that the induced map $U \to \mathrm{Spa}(A, A^+)$ is a homeomorphism, and the map of (pre)sheaves $\mathcal{O}_{\mathrm{Spa}(A,A^+)} \to \mathcal{O}_U$ is an isomorphism locally on U.[2] Then

$$(A, A^+) \cong (H^0(U, \mathcal{O}_X), H^0(U, \mathcal{O}_X^+))$$

is an affinoid perfectoid K-algebra.

---

[2] This is saying that U is an affinoid adic space in the sense of [**31**].

COROLLARY 2.18. *Assume Conjecture 2.16. Let X be an affinoid perfectoid space over K, and let Y be an affinoid noetherian adic space over K, and $f : X \to Y$ a map of adic spaces over K. Let V be another affinoid noetherian adic space over K, with an étale map $g : V \to Y$. Then $U = V \times_Y X \to X$ is étale, and U is affinoid perfectoid.*

REMARK 2.19. As the proof shows, one can also compute the global sections of U in the expected way.

PROOF. Let $X = \mathrm{Spa}(A, A^+)$, $Y = \mathrm{Spa}(B, B^+)$, $V = \mathrm{Spa}(C, C^+)$. Then let $D = A \otimes_B C$, with the topology making the image of $A_0 \otimes_{B_0} C_0$ a ring of definition, with ideal of definition the image of $I \otimes J$, where $A_0 \subset A$, $B_0 \subset B$, $C_0 \subset C$ are rings of definition, and $I \subset A_0$, $J \subset C_0$ are ideals of definition. Let $D^+ \subset D$ be the integral closure of the image of $A^+ \otimes_{B^+} C^+$. Then the completion of (D, D^+) is an affinoid K-algebra, and it satisfies the hypothesis of Conjecture 2.16. For this, use that if $g : V \to Y$ is a composition of rational subsets and finite étale maps, then the completion of (D, D^+) is a perfectoid affinoid K-algebra, with $\mathrm{Spa}(D, D^+) = U \subset X$ (as follows from the definition of the structure sheaf on rational subsets, resp. [30], Theorem 7.9 (iii)). $\qquad\qquad\square$

Let us also note some subtleties related to saying that 'an inverse limit of adic spaces is perfectoid'. For this, we recall a definition from [31] in the case of interest here.

DEFINITION 2.20. Let K be a perfectoid field, and let X_i, $i \in I$, be a cofiltered inverse system of locally noetherian adic spaces over K, with qcqs transition maps. Let X be a perfectoid space over K with a compatible system of maps $X \to X_i$, $i \in I$. Write

$$X \sim \varprojlim_i X_i$$

if $|X| \cong \varprojlim_i |X_i|$ is a homeomorphism, and there exists a cover of X by open perfectoid affinoid $U = \mathrm{Spa}(R, R^+)$ for which the map

$$\varinjlim_{\mathrm{Spa}(R_i, R_i^+) \subset X_i} R_i \to R$$

has dense image, where the direct limit runs over all

$$\mathrm{Spa}(R_i, R_i^+) \subset X_i$$

over which $\mathrm{Spa}(R, R^+) \to X$ factors.

REMARK 2.21. In [31, Definition 2.4.1], this notion is introduced for general adic spaces (in the sense of [31, Definition 2.1.5]). The elementary properties of the following proposition also hold in that generality. It also follows from the arguments of the following proposition that the given definition is equivalent to the one in [31] in this situation: Note that here we require the open affinoid $U = \mathrm{Spa}(R, R^+)$ to come from a perfectoid affinoid K-algebra (R, R^+), which would a priori lead to a stronger notion.

Let us say that a perfectoid affinoid open subset $\mathrm{Spa}(R, R^+) \subset X$ is good if

$$\varinjlim_{\mathrm{Spa}(R_i, R_i^+) \subset X_i} R_i \to R$$

has dense image, and

$$|\operatorname{Spa}(R, R^+)| \cong \varprojlim_{\operatorname{Spa}(R_i, R_i^+) \subset X_i} |\operatorname{Spa}(R_i, R_i^+)|$$

is a homeomorphism.

PROPOSITION 2.22. (i) *If* $X \sim \varprojlim_i X_i$, *then there exists a cover of* X *by perfectoid affinoid open* $U = \operatorname{Spa}(R, R^+) \subset X$ *which are good.*

(ii) *If* $U = \operatorname{Spa}(R, R^+) \subset X$ *is good, then any rational subset of* U *is good.*

PROOF. (i) Take any perfectoid affinoid $U = \operatorname{Spa}(R, R^+) \subset X$ such that

$$\varinjlim_{\operatorname{Spa}(R_i, R_i^+) \subset X_i} R_i \to R$$

has dense image. Clearly,

$$|U| \subset \varprojlim_{\operatorname{Spa}(R_i, R_i^+) \subset X_i} |\operatorname{Spa}(R_i, R_i^+)| (\subset |X|)$$

is open. Therefore, there is some qcqs open subset $U_i \subset \operatorname{Spa}(R_i, R_i^+)$ for some i, such that U is the preimage of U_i. We may cover U_i by rational subsets $U_i' \subset \operatorname{Spa}(R_i, R_i^+)$; let $U' \subset U$ be the preimage, which is a rational subset of U. Then

$$|U'| = \varprojlim_{\operatorname{Spa}(R_i', R_i'^+) \subset X_i} |\operatorname{Spa}(R_i', R_i'^+)| \,,$$

where now the inverse limit runs over all $\operatorname{Spa}(R_i', R_i'^+) \subset X_i$ over which $U' \to X_i$ factors. Moreover, the density statement is preserved (by the definition of the structure presheaf on rational subsets), so U' is good.

(ii) A rational subset of U comes as the preimage of a rational subset of some $\operatorname{Spa}(R_i, R_i^+)$; then use the argument from (i).

$\square$

PROPOSITION 2.23 ([**31**, Proposition 2.4.5]). *In the situation of Definition 2.20, the perfectoid space* X *represents the functor*

$$Y \mapsto \varprojlim_i \operatorname{Hom}(Y, X_i)$$

on perfectoid spaces over K. *In particular, it is unique up to unique isomorphism.*

For the next proposition, one needs a slight variant of Conjecture 2.16.

CONJECTURE 2.24. *Let* K *be some perfectoid field, and* (A, A^+) *a complete affinoid* K-*algebra for which* A^+ *has the p-adic topology. Assume that there is a covering of* $X = \operatorname{Spa}(A, A^+)$ *by rational subsets* $U_i \subset X$ *for which* $\widehat{\mathcal{O}_X(U_i)}$ *is a perfectoid* K-*algebra, where the completion is taken with respect to the topology giving* $\mathcal{O}_X^+(U_i)$ *the p-adic topology. Then* (A, A^+) *is a perfectoid affinoid* K-*algebra.*

REMARK 2.25. It is not clear whether the natural topology on $\mathcal{O}_X^+(U_i)$ is the p-adic topology, even if A^+ has the p-adic topology. If (A, A^+) is perfectoid, this is true.

PROPOSITION 2.26. *Assume Conjecture 2.24. Assume that all* $X_i = \operatorname{Spa}(R_i, R_i^+)$ *are affinoid, and that* $X \sim \varprojlim_i X_i$. *Then* X *is good, i.e.* $X = \operatorname{Spa}(R, R^+)$ *is affinoid perfectoid and* $\varinjlim_i R_i \to R$ *has dense image.*

PROOF. As before: One knows the result after localizing to rational subsets, so Conjecture 2.24 gives the result (using the completion of the direct limit of the (R_i, R_i^+), with the p-adic topology on $\varinjlim R_i^+$, as (A, A^+)). Note that we have to enforce artificially the p-adic topology on A^+, and we have to do the same on rational subsets, which is the reason that we need Conjecture 2.24 in place of Conjecture 2.16. $\qquad\square$

3. p-adic Hodge theory

3.1. Introduction. This introduction is essentially identical to the introduction to [**29**].

In the paper [**29**], we started to investigate to what extent p-adic comparison theorems stay true for rigid-analytic varieties. Up to now, such comparison isomorphisms were mostly studied for schemes over p-adic fields, but we show there that the whole theory extends naturally to rigid-analytic varieties over p-adic fields. This is of course in analogy with classical Hodge theory, which most naturally is formulated in terms of complex-analytic spaces.

Several difficulties have to be overcome to make this work. The first is that finiteness of p-adic étale cohomology is not known for rigid-analytic varieties over p-adic fields. In fact, it is false if one does not make a restriction to the proper case. However, we prove that for proper smooth rigid-analytic varieties, finiteness of p-adic étale cohomology holds.

THEOREM 3.1. *Let C be a complete algebraically closed extension of $\mathbb{Q}_p$, let X/C be a proper smooth rigid-analytic variety, and let $\mathbb{L}$ be an $\mathbb{F}_p$-local system on $X_{\text{ét}}$. Then $H^i(X_{\text{ét}}, \mathbb{L})$ is a finite-dimensional $\mathbb{F}_p$-vector space for all $i \geq 0$, which vanishes for $i > 2\dim X$.*

The properness assumption is crucial here; the smoothness assumption is in fact unnecessary, and an artefact of the proof – using resolution of singularities, one can deduce the result for general proper rigid-analytic varieties, see Theorem 3.17 below. We note that in the smooth case, it would be interesting to prove Poincaré duality.

Let us first explain our proof of this theorem. We build upon Faltings's theory of almost étale extensions, amplified by the theory of perfectoid spaces. One important difficulty in p-adic Hodge theory as compared to classical Hodge theory is that the local structure of rigid-analytic varieties is very complicated; small open subsets still have a large étale fundamental group. We introduce the pro-étale site $X_{\text{proét}}$ whose open subsets are roughly of the form $V \to U \to X$, where $U \to X$ is some étale morphism, and $V \to U$ is an inverse limit of finite étale maps. Then the local structure of X in the pro-étale topology is simpler, namely, it is locally perfectoid. This amounts to extracting lots of p-power roots of units in the tower $V \to U$. We note that the idea to extract many p-power roots is common to all known proofs of comparison theorems in p-adic Hodge theory.

The following result gives further justification to the definition of pro-étale site.

THEOREM 3.2. *Let X be a connected affinoid rigid-analytic variety over C. Then X is a $K(\pi, 1)$ for p-torsion coefficients, i.e. for all p-torsion local systems $\mathbb{L}$ on X, the natural map*

$$H^i_{\text{cont}}(\pi_1(X, x), \mathbb{L}_x) \to H^i(X_{\text{ét}}, \mathbb{L})$$

is an isomorphism. Here, $x \in X(C)$ is a base point, and $\pi_1(X, x)$ denotes the profinite étale fundamental group.

We note that we assume only that X is affinoid; no smallness or nonsingularity hypothesis is necessary for this result. This theorem implies that X is 'locally contractible' in the pro-étale site, at least for p-torsion local systems.

Now, on affinoid perfectoid subsets U, one knows that $H^i(U_{\text{ét}}, \mathcal{O}_X^+/p)$ is almost zero for $i > 0$, where $\mathcal{O}_X^+ \subset \mathcal{O}_X$ is the subsheaf of functions of absolute value ≤ 1 everywhere. This should be seen as the basic finiteness result, and is related to Faltings's almost purity theorem. Starting from this and a suitable cover of X by affinoid perfectoid subsets in $X_{\text{proét}}$, one can deduce that $H^i(X_{\text{ét}}, \mathcal{O}_X^+/p)$ is almost finitely generated over $\mathcal{O}_K$. At this point, one uses that X is proper, and in fact the proof of this finiteness result is inspired by the proof of finiteness of coherent cohomology of proper rigid-analytic varieties, as given by Kiehl, [**20**]. Then one deduces finiteness results for the $\mathbb{F}_p$-cohomology by using a variant of the Artin-Schreier sequence

$$0 \to \mathbb{F}_p \to \mathcal{O}_X^+/p \to \mathcal{O}_X^+/p \to 0 \ .$$

In fact, the proof shows at the same time the following result, which is closely related to §3, Theorem 8, of [**10**].

THEOREM 3.3. *In the situation of Theorem 3.1, there is an almost isomorphism of $\mathcal{O}_C$-modules for all $i \geq 0$,*

$$H^i(X_{\text{ét}}, \mathbb{L}) \otimes \mathcal{O}_C/p \to H^i(X_{\text{ét}}, \mathbb{L} \otimes \mathcal{O}_X^+/p) \ .$$

More generally, assume that $f : X \to Y$ is a proper smooth morphism of rigid-analytic varieties over C, and $\mathbb{L}$ is an $\mathbb{F}_p$-local system on $X_{\text{ét}}$. Then there is an almost isomorphism for all $i \geq 0$,

$$(R^i f_{\text{ét}*}\mathbb{L}) \otimes \mathcal{O}_Y^+/p \to R^i f_{\text{ét}*}(\mathbb{L} \otimes \mathcal{O}_X^+/p) \ .$$

REMARK 3.4. The relative case was already considered in an appendix to [**10**]: Under the assumption that X, Y and f are algebraic and have suitable integral models, this is §6, Theorem 6, of [**10**]. In our approach, it is a direct corollary of the absolute version.

In a sense, this can be regarded as a primitive version of a comparison theorem. It has the very interesting (and apparently paradoxical) feature that on the right-hand side, one has a 'coherent cohomology group modulo p', but the cohomology is computed on the generic fibre. That this group behaves well relies on the fact that perfectoid spaces have a canonical 'almost integral' structure. Moreover, it implies the following strange property of

$$H^i(X_{\text{ét}}, \mathcal{O}_X^+) \ :$$

After inverting p, these groups are usual coherent cohomology $H^i(X_{\text{ét}}, \mathcal{O}_X) = H^i(X_{\text{an}}, \mathcal{O}_X)$, but after modding out p, one gets étale cohomology. Thus, these integral cohomology groups build a bridge between étale and coherent cohomology.

Although it should be possible to deduce (log-)crystalline comparison theorems from here, we did only the de Rham case. For this, we introduced sheaves on $X_{\text{proét}}$, which we call period sheaves, as their values on pro-étale covers of X give period rings. Among them is the sheaf $\mathbb{B}_{\text{dR}}^+$, which is the relative version of Fontaine's ring

B_{dR}^+. Any lisse $\mathbb{Z}_p$-sheaf $\mathbb{L}$ on $X_{\text{ét}}$ gives rise to a $\mathbb{B}_{\mathrm{dR}}^+$-local system $\mathbb{M}$ on $X_{\text{proét}}$, and it is a formal consequence of Theorem 3.3 that

$$(1) \qquad H^i_{\text{ét}}(X, \mathbb{L}) \otimes_{\mathbb{Z}_p} B_{\mathrm{dR}}^+ \cong H^i(X_{\text{proét}}, \mathbb{M}) \ .$$

We want to compare this to de Rham cohomology. For this, we first relate filtered modules with integrable connection to $\mathbb{B}_{\mathrm{dR}}^+$-local systems.

THEOREM 3.5. *Let X be a smooth rigid-analytic variety over k, where k is a complete discretely valued nonarchimedean extension of $\mathbb{Q}_p$ with perfect residue field. Then there is a fully faithful functor from the category of filtered $\mathcal{O}_X$-modules with an integrable connection satisfying Griffiths transversality, to the category of $\mathbb{B}_{\mathrm{dR}}^+$-local systems.*

The proof makes use of the period rings introduced in Brinon's book [**5**], and relies on some of the computations of Galois cohomology groups done there. We say that a lisse $\mathbb{Z}_p$-sheaf $\mathbb{L}$ is de Rham if the associated $\mathbb{B}_{\mathrm{dR}}^+$-local system $\mathbb{M}$ lies in the essential image of this functor. We get the following comparison result.

THEOREM 3.6. *Let k be a discretely valued complete nonarchimedean extension of $\mathbb{Q}_p$ with perfect residue field κ, and algebraic closure $\bar{k}$, and let X be a proper smooth rigid-analytic variety over k. For any lisse $\mathbb{Z}_p$-sheaf $\mathbb{L}$ on $X_{\text{ét}}$ with associated $\mathbb{B}_{\mathrm{dR}}^+$-local system $\mathbb{M}$, we have a $\mathrm{Gal}(\bar{k}/k)$-equivariant isomorphism*

$$H^i_{\text{ét}}(X_{\bar{k}}, \mathbb{L}) \otimes_{\mathbb{Z}_p} B_{\mathrm{dR}}^+ \cong H^i(X_{\bar{k}, \text{proét}}, \mathbb{M}) \ .$$

If $\mathbb{L}$ is de Rham, with associated filtered module with integrable connection $(\mathcal{E}, \nabla, \mathrm{Fil}^\bullet)$, then the Hodge-de Rham spectral sequence

$$H^{i-j,j}_{\mathrm{Hodge}}(X, \mathcal{E}) \Rightarrow H^i_{\mathrm{dR}}(X, \mathcal{E})$$

degenerates. Moreover, $H^i_{\text{ét}}(X_{\bar{k}}, \mathbb{L})$ is a de Rham representation of $\mathrm{Gal}(\bar{k}/k)$ with associated filtered k-vector space $H^i_{\mathrm{dR}}(X, \mathcal{E})$. In particular, there is also a $\mathrm{Gal}(\bar{k}/k)$-equivariant isomorphism

$$H^i_{\text{ét}}(X_{\bar{k}}, \mathbb{L}) \otimes_{\mathbb{Z}_p} \hat{\bar{k}} \cong \bigoplus_j H^{i-j,j}_{\mathrm{Hodge}}(X, \mathcal{E}) \otimes_k \hat{\bar{k}}(-j) \ .$$

REMARK 3.7. We define the Hodge cohomology as the hypercohomology of the associated gradeds of the de Rham complex of $\mathcal{E}$, with the filtration induced from $\mathrm{Fil}^\bullet$.

In particular, we get the following corollary, which answers a question of Tate, [**33**], Remark on p.180.

COROLLARY 3.8. *For any proper smooth rigid-analytic variety X over k, the Hodge-de Rham spectral sequence*

$$H^j(X, \Omega^i_X) \Rightarrow H^{i+j}_{\mathrm{dR}}(X)$$

degenerates, there is a Hodge-Tate decomposition

$$H^i_{\text{ét}}(X_{\bar{k}}, \mathbb{Q}_p) \otimes_{\mathbb{Q}_p} \hat{\bar{k}} \cong \bigoplus_{j=0}^{i} H^{i-j}(X, \Omega^j_X) \otimes_k \hat{\bar{k}}(-j) \ ,$$

and the p-adic étale cohomology $H^i_{\text{ét}}(X_{\bar{k}}, \mathbb{Q}_p)$ is de Rham, with associated filtered k-vector space $H^i_{\mathrm{dR}}(X)$.

Interestingly, no 'Kähler' assumption is necessary for this result in the p-adic case as compared to classical Hodge theory. In particular, one gets degeneration for all proper smooth varieties over fields of characteristic 0 without using Chow's lemma.

Examples of non-algebraic proper smooth rigid-analytic varieties can be constructed by starting from a proper smooth variety in characteristic p, and taking a formal, non-algebraizable, lift to characteristic 0. This can be done for example for abelian varieties or K3 surfaces. More generally, there is the theory of abeloid varieties, which are 'non-algebraic abelian rigid-analytic varieties', roughly, cf. [22].

There are also some non-Kähler compact complex manifolds over $\mathbb{C}$ which have p-adic analogues: Fortunately, only those for which Hodge-de Rham degeneration holds. An example is the Hopf surface. Over a p-adic field, this can be defined as follows. Fix an element $q \in k$ with $|q| < 1$, and let

$$X = (\mathbb{A}^2 \setminus \{(0,0)\})/q^{\mathbb{Z}} \, ,$$

where q acts by diagonal multiplication. It is easy to see that X is proper and smooth. It has $H^0(X, \Omega^1_X) = 0$, $H^1(X, \mathcal{O}_X) = k$ (so Hodge symmetry fails!), and $H^1_{\text{ét}}(X, \mathbb{Z}_\ell) = \mathbb{Z}_\ell$ for any prime number ℓ (including $\ell = p$). In particular, the weight-monodromy conjecture fails badly for this non-algebraic variety. However, one may formulate the following conjecture on independence of ℓ:

CONJECTURE 3.9. *Let X be a proper smooth rigid-analytic variety over a finite extension k of $\mathbb{Q}_p$. Then the Weil-Deligne representation associated to $H^i_{\text{ét}}(X_{\bar{k}}, \mathbb{Q}_\ell)$ is independent of ℓ (including $\ell = p$).*

For $\ell \neq p$, we take the usual recipee, and for $\ell = p$, we use Fontaine's recipee, using that $H^i_{\text{ét}}(X_{\bar{k}}, \mathbb{Q}_\ell)$ is de Rham (and thus potentially semistable).

Theorem 3.6 also has the following consequence, which was conjectured by Schneider, cf. [26], p.633.

COROLLARY 3.10. *Let k be a finite extension of $\mathbb{Q}_p$, let $X = \Omega^n_k$ be Drinfeld's upper half-space, which is the complement of all k-rational hyperplanes in $\mathbb{P}^{n-1}_k$, and let $\Gamma \subset \mathrm{PGL}_n(k)$ be a discrete cocompact subgroup acting without fixed points on Ω^n_k. One gets the quotient $X_\Gamma = X/\Gamma$, which is a proper smooth rigid-analytic variety over k. Let M be a representation of Γ on a finite-dimensional k-vector space, such that M admits a Γ-invariant $\mathcal{O}_k$-lattice. It gives rise to a local system $\mathcal{M}_\Gamma$ of k-vector spaces on X_Γ. Then the twisted Hodge-de Rham spectral sequence*

$$H^j(X_\Gamma, \Omega^i_{X_\Gamma} \otimes \mathcal{M}_\Gamma) \Rightarrow H^{i+j}_{\mathrm{dR}}(X_\Gamma, \mathcal{O}_{X_\Gamma} \otimes \mathcal{M}_\Gamma)$$

degenerates.

The proof of Theorem 3.6 follows the ideas of Andreatta and Iovita, [1], in the crystalline case. One uses a version of the Poincaré lemma, which says here that one has an exact sequence of sheaves over $X_{\text{proét}}$,

$$0 \to \mathbb{B}^+_{\mathrm{dR}} \to \mathcal{O}\mathbb{B}^+_{\mathrm{dR}} \xrightarrow{\nabla} \mathcal{O}\mathbb{B}^+_{\mathrm{dR}} \otimes_{\mathcal{O}_X} \Omega^1_X \xrightarrow{\nabla} \dots \, ,$$

where we use slightly nonstandard notation. In [5] and [1], $\mathbb{B}^+_{\mathrm{dR}}$ would be called $\mathbb{B}^{\nabla+}_{\mathrm{dR}}$, and $\mathcal{O}\mathbb{B}^+_{\mathrm{dR}}$ would be called $\mathbb{B}^+_{\mathrm{dR}}$. This choice of notation is used because many sources do not consider sheaves like $\mathcal{O}\mathbb{B}^+_{\mathrm{dR}}$, and agree with our notation in writing $\mathbb{B}^+_{\mathrm{dR}}$ for the sheaf that is sometimes called $\mathbb{B}^{\nabla+}_{\mathrm{dR}}$.

Given this Poincaré lemma, it only remains to calculate the cohomology of $\mathcal{O}\mathbb{B}^+_{\mathrm{dR}}$, which turns out to be given by coherent cohomology through some explicit calculation. This finishes the proof of Theorem 3.6. We note that this proof is direct: All desired isomorphisms are proved by a direct argument, and not by producing a map between two cohomology theories and then proving that it has to be an isomorphism by abstract arguments. In fact, such arguments would not be available for us, as results like Poincaré duality are not known for the p-adic étale cohomology of rigid-analytic varieties over p-adic fields. It also turns out that our methods are flexible enough to handle the relative case, and our results imply directly the corresponding results for proper smooth algebraic varieties, by suitable GAGA results. This gives for example the following result. We should note that this is the first general relative de Rham comparison result, even in the algebraic case.

THEOREM 3.11. *Let k be a discretely valued complete nonarchimedean extension of $\mathbb{Q}_p$ with perfect residue field κ, and let $f : X \to Y$ be a proper smooth morphism of smooth rigid-analytic varieties over k. Let $\mathbb{L}$ be a lisse $\mathbb{Z}_p$-sheaf on $X_{\text{ét}}$ which is de Rham, with associated filtered module with integrable connection $(\mathcal{E}, \nabla, \mathrm{Fil}^\bullet)$. Assume that $R^i f_{\text{ét}*}\mathbb{L}$ is a lisse $\mathbb{Z}_p$-sheaf on $Y_{\text{ét}}$; this holds true, for example, if the situation comes as the analytification of algebraic objects.*

Then $R^i f_{\text{ét}*}\mathbb{L}$ is de Rham, with associated filtered module with integrable connection given by $R^i f_{\mathrm{dR}*}(\mathcal{E}, \nabla, \mathrm{Fil}^\bullet)$.

We note that we make use of the full strength of the theory of perfectoid spaces. Apart from this, our argument is rather elementary and self-contained, making use of little more than basic rigid-analytic geometry, which we reformulate in terms of adic spaces, and basic almost mathematics. In particular, we work entirely on the generic fibre. This eliminates in particular any assumptions on the reduction type of our variety, and we do not need any version of de Jong's alterations, neither do we need log structures. The introduction of the pro-étale site makes all constructions functorial, and it also eliminates the need to talk about formal projective or formal inductive systems of sheaves, as was done e.g. in [**10**], [**1**]: All period sheaves are honest sheaves on the pro-étale site.

3.2. A comparison result for constructible coefficients. The methods of [**29**] have some consequences that were not included there. First, we explain how to deduce a comparison result in the style of Theorem 3.3 with constructible coefficients.

Let C be a complete algebraically closed extension of $\mathbb{Q}_p$, with a fixed open valuation subring $C^+ \subset C$ (e.g. $C^+ = \mathcal{O}_C$), and let $f : X \to Y$ be a proper map of schemes of finite type over C, with associated adic spaces $f^{\mathrm{ad}} : X^{\mathrm{ad}} \to Y^{\mathrm{ad}}$ over $\mathrm{Spa}(C, C^+)$. Let $\mathbb{L}$ be a constructible $\mathbb{F}_p$-sheaf on X, with pullback $\mathbb{L}^{\mathrm{ad}}$ to X^{ad}. Recall the following result of Huber, [**17**], Theorem 3.7.2.

THEOREM 3.12. *For all $i \geq 0$, $(R^i f_{\text{ét}*}\mathbb{L})^{\mathrm{ad}} \xrightarrow{\cong} R^i f^{\mathrm{ad}}_{\text{ét}*}\mathbb{L}^{\mathrm{ad}}$.*

In this section, we prove the following constructible version of Theorem 3.3.

THEOREM 3.13. *For all $i \geq 0$, the map $(R^i f^{\mathrm{ad}}_{\text{ét}*}\mathbb{L}^{\mathrm{ad}}) \otimes \mathcal{O}^+_{Y^{\mathrm{ad}}}/p \to R^i f^{\mathrm{ad}}_{\text{ét}*}(\mathbb{L}^{\mathrm{ad}} \otimes \mathcal{O}^+_{X^{\mathrm{ad}}}/p)$ is an almost isomorphism.*

We remark that if $Y = \mathrm{Spec}\, C$ (i.e., in the absolute context), this says that

$$H^i(X_{\text{ét}}, \mathbb{L}) \otimes C^+/p \cong H^i(X^{\mathrm{ad}}_{\text{ét}}, \mathbb{L}^{\mathrm{ad}}) \otimes C^+/p \to H^i(X^{\mathrm{ad}}_{\text{ét}}, \mathbb{L}^{\mathrm{ad}} \otimes \mathcal{O}^+_{X^{\mathrm{ad}}}/p)$$

is an almost isomorphism. We need a simple lemma.

Lemma 3.14. *Let X be a locally noetherian adic space over $\mathrm{Spa}(\mathbb{Q}_p, \mathbb{Z}_p)$, with closed subspace $i : Z \hookrightarrow X$.*

(i) The map $i^*\mathcal{O}_X^+/p \to \mathcal{O}_Z^+/p$ is an isomorphism (on Z_{an}, and on $Z_{\mathrm{ét}}$).

(ii) For any $\mathbb{F}_p$-sheaf F on $Z_{\mathrm{ét}}$,

$$i_*F \otimes \mathcal{O}_X^+/p \xrightarrow{\cong} i_*(F \otimes \mathcal{O}_Z^+/p) \ .$$

Proof. (i) Let $\overline{f} \in \mathcal{O}_Z^+/p$ be a section; take a lift $f \in \mathcal{O}_Z^+$; by approximation, we may assume that f is the image of some $g \in \mathcal{O}_X$. The locus $|g| \leq 1$ in X is open; thus, we may assume $g \in \mathcal{O}_X^+$. This shows that $i^*\mathcal{O}_X^+/p \to \mathcal{O}_Z^+/p$ is surjective. Moreover, if $g \in \mathcal{O}_X^+$ that becomes divisible by p on Z, look at the open locus $|g| \leq |p|$ in X: Again, it contains Z, which shows that g becomes 0 in $i^*\mathcal{O}_X^+/p$.

(ii) We check fibres at all geometric points. At fibres outside Z, both are zero. At fibres in Z, it follows from (i), noting that we may ignore i_* then, and that taking fibres commutes with tensor products. $\square$

We recall from Theorem 3.3 that Theorem 3.13 is true when X is smooth and $\mathbb{L}$ is trivial (or just locally constant). From here, we get another base case.

Lemma 3.15. *Assume that X is smooth, and let $D = \bigcup_a D_a \subset X$, $a \in I$, be a simple normal crossings divisor with smooth irreducible components $D_a \subset X$. Let $U = X \setminus D$, with open embedding $j : U \to X$. Then Theorem 3.13 holds true for $\mathbb{L} = j_!\mathbb{F}_p$ and $Y = \mathrm{Spec}\, C$ (i.e., in the absolute context).*

Proof. For $J \subset I$, let $D_J = \bigcap_{a \in J} D_a$, and set $D_\emptyset = X$. Then all D_J are proper and smooth. We have the closed embeddings $i_J : D_J \to X$, and a long exact sequence

$$0 \to j_!\mathbb{F}_p \to \mathbb{F}_p \to \bigoplus_a i_{a*}\mathbb{F}_p \to \cdots \to \bigoplus_{|J|=k} i_{J*}\mathbb{F}_p \to \cdots \to i_{I*}\mathbb{F}_p \to 0$$

of sheaves on $X_{\mathrm{ét}}$. By pullback, we get a similar exact sequence on $X_{\mathrm{ét}}^{\mathrm{ad}}$. Now, we tensor with $\mathcal{O}_{X^{\mathrm{ad}}}^+/p$ and get a similar long exact sequence

$$0 \to j_!\mathbb{F}_p \otimes \mathcal{O}_{X^{\mathrm{ad}}}^+/p \to \mathcal{O}_{X^{\mathrm{ad}}}^+/p \to \bigoplus_a i_{a*}\mathbb{F}_p \otimes \mathcal{O}_{X^{\mathrm{ad}}}^+/p \to \cdots \ .$$

But $i_{J*}\mathbb{F}_p \otimes \mathcal{O}_{X^{\mathrm{ad}}}^+/p = i_{J*}\mathcal{O}_{D_J^{\mathrm{ad}}}^+/p$. Thus, the lemma follows upon applying the result for all D_J, with trivial coefficients. $\square$

Lemma 3.16. *Let $Y = \mathrm{Spec}\, C$, and X proper over Y. Let $j : U \hookrightarrow X$ be a smooth dense open subscheme. Then Theorem 3.13 holds true for $\mathbb{L} = j_!\mathbb{F}_p$.*

Proof. Take a resolution of singularities $f : \tilde{X} \to X$, which is an isomorphism above U, and such that the boundary $D = \tilde{X} \setminus U \subset \tilde{X}$ is a divisor with simple normal crossings. Let $\tilde{j} : U \hookrightarrow \tilde{X}$ denote the lifted embedding. The result follows from the previous lemma, together with the simple observation

$$j_!\mathbb{F}_p \otimes \mathcal{O}_{X^{\mathrm{ad}}}^+/p \cong Rf_{\mathrm{ét}*}^{\mathrm{ad}}(\tilde{j}_!\mathbb{F}_p \otimes \mathcal{O}_{\tilde{X}^{\mathrm{ad}}}^+/p) \ .$$

This may be checked on fibres at points. Over U, it is clear, and outside U, everything vanishes. (Use Proposition 2.6.1 of [**17**] to compute the fibre of the pushforward as the cohomology of the fibre.) $\square$

PROOF. *(of Theorem 3.13.)* As in [**29**], proof of Corollary 5.11, the relative version reduces immediately to the absolute version, so we may assume that $Y = \operatorname{Spec} C$, $Y^{\mathrm{ad}} = \operatorname{Spa}(C, C^+)$. (This reduction step is the reason that we allow general open valuation rings $C^+ \subset C$ from the start.)

We argue by induction on $\dim X$. So assume the statement is known in dimension $< n$, let X be of dimension n, and let $\mathbb{L}$ be some constructible $\mathbb{F}_p$-sheaf on X. We may assume that X is reduced. Then there is a smooth dense open subscheme $j : U \hookrightarrow X$ such that $j^*\mathbb{L}$ is locally constant; let $i : Z \hookrightarrow X$ be the closed complement, so that $\dim Z < n$. We have a short exact sequence

$$0 \to j_! j^*\mathbb{L} \to \mathbb{L} \to i_* i^*\mathbb{L} \to 0 \ ,$$

which gives rise to a short exact sequence

$$0 \to j_! j^*\mathbb{L}^{\mathrm{ad}} \otimes \mathcal{O}^+_{X^{\mathrm{ad}}}/p \to \mathbb{L}^{\mathrm{ad}} \otimes \mathcal{O}^+_{X^{\mathrm{ad}}}/p \to (i_* i^*\mathbb{L}^{\mathrm{ad}}) \otimes \mathcal{O}^+_{X^{\mathrm{ad}}}/p \to 0 \ .$$

But $(i_* i^*\mathbb{L}^{\mathrm{ad}}) \otimes \mathcal{O}^+_{X^{\mathrm{ad}}}/p = i_*(i^*\mathbb{L}^{\mathrm{ad}} \otimes \mathcal{O}^+_{Z^{\mathrm{ad}}}/p)$, so the isomorphism for the right-hand term follows by induction from the assertion for Z and $i^*\mathbb{L}$. Thus, we may assume that there is a smooth dense open subscheme $j : U \hookrightarrow X$ and a locally constant sheaf F on U such that $\mathbb{L} = j_! F$.

The case that F is trivial has been handled in Lemma 3.15. In particular, this handles the case $n = 0$ of relative dimension 0.

In the general case, let $V \to U$ be a finite étale Galois cover over which F becomes trivial, and let $Y \to X$ be the normalization of X in V. Let Y_k be the k-fold fibre product of Y over X, with open subset $V_k \subset Y_k$, which is the k-fold fibre product of V over U. Let $j_k : V_k \to Y_k$ be the open embedding, and $f_k : Y_k \to X$, $g_k : V_k \to U$ the projections. We have a resolution of $j_! F$ on $X_{\text{ét}}$:

$$0 \to j_! F \to f_{1*} j_{1!} g_1^* F \to f_{2*} j_{2!} g_2^* F \to \dots \ .$$

This induces a similar resolution on $X^{\mathrm{ad}}_{\text{ét}}$. Similarly, we have a resolution of $(j_! F \otimes \mathcal{O}^+_{X^{\mathrm{ad}}}/p)^a$ on $X^{\mathrm{ad}}_{\text{ét}}$:

$$0 \to ((j_! F)^{\mathrm{ad}} \otimes \mathcal{O}^+_{X^{\mathrm{ad}}}/p)^a \to f^{\mathrm{ad}}_{1*}((j_{1!} g_1^* F)^{\mathrm{ad}} \otimes \mathcal{O}^+_{Y_1^{\mathrm{ad}}}/p)^a \to f^{\mathrm{ad}}_{2*}((j_{2!} g_2^* F)^{\mathrm{ad}} \otimes \mathcal{O}^+_{Y_2^{\mathrm{ad}}}/p)^a \to \dots :$$

To see this, we may check on fibres. Away from U, everything is 0. At a geometric point $\bar{x} \to U$ with values in (L, L^+), and fibre $Y_{k\bar{x}} \subset Y_k$, the sequence identifies with

$$0 \to F_{\bar{x}} \otimes L^{+a}/p \to H^0(Y_{1\bar{x}}, F_{\bar{x}}) \otimes L^{+a}/p \to H^0(Y_{2\bar{x}}, F_{\bar{x}}) \otimes L^{+a}/p \to \dots$$

(by the result in relative dimension 0). The exactness of this sequence is just the exactness of

$$0 \to j_! F \to f_{1*} j_{1!} g_1^* F \to f_{2*} j_{2!} g_2^* F \to \dots$$

at $\bar{x}$, tensored with L^{+a}/p.

Finally, we get the result by applying it for all $j_{k!} g_k^* F$ on Y_k, noting that $g_k^* F$ is trivial. $\qquad\square$

At this point, let us also remark that by using resolution of singularities for rigid-analytic varieties[3], one gets Theorem 3.1 for non-smooth spaces by the same argument as above.

---

[3]See the paper by Bierstone-Milman [**2**], situation (0.1) (2), where they state that it works for (good) Berkovich spaces. But proper rigid-analytic varieties, proper adic spaces and proper Berkovich spaces are canonically equivalent, and proper Berkovich spaces are good by [**34**], so one may use their result.

THEOREM 3.17. *Let C be an algebraically closed complete extension of $\mathbb{Q}_p$, and let X/C be a proper rigid-analytic variety. Then $H^i(X_{\text{ét}}, \mathbb{F}_p)$ is a finite-dimensional $\mathbb{F}_p$-vector space for all $i \geq 0$, which vanishes for $i > 2\dim X$. Moreover, there is an almost isomorphism*

$$H^i(X_{\text{ét}}, \mathbb{F}_p) \otimes_{\mathbb{F}_p} \mathcal{O}_C/p \to H^i(X_{\text{ét}}, \mathcal{O}_X^+/p)$$

for all $i \geq 0$.

PROOF. We use the same arguments as above. We claim more generally that if $U = X \setminus Z$ is the complement of a Zariski closed subset Z in a proper rigid-analytic variety X, then $H^i(X_{\text{ét}}, j_!\mathbb{F}_p)$ is a finite-dimensional $\mathbb{F}_p$-vector space, which vanishes for $i > 2\dim X$, and with a similar almost isomorphism. By induction on $\dim X$, the result is known for Z, and the result for X is equivalent to the result for U. Given X, we may thus assume that U is smooth. By resolution of singularities, we may then assume that X is smooth, and $Z \subset X$ is a divisor with normal crossings (cf. proof of Lemma 3.16). In that case, one argues as in the proof of Lemma 3.15. $\square$

3.3. The Hodge-Tate spectral sequence. Let C be an algebraically closed complete extension of $\mathbb{Q}_p$, and let X/C be a proper smooth rigid-analytic variety. There is a general Hodge-Tate spectral sequence. It is instructive to compare it to the Hodge-de Rham spectral sequence:

THEOREM 3.18. *There is a Hodge-de Rham spectral sequence*

$$E_1^{ij} = H^j(X, \Omega_X^i) \Rightarrow H_{\text{dR}}^{i+j}(X) .$$

REMARK 3.19. If X is a scheme, or X is defined over a discretely valued subfield $K \subset C$ with perfect residue field, then this sequence degenerates (by the Lefschetz principle, resp. our result stated above). We conjecture that it degenerates in general; however, our methods are not sufficient to prove this.[4] Assuming that it degenerates, one gets a decreasing Hodge-de Rham filtration $\text{Fil}^\bullet H_{\text{dR}}^i(X)$, with

$$\text{Fil}^q H_{\text{dR}}^i(X)/ \text{Fil}^{q+1} H_{\text{dR}}^i(X) = H^{i-q}(X, \Omega_X^q) .$$

THEOREM 3.20. *There is a Hodge-Tate spectral sequence*

$$E_2^{ij} = H^i(X, \Omega_X^j)(-j) \Rightarrow H_{\text{ét}}^{i+j}(X, \mathbb{Q}_p) \otimes_{\mathbb{Q}_p} C .$$

REMARK 3.21. Again, if X is a scheme, or X is defined over a discretely valued subfield $K \subset C$ with perfect residue field (more generally, if $C(-j)^{\text{Gal}(\bar{K}/K)} = 0$ for $j \neq 0$), then this spectral sequence degenerates. For schemes, this follows by a dimension count (noting that by the Lefschetz principle, étale and de Rham cohomology have the same dimension); in the other case, it follows because the differentials are $\text{Gal}(\bar{K}/K)$-equivariant. Note that in the case of schemes, one could also use a spreading-out argument to reduce to the case of a discretely valued subfield $K \subset C$ with perfect residue field; the same argument could be applied in the Hodge-de Rham case. This gives a purely p-adic proof of these results.

In case the Hodge-Tate spectral sequence degenerates, one gets a decreasing Hodge-Tate filtration $\text{Fil}^\bullet(H_{\text{ét}}^i(X, \mathbb{Q}_p) \otimes_{\mathbb{Q}_p} C)$, with

$$\text{Fil}^q(H_{\text{ét}}^i(X, \mathbb{Q}_p) \otimes_{\mathbb{Q}_p} C)/ \text{Fil}^{q+1}(H_{\text{ét}}^i(X, \mathbb{Q}_p) \otimes_{\mathbb{Q}_p} C) = H^q(X, \Omega_X^{i-q})(q - i) .$$

---

[4]O. Gabber explained to us that X is always defined over a rigid space of finite type over $\mathbb{Q}_p$, implying degeneration in general.

REMARK 3.22. The Hodge-Tate spectral sequence does not have a direct analogue over the complex numbers $\mathbb{C}$. Note that in the p-adic case, $H^i_{\text{ét}}(X, \mathbb{Q}_p) \otimes_{\mathbb{Q}_p} C$ is not canonically isomorphic to $H^i_{\text{dR}}(X)$; thus the two spectral sequences do not converge to the same cohomology groups. Over $\mathbb{C}$, the Hodge-de Rham filtration is canonically split; in the p-adic case, however, both the Hodge-de Rham and the Hodge-Tate filtration are not canonically split.

PROOF. *(of Theorem 3.20.)* We use our results from [**29**]. First, Theorem 3.3 says that

$$H^i(X_{\text{ét}}, \mathbb{F}_p) \otimes_{\mathbb{F}_p} \mathcal{O}_C/p \to H^i(X_{\text{ét}}, \mathcal{O}_X^+/p)$$

is an almost isomorphism. By induction on n, we find that

$$H^i(X_{\text{ét}}, \mathbb{Z}/p^n\mathbb{Z}) \otimes_{\mathbb{Z}/p^n\mathbb{Z}} \mathcal{O}_C/p^n \to H^i(X_{\text{ét}}, \mathcal{O}_X^+/p^n)$$

is an almost isomorphism. Passing to the inverse limit over n (cf. [**29**, Lemma 3.18]), we get that

$$H^i(X_{\text{proét}}, \hat{\mathbb{Z}}_p) \otimes_{\mathbb{Z}_p} \mathcal{O}_C \to H^i(X_{\text{proét}}, \hat{\mathcal{O}}_X^+)$$

is an almost isomorphism; also, $H^i(X_{\text{proét}}, \hat{\mathbb{Z}}_p) = H^i_{\text{ét}}(X, \mathbb{Z}_p)$, with the usual definition of the right-hand side. Inverting p, we get an isomorphism

$$H^i_{\text{ét}}(X, \mathbb{Q}_p) \otimes_{\mathbb{Q}_p} C \cong H^i(X_{\text{proét}}, \hat{\mathcal{O}}_X) \ .$$

It remains to compute the right-hand side. For this, we use the projection $\nu : X_{\text{proét}} \to X_{\text{ét}}$, and the spectral sequence

$$E_2^{ij} = H^i(X_{\text{ét}}, R^j\nu_*\hat{\mathcal{O}}_X) \Rightarrow H^{i+j}(X_{\text{proét}}, \hat{\mathcal{O}}_X) = H^{i+j}_{\text{ét}}(X, \mathbb{Q}_p) \otimes_{\mathbb{Q}_p} C \ ;$$

this reduces us to the next proposition. $\square$

PROPOSITION 3.23. *Let X/C be a smooth adic space. Let $\nu : X_{\text{proét}} \to X_{\text{ét}}$ be the projection. Then $\mathcal{O}_{X_{\text{ét}}} \xrightarrow{\cong} \nu_*\hat{\mathcal{O}}_X$, and there is a natural isomorphism $\Omega^1_{X_{\text{ét}}}(-1) \xrightarrow{\cong} R^1\nu_*\hat{\mathcal{O}}_X$; it induces isomorphisms (via taking the exterior power, and cup products)*

$$\Omega^j_{X_{\text{ét}}}(-j) \xrightarrow{\cong} R^j\nu_*\hat{\mathcal{O}}_X$$

for all $j \geq 0$.

PROOF. First, we prove that $\mathcal{E} = R^1\nu_*\hat{\mathcal{O}}_X$ is a locally free $\mathcal{O}_{X_{\text{ét}}}$-module of rank $\dim X$, such that

$$\bigwedge^j \mathcal{E} \cong R^j\nu_*\hat{\mathcal{O}}_X$$

for all $j \geq 0$. This can be checked locally, so we can assume that there is an étale map $X \to \mathbb{T}^n$ that factors as a composite of rational embeddings and finite étale maps, where

$$\mathbb{T}^n = \text{Spa}(C\langle T_1^{\pm 1}, \ldots, T_n^{\pm 1}\rangle, \mathcal{O}_C\langle T_1^{\pm 1}, \ldots, T_n^{\pm 1}\rangle)$$

is the n-dimensional torus. We have the natural pro-finite étale cover

$$\tilde{\mathbb{T}}^n = \text{Spa}(C\langle T_1^{\pm 1/p^\infty}, \ldots, T_n^{\pm 1/p^\infty}, \mathcal{O}_C\langle T_1^{\pm 1/p^\infty}, \ldots, T_n^{1/p^\infty}\rangle) \ .$$

By pullback, it induces a pro-finite étale cover $\tilde{X} \to X$, and one has

$$H^i(X_{\text{proét}}, \hat{\mathcal{O}}_X) = H^i_{\text{cont}}(\mathbb{Z}_p^n, \mathcal{O}_{\tilde{X}}(\tilde{X})) \ .$$

Thus, the statement follows from [**29**, Lemma 4.5, Lemma 5.5]: First, Lemma 4.5 shows that

$$\mathcal{O}_{\tilde{X}}(\tilde{X}) = \mathcal{O}_X(X)\hat{\otimes}_{C\langle T_1^{\pm 1},\ldots,T_n^{\pm 1}\rangle} C\langle T_1^{\pm 1/p^\infty},\ldots,T_n^{\pm 1/p^\infty}\rangle \,,$$

and then Lemma 5.5 computes this group.

In particular, we find that $\mathcal{O}_{X_{\text{ét}}} \xrightarrow{\cong} \nu_*\hat{\mathcal{O}}_X$. It remains to prove that $\mathcal{E} \cong \Omega^1_{X_{\text{ét}}}(-1)$. For this, we prove the following lemma. $\qquad\square$

LEMMA 3.24. *Consider the exact sequence*

$$0 \to \hat{\mathbb{Z}}_p(1) \to \varprojlim_{\times p} \mathcal{O}_X^\times \to \mathcal{O}_X^\times \to 0$$

on $X_{\text{proét}}$, where $\hat{\mathbb{Z}}_p(1) = \varprojlim \mu_{p^n}$. It induces a boundary map

$$\mathcal{O}_{X_{\text{ét}}}^\times = \nu_*\mathcal{O}_X^\times \to R^1\nu_*\hat{\mathbb{Z}}_p(1) \,.$$

There is a unique $\mathcal{O}_{X_{\text{ét}}}$-linear map $\Omega^1_{X_{\text{ét}}} \to \mathcal{E} = R^1\nu_*\hat{\mathcal{O}}_X(1)$ such that the diagram

$$
\begin{array}{ccc}
\mathcal{O}_{X_{\text{ét}}}^\times & \longrightarrow & R^1\nu_*\hat{\mathbb{Z}}_p(1) \\
\downarrow{\scriptstyle d\log} & & \downarrow \\
\Omega^1_{X_{\text{ét}}} & \longrightarrow & R^1\nu_*\hat{\mathcal{O}}_X(1)
\end{array}
$$

commutes. This map is an isomorphism.

PROOF. The assertion is local, so we can again assume that $X \to \mathbb{T}^n$ is a composition of finite étale maps and rational subsets; in particular, $X = \operatorname{Spa}(R, R^\circ)$ is affinoid. Let $T_1,\ldots,T_n$ be the coordinates on $\mathbb{T}^n$; then $d\log(T_i) \in \Omega^1_{X_{\text{ét}}}$ are an $\mathcal{O}_{X_{\text{ét}}}$-basis. Their images are prescribed uniquely by the requirement of the lemma, thus the map is unique if it exists. Moreover, the local computation shows that this map is an isomorphism.

It remains to prove existence. For this, we have to check that the diagram

$$
\begin{array}{ccc}
R^\times = H^0(X_{\text{proét}}, \mathcal{O}_X^\times) & \longrightarrow & H^1(X_{\text{proét}}, \hat{\mathbb{Z}}_p(1)) \\
\downarrow{\scriptstyle d\log} & & \downarrow \\
\Omega^1_{R/C} = H^0(X_{\text{proét}}, \Omega^1_X) & \longrightarrow & H^1(X_{\text{proét}}, \hat{\mathcal{O}}_X(1))
\end{array}
$$

commutes, where the lower map is defined to be the one pinned down by the $d\log(T_i)$. First, by the local computation, $H^1(X_{\text{proét}}, \hat{\mathcal{O}}_X(1))$ is a finite free R-module. In particular, it carries a natural topology. We claim that via both maps, the image of $R^{\circ\times}$ in $H^1(X_{\text{proét}}, \hat{\mathcal{O}}_X(1))$ is bounded (the commutativity of the diagram implies a posteriori the fact that the image of $R^\times$ is bounded). For the map over the upper right corner, this follows by observing that it factors over $H^1(X_{\text{proét}}, \hat{\mathcal{O}}_X^+(1))$. For the map over the lower left corner, observe that on $R^{\circ\times}$, $d\log$ factors over $\Omega^1_{R^\circ/\mathcal{O}_C}$, where the latter is an R°-module of finite type (as R° is topologically finitely generated by [**3**]).

Now write $C = \varinjlim A_i$ as the filtered direct limit of $\mathbb{Q}_p$-algebras A_i of finite type. By resolution of singularities, we may assume that all A_i are smooth. We may thus find formally smooth $\mathbb{Q}_p$-algebras B_i topologically of finite type with $C = \widehat{\varinjlim B_i}$.

We may also assume that there are étale maps $Y_i = \mathrm{Spa}(B_i, B_i^\circ) \to \mathbb{T}^{n_i}$ for some n_i. For i large enough, X comes as base extension from $X_i \to Y_i$ (use that finite étale covers and rational subsets on an inverse limits of adic spaces come from a finite level). Let $X_i = \mathrm{Spa}(R_i, R_i^\circ)$, where now R_i is topologically of finite type over $\mathbb{Q}_p$. Moreover, given any $f \in R^\times$ and $m \geq 1$, one can find an i and some $f_i \in R_i^\times$ such that $\frac{f}{f_i} \in 1 + p^m R^\circ$. By the boundedness statement on the image of $R^{\circ\times}$, and because elements of $1 + p^m R^\circ$ are p^{m-2}-th powers of elements of $1 + p^2 R^\circ \subset R^{\circ\times}$ (say $m \geq 2$), it suffices to prove the statement for $f = f_i$.

Now we recall Faltings's extension from [**29**, Corollary 6.14]; this is an exact sequence

$$0 \to \hat{\mathcal{O}}_{X_i}(1) \to \mathcal{F}_i \to \hat{\mathcal{O}}_{X_i} \otimes_{\mathcal{O}_{X_i}} \Omega^1_{X_i} \to 0$$

of $\hat{\mathcal{O}}_{X_i}$-modules on $(X_i)_{\mathrm{pro\acute{e}t}}$. Moreover, there is an exact commutative diagram

$$
\begin{array}{ccccccccc}
0 & \longrightarrow & \hat{\mathbb{Z}}_p(1) & \longrightarrow & \varprojlim_{\times p} \mathcal{O}^\times_{X_i} & \longrightarrow & \mathcal{O}^\times_{X_i} & \longrightarrow & 0 \\
& & \downarrow & & \downarrow & & \downarrow{\scriptstyle d\log} & & \\
0 & \longrightarrow & \hat{\mathcal{O}}_{X_i}(1) & \longrightarrow & \mathcal{F}_i & \longrightarrow & \hat{\mathcal{O}}_{X_i} \otimes_{\mathcal{O}_{X_i}} \Omega^1_{X_i} & \longrightarrow & 0
\end{array}
$$

on $(X_i)_{\mathrm{pro\acute{e}t}}$. Here, an element of $\varprojlim_{\times p} \mathcal{O}^\times_{X_i}$ gives an element $U \in \hat{\mathcal{O}}^{\flat\times}_{X_i}$, and an element $V \in \mathcal{O}^\times_{X_i}$; then $\log(V/[U]) \in \mathcal{O}\mathbb{B}^+_{\mathrm{dR},X_i}$ lies in the kernel of Θ, and defines an element of $\mathcal{F}_i$; this gives the middle vertical map. Commutativity is a direct check; note that $d[U] = 0$. Applying g_i^* (where $g_i : X_{\mathrm{pro\acute{e}t}} \to (X_i)_{\mathrm{pro\acute{e}t}}$) to the diagram and tensoring the lower sequence with $\hat{\mathcal{O}}_X$ over $g_i^* \hat{\mathcal{O}}_{X_i}$ gives an exact commutative diagram

$$
\begin{array}{ccccccccc}
0 & \longrightarrow & g_i^* \hat{\mathbb{Z}}_p(1) & \longrightarrow & g_i^* \varprojlim_{\times p} \mathcal{O}^\times_{X_i} & \longrightarrow & g_i^* \mathcal{O}^\times_{X_i} & \longrightarrow & 0 \\
& & \downarrow & & \downarrow & & \downarrow & & \\
0 & \longrightarrow & \hat{\mathcal{O}}_X(1) & \longrightarrow & \mathcal{F}_i' & \longrightarrow & \hat{\mathcal{O}}_X \otimes_{\mathcal{O}_X} (\Omega^1_X)' & \longrightarrow & 0
\end{array}
$$

on $X_{\mathrm{pro\acute{e}t}}$, where $\mathcal{F}_i'$ is some sheaf of $\hat{\mathcal{O}}_X$-modules, and $(\Omega^1_X)' = g_i^* \Omega^1_{X_i} \otimes_{g_i^* \mathcal{O}_{X_i}} \mathcal{O}_X$ is a free $\mathcal{O}_X$-module with a projection $(\Omega^1_X)' \to \Omega^1_X$. The kernel comes from $\Omega^1_{Y_i}$, and is generated by the image of $\mathcal{O}^\times_{Y_i}$ via $d\log$ (using that Y_i is étale over $\mathbb{T}^{n_i}$). Now look at the associated commutative diagram of boundary maps:

$$
\begin{array}{ccc}
H^0(X_{\mathrm{pro\acute{e}t}}, g_i^* \mathcal{O}^\times_{X_i}) & \longrightarrow & H^1(X_{\mathrm{pro\acute{e}t}}, g_i^* \hat{\mathbb{Z}}_p(1)) \\
\downarrow & & \downarrow \\
H^0(X_{\mathrm{pro\acute{e}t}}, \hat{\mathcal{O}}_X \otimes_{\mathcal{O}_X} (\Omega^1_X)') & \longrightarrow & H^1(X_{\mathrm{pro\acute{e}t}}, \hat{\mathcal{O}}_X(1))
\end{array}
$$

Note also that

$$
\begin{array}{ccc}
H^0(X_{\mathrm{pro\acute{e}t}}, g_i^* \mathcal{O}^\times_{X_i}) & \longrightarrow & H^1(X_{\mathrm{pro\acute{e}t}}, g_i^* \hat{\mathbb{Z}}_p(1)) \\
\downarrow & & \downarrow \\
R^\times & \longrightarrow & H^1(X_{\mathrm{pro\acute{e}t}}, \hat{\mathbb{Z}}_p(1))
\end{array}
$$

commutes, as one can relate the two exact sequences defining the boundary maps. Using the two diagrams together, one finds that the map

$$H^0(X_{\text{proét}}, \hat{\mathcal{O}}_X \otimes_{\mathcal{O}_X} (\Omega^1_X)') \to H^1(X_{\text{proét}}, \hat{\mathcal{O}}_X(1))$$

defined by the $\hat{\mathcal{O}}_X$-linear extension $\mathcal{F}'_i$ is $\hat{\mathcal{O}}_X$-linear, and maps $d\log(T_i)$ to the correct elements. Moreover, it factors over

$$\Omega^1_{R/C} = H^0(X_{\text{proét}}, \hat{\mathcal{O}}_X \otimes_{\mathcal{O}_X} \Omega^1_X) \ :$$

For this, we have to check that it kills all $d\log(h)$, where h is one of the coordinates of $\mathbb{T}^{n_i}$, pulled back to Y_i. But then h gives an element of C, and it has in C a sequence of p-power roots; it follows that h lifts to $H^0(X_{\text{proét}}, g_i^* \varprojlim_{\times p} \mathcal{O}^\times_{X_i})$, so that it vanishes under the boundary map.

It follows that

$$H^0(X_{\text{proét}}, \hat{\mathcal{O}}_X \otimes_{\mathcal{O}_X} (\Omega^1_X)') \to H^1(X_{\text{proét}}, \hat{\mathcal{O}}_X(1))$$

is the composite of the projection

$$H^0(X_{\text{proét}}, \hat{\mathcal{O}}_X \otimes_{\mathcal{O}_X} (\Omega^1_X)') \to \Omega^1_{R/C} = H^0(X_{\text{proét}}, \hat{\mathcal{O}}_X \otimes_{\mathcal{O}_X} \Omega^1_X)$$

with the map

$$\Omega^1_{R/C} \to H^1(X_{\text{proét}}, \hat{\mathcal{O}}_X(1))$$

pinned down by the $d\log(T_i)$. Now, use the commutativity of the diagrams for $f_i \in H^0(X_{\text{proét}}, g_i^* \mathcal{O}^\times_{X_i})$ to conclude. $\qquad\square$

4. A p-adic analogue of Riemann's classification of complex abelian varieties

In this section, we explain a result on p-divisible groups over $\mathcal{O}_C$, where C is an algebraically closed complete extension of $\mathbb{Q}_p$, proved in joint work with J. Weinstein, [**31**]. Although in [**31**], perfectoid spaces are used at several points, the material of this section is independent of the theory of perfectoid spaces, except for Subsection 4.2, which is however not needed to state and prove the main result discussed below. Thus, the reader may prefer to skip Subsection 4.2, except for reading the statement of Fargues's theorem 4.13.

4.1. Riemann's theorem. Let us first recall the classical theory over the complex numbers $\mathbb{C}$.

DEFINITION 4.1. A complex torus is a connected compact complex Lie group T.

LEMMA 4.2. *A complex torus T is commutative.*

PROOF. Let $\mathcal{O}_{T,0}$ be the local ring at 0, with maximal ideal $\mathfrak{m}$. The adjoint action

$$T \to \mathrm{GL}(\mathcal{O}_{T,0}/\mathfrak{m}^n)$$

is trivial, as T is compact and connected, and the right-hand side is affine. $\qquad\square$

Let $\mathfrak{t}$ be the Lie algebra of T; it is a finite-dimensional $\mathbb{C}$-vector space, $\mathfrak{t} \cong \mathbb{C}^g$. The exponential map

$$\exp : \mathfrak{t} \to T$$

makes $\mathfrak{t}$ the universal covering of T, and if we let $\Lambda = \ker(\exp) \subset \mathfrak{t}$, then $T = \mathfrak{t}/\Lambda$. Thus, $\Lambda \subset \mathfrak{t}$ is a discrete cocompact subgroup, i.e. a lattice $\Lambda \cong \mathbb{Z}^{2g}$. Thus, we arrive at the classification of complex tori.

PROPOSITION 4.3. *The category of complex tori is equivalent to the category of pairs $(\mathfrak{t}, \Lambda)$, where $\mathfrak{t}$ is a finite-dimensional $\mathbb{C}$-vector space, and $\Lambda \subset \mathfrak{t}$ is a lattice.*

One can reformulate this classification in terms of Hodge structures.

DEFINITION 4.4. (i) A $\mathbb{Z}$-Hodge structure of weight -1 is a finite free $\mathbb{Z}$-module Λ together with a $\mathbb{C}$-subvectorspace $V \subset \Lambda \otimes_{\mathbb{Z}} \mathbb{C}$, such that $V \oplus \bar{V} \xrightarrow{\cong} \Lambda \otimes_{\mathbb{Z}} \mathbb{C}$.

(ii) A polarization on a $\mathbb{Z}$-Hodge structure (Λ, V) of weight -1 is an alternating form

$$\psi : \Lambda \otimes \Lambda \to 2\pi i \mathbb{Z}$$

such that $\psi(x, Cy)$ is a symmetric positive definite form on $\Lambda \otimes_{\mathbb{Z}} \mathbb{R}$, where C is Weil's operator on $\Lambda \otimes_{\mathbb{Z}} \mathbb{C} \cong V \oplus \bar{V}$, acting as i on V, and as $-i$ on $\bar{V}$.

We note that it follows that complex tori are equivalent to $\mathbb{Z}$-Hodge structures of weight -1, via mapping $(\mathfrak{t}, \Lambda)$ to (Λ, V), with $V = \ker(\Lambda \otimes \mathbb{C} \to \mathfrak{t})$.

DEFINITION 4.5. A complex abelian variety is a projective complex torus.

THEOREM 4.6 (Riemann). *The category of complex abelian varieties is equivalent to the category of polarizable $\mathbb{Z}$-Hodge structures of weight -1.*

Let us call this theorem, stating an abstract equivalence between some geometric objects (abelian varieties) with some Hodge-theoretic data, the 'Hodge-theoretic perspective'. In this case over $\mathbb{C}$, we have seen that this equivalence has a very direct geometric meaning, which we call 'the geometric perspective': All complex tori of dimension g have the same universal cover $\mathbb{C}^g$, and thus are of the form $\mathbb{C}^g/\Lambda$ for a lattice $\Lambda \subset \mathbb{C}^g$. When can one form the quotient $\mathbb{C}^g/\Lambda$? Always as a complex manifold, sometimes (as determined by Riemann) as an algebraic variety.

4.2. The Hodge-Tate sequence for abelian varieties and p-divisible groups. Let C be an algebraically closed complete extension of $\mathbb{Q}_p$. Let X/C be a proper smooth scheme. We recall the following results.

THEOREM 4.7. *There is a Hodge-de Rham spectral sequence*

$$E_1^{ij} = H^j(X, \Omega_X^i) \Rightarrow H_{\mathrm{dR}}^{i+j}(X) \ .$$

It degenerates at E_1. One gets a decreasing Hodge-de Rham filtration $\mathrm{Fil}^\bullet H_{\mathrm{dR}}^i(X)$, with

$$\mathrm{Fil}^q H_{\mathrm{dR}}^i(X) / \mathrm{Fil}^{q+1} H_{\mathrm{dR}}^i(X) = H^{i-q}(X, \Omega_X^q) \ .$$

THEOREM 4.8. *There is a Hodge-Tate spectral sequence*

$$E_2^{ij} = H^i(X, \Omega_X^j)(-j) \Rightarrow H_{\text{ét}}^{i+j}(X, \mathbb{Q}_p) \otimes_{\mathbb{Q}_p} C \ .$$

It degenerates at E_2. One gets a decreasing Hodge-Tate filtration $\mathrm{Fil}^\bullet(H_{\text{ét}}^i(X, \mathbb{Q}_p) \otimes_{\mathbb{Q}_p} C)$, with

$$\mathrm{Fil}^q(H_{\text{ét}}^i(X, \mathbb{Q}_p) \otimes_{\mathbb{Q}_p} C) / \mathrm{Fil}^{q+1}(H_{\text{ét}}^i(X, \mathbb{Q}_p) \otimes_{\mathbb{Q}_p} C) = H^q(X, \Omega_X^{i-q})(q - i) \ .$$

EXAMPLE 4.9. Let A/C be an abelian variety, with universal vector extension $EA \to A$, and p-adic Tate module Λ. Recall that $\operatorname{Lie} EA$ is dual to $H^1_{\mathrm{dR}}(A)$, and Λ is dual to $H^1_{\mathrm{ét}}(A, \mathbb{Z}_p)$. One has two short exact sequences (where A^* is the dual abelian variety)

$$0 \to (\operatorname{Lie} A^*)^* \to \operatorname{Lie} EA \to \operatorname{Lie} A \to 0 \ ,$$

$$0 \to (\operatorname{Lie} A)(1) \to \Lambda \otimes_{\mathbb{Z}_p} C \to (\operatorname{Lie} A^*)^* \to 0 \ .$$

One has the following compatibility of the Hodge-Tate sequence with duality.

PROPOSITION 4.10. *Let A/C be an abelian variety, and let A^*/C be the dual abelian variety. Identify $H^0(A, \Omega^1_A) = (\operatorname{Lie} A)^*$, $H^1(A, \mathcal{O}_A) = \operatorname{Lie} A^*$, and similarly for A^*. The two sequences*

$$0 \to \operatorname{Lie} A^* \to H^1_{\mathrm{ét}}(A, \mathbb{Z}_p) \otimes_{\mathbb{Z}_p} C \to (\operatorname{Lie} A)^*(-1) \to 0 \ ,$$

$$0 \to \operatorname{Lie} A \to H^1_{\mathrm{ét}}(A^*, \mathbb{Z}_p) \otimes_{\mathbb{Z}_p} C \to (\operatorname{Lie} A^*)^*(-1) \to 0$$

are dual to each other under the Weil pairing

$$H^1_{\mathrm{ét}}(A, \mathbb{Z}_p) \otimes_{\mathbb{Z}_p} H^1_{\mathrm{ét}}(A^*, \mathbb{Z}_p) \to \mathbb{Z}_p(-1) \ .$$

PROOF. Let $\mathfrak{P}$ be the Poincaré bundle on $A \times A^*$. Unraveling definitions, the proposition follows from the following compatibility for the first Chern class $c_1(\mathfrak{P})$ on $A \times A^*$: $c_1^{\mathrm{dR}}(\mathfrak{P})$ encodes the duality between $H^1(A, \mathcal{O}_A)$ and $H^0(A^*, \Omega^1_{A^*})$, whereas $c_1^{\mathrm{ét}}(\mathfrak{P})$ encodes the Weil pairing. $\square$

LEMMA 4.11. *Let X/C be a proper smooth scheme, and $\mathcal{L}$ a line bundle on X; regard $\mathcal{L}$ as an element of $H^1(X, \mathcal{O}_X^\times)$. Define*

$$c_1^{\mathrm{dR}}(\mathcal{L}) \in H^1(X, \Omega^1_X)$$

as the image of $\mathcal{L} \in H^1(X, \mathcal{O}_X^\times)$ under $d\log : \mathcal{O}_X^\times \to \Omega^1_X$. Define

$$c_1^{\mathrm{ét}}(\mathcal{L}) \in H^2(X_{\mathrm{proét}}, \hat{\mathbb{Z}}_p(1)) = H^2_{\mathrm{ét}}(X, \mathbb{Z}_p)(1)$$

as the image of $\mathcal{L} \in H^1(X, \mathcal{O}_X^\times) = H^1(X_{\mathrm{proét}}, \mathcal{O}_X^\times)$ under the boundary map associated to the short exact sequence

$$0 \to \hat{\mathbb{Z}}_p(1) \to \varprojlim_{\times p} \mathcal{O}_X^\times \to \mathcal{O}_X^\times \to 0$$

on $X_{\mathrm{proét}}$. Then $c_1^{\mathrm{ét}}(\mathcal{L}) \otimes 1 \in H^2_{\mathrm{ét}}(X, \mathbb{Z}_p) \otimes_{\mathbb{Z}_p} C(1)$ lies in $\mathrm{Fil}^1(H^2_{\mathrm{ét}}(X, \mathbb{Z}_p) \otimes_{\mathbb{Z}_p} C)(1)$, and maps to

$$c_1^{\mathrm{dR}}(\mathcal{L}) \in H^1(X, \Omega^1_X) = \mathrm{gr}^1(H^2_{\mathrm{ét}}(X, \mathbb{Z}_p) \otimes_{\mathbb{Z}_p} C)(1) \ .$$

REMARK 4.12. The statement is true more generally for X/C a proper smooth rigid-analytic variety for which the Hodge-Tate spectral sequence degenerates.

PROOF. One has a map $\mathcal{O}_X^\times \to \hat{\mathbb{Z}}_p(1)[1] \to \hat{\mathcal{O}}_X(1)[1]$ in the derived category of sheaves on $X_{\mathrm{proét}}$. The associated map of Leray spectral sequences shows that $c_1^{\mathrm{ét}}(\mathcal{L})$ lies in

$$\mathrm{Fil}^1(H^2_{\mathrm{ét}}(X, \mathbb{Z}_p) \otimes_{\mathbb{Z}_p} C)(1) = \mathrm{Fil}^1(H^2(X_{\mathrm{proét}}, \hat{\mathcal{O}}_X(1))) \ ;$$

note that $R\nu_* \mathcal{O}_X^\times = \mathcal{O}_{X_{\text{ét}}}^\times$ by [**29**, Corollary 3.17 (i)]. The final statement follows from the commutative diagram

$$
\begin{array}{ccc}
\mathcal{O}_{X_{\text{ét}}}^\times & \longrightarrow & R^1\nu_* \hat{\mathbb{Z}}_p(1) \\
\downarrow {\scriptstyle d\log} & & \downarrow \\
\Omega^1_{X_{\text{ét}}} & \overset{\cong}{\longrightarrow} & R^1\nu_* \hat{\mathcal{O}}_X(1)
\end{array}
$$

upon applying $H^1(X_{\text{ét}}, -)$. $\square$

Now assume that A has good reduction, i.e. we have an abelian variety $A/\mathcal{O}_C$, where $\mathcal{O}_C \subset C$ is the ring of integers. Then we can describe everything in terms of the p-divisible group $G = A[p^\infty]$. Indeed, we have the universal vector extension $EG \to G$, and the p-adic Tate module Λ of G. One has the short exact sequence of finite free $\mathcal{O}_C$-modules

$$0 \to (\operatorname{Lie} G^*)^* \to \operatorname{Lie} EG \to \operatorname{Lie} G \to 0 \ ,$$

where G^* denotes the Serre dual p-divisible group.

THEOREM 4.13 (Fargues, [**14**, Ch.2,App.C]). *There is a complex of finite free* $\mathcal{O}_C$*-modules*

$$0 \to (\operatorname{Lie} G)(1) \overset{\alpha_{G^*}^*(1)}{\longrightarrow} \Lambda \otimes_{\mathbb{Z}_p} \mathcal{O}_C \overset{\alpha_G}{\longrightarrow} (\operatorname{Lie} G^*)^* \to 0 \ .$$

Its cohomology groups are killed by $p^{1/(p-1)}$.

REMARK 4.14. Fargues uses the following direct definition of α_G. Take any $\lambda \in \Lambda$. Thus,

$$\lambda \in \varprojlim_n G[p^n](C) = \varprojlim_n G[p^n](\mathcal{O}_C) = \operatorname{Hom}_{\mathcal{O}_C}(\mathbb{Q}_p/\mathbb{Z}_p, G) \ .$$

Thus, by duality, one gets a map $G^* \to \mu_{p^\infty}$, which on Lie algebras gives map $\operatorname{Lie} G^* \to \operatorname{Lie} \mu_{p^\infty} \cong \mathcal{O}_C$; i.e. we get an element of $(\operatorname{Lie} G^*)^*$.

PROPOSITION 4.15. *Let* $A/\mathcal{O}_C$ *be an abelian variety,* $G = A[p^\infty]$, *with Tate module* Λ. *The two Hodge-Tate sequences (the first from Theorem 3.20, the second from Fargues's theorem)*

$$0 \to \operatorname{Lie} A^* \otimes_{\mathcal{O}_C} C \to H^1_{\text{ét}}(A_C, \mathbb{Z}_p) \otimes_{\mathbb{Z}_p} C \to (\operatorname{Lie} A)^* \otimes_{\mathcal{O}_C} C(-1) \to 0 \ ,$$

$$0 \to \operatorname{Lie} G \otimes_{\mathcal{O}_C} C(1) \to \Lambda \otimes_{\mathbb{Z}_p} C \to (\operatorname{Lie} G^*)^* \otimes_{\mathcal{O}_C} C \to 0$$

are dual to each other.

PROOF. By definition, the second exact sequence is compatible with duality $G \mapsto G^*$. By Proposition 4.10, the first exact sequence is compatible with duality $A \mapsto A^*$; as $A^*[p^\infty] = G^*$, we are reduced to checking that

$$H^1_{\text{ét}}(A_C, \mathbb{Z}_p) \otimes_{\mathbb{Z}_p} C(1) \to (\operatorname{Lie} A)^* \otimes_{\mathcal{O}_C} C$$

agrees with

$$\Lambda^* \otimes_{\mathbb{Z}_p} C(1) \to (\operatorname{Lie} G)^* \otimes_{\mathcal{O}_C} C \ .$$

Recall that the first map is defined as the composite

$$H^1_{\text{ét}}(A_C, \mathbb{Z}_p) \otimes_{\mathbb{Z}_p} C(1) \cong H^1(A_{C,\text{proét}}, \hat{\mathcal{O}}_A)(1)$$

$$\to H^0(A_{C,\text{ét}}, R^1\nu_* \hat{\mathcal{O}}_A)(1) \cong H^0(A_{C,\text{ét}}, \Omega^1_A) = (\operatorname{Lie} A)^* \otimes_{\mathcal{O}_C} C \ .$$

Consider G as a formal scheme over $\operatorname{Spf}\mathcal{O}_C$, and let G_η be its generic fibre. Then $G_\eta \subset A_C$ is a rigid-analytic open subset. Fix an element $\lambda \in \Lambda^*(1)$; it corresponds to a map $\mathbb{Q}_p/\mathbb{Z}_p \to G^*$ of p-divisible groups over $\mathcal{O}_C$, thus to a morphism $G \to G' = \mu_{p^\infty}$ over $\mathcal{O}_C$. One generic fibres, it induces a morphism $G_\eta \to G'_\eta$. Note that the latter is just an open unit ball. We get a commutative diagram

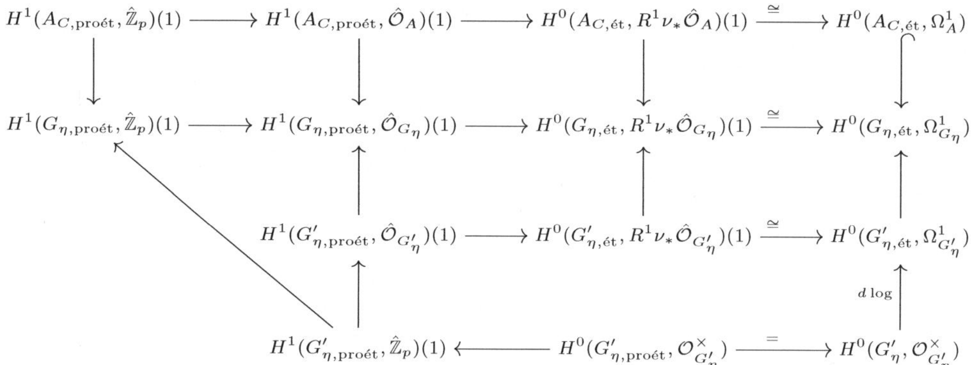

Identify $G' = \mu_{p^\infty} = \operatorname{Spf}\mathbb{Z}_p[[T]]$ in the usual way; now look at what happens to the element $1 + T$ as an element of the lower-right group. Under $d\log$, it maps to the standard basis element of $\operatorname{Lie}G' \otimes_{\mathcal{O}_C} C$, thus the image in

$$H^0(G_{\eta,\text{ét}}, \Omega^1_{G_\eta}) = (\operatorname{Lie}G)^* \otimes_{\mathcal{O}_C} H^0(G_{\eta,\text{ét}}, \mathcal{O}_{G_{\eta,\text{ét}}})$$

is given by $\alpha_{G^*}(\lambda) \otimes 1$.

On the other hand, mapping $1 + T$ around the lower-left hand corner, we use Kummer theory first. Note that the $H^1(-, \hat{\mathbb{Z}}_p)(1)$-groups on the left classify $\mathbb{Z}_p(1)$-covers. Extracting a sequence of p-power roots of $1 + T$ amounts exactly to the tower corresponding to multiplication by p-powers on G'_η. On the other hand, $\lambda \in \Lambda^*(1) = H^1(A_{C,\text{proét}}, \hat{\mathbb{Z}}_p)(1)$ gives a similar tower over A_C; the two towers become equal after restriction to G_η. Now the result follows from the commutativity of the diagram, and the injectivity of the vertical upper right map. $\qquad\square$

4.3. A p-adic analogue: The Hodge-theoretic perspective. The preceding discussion gives a functor from p-divisible groups over $\mathcal{O}_C$ to pairs (Λ, W), where Λ is a finite free $\mathbb{Z}_p$-module, and $W \subset \Lambda \otimes_{\mathbb{Z}_p} C$ is a C-subvectorspace.

THEOREM 4.16 ([**31**, Theorem 5.2.1]). *This functor is an equivalence of categories.*

Let us make a series of remarks.

REMARK 4.17. (i) In an unpublished manuscript, Fargues ([**12**]) had previously proved fully faithfulness, and more.

(ii) This is the first instance of a classification of p-divisible groups in terms of linear algebra (instead of σ-linear algebra, as usual in Dieudonné theory).

(iii) Let κ be the residue field of $\mathcal{O}_C$. Then, by reduction to the special fibre, one has a functor from p-divisible groups over $\mathcal{O}_C$ to p-divisible groups over κ. Recall that the latter are classified by Dieudonné modules (M, F, V), where M is a finite free $W(\kappa)$-module, $F : M \to M$ is a σ-linear map, and $V : M \to M$ is a σ^{-1}-linear map, such that $FV = VF = p$. Here, $\sigma : W(\kappa) \to W(\kappa)$ is the lift of Frobenius on κ.

Using these equivalences of categories, one gets a functor $(\Lambda, W) \mapsto (M, F, V)$, which the author does not know how to describe. Describing this functor amounts to an integral comparison between the étale and crystalline cohomology of p-divisible groups. However, in the current situation over C, there is no Galois action on the Tate module, which would usually be used in Fontaine's theory.[5]

(iv) If $0 \to G_1 \to G_2 \to G_3 \to 0$ is an exact sequence of p-divisible groups over $\mathcal{O}_C$, then the corresponding sequences $0 \to W_1 \to W_2 \to W_3 \to 0$ and $0 \to \Lambda_1 \to \Lambda_2 \to \Lambda_3 \to 0$ are exact. However, the converse is not true. For example, take G_2 of height 2 and dimension 1, with supersingular special fibre. Then there exists a complex

$$0 \to \mathbb{Q}_p/\mathbb{Z}_p \to G_2 \to \mu_{p^\infty} \to 0$$

that becomes exact on Λ's and W's.

(v) Given (Λ, W), one gets a p-divisible group G over $\mathcal{O}_C$ by the theorem, and thus an $\mathcal{O}_C$-lattice

$$W^\circ = (\mathrm{Lie}\, G)(1) \subset W = (\mathrm{Lie}(G)(1)) \otimes_{\mathcal{O}_C} C \subset \Lambda \otimes_{\mathbb{Z}_p} C\ .$$

Can one describe W° directly in terms of (Λ, W)? We note that there is a second natural $\mathcal{O}_C$-lattice $(W^\circ)' = (\Lambda \otimes \mathcal{O}_C) \cap W$. Then Fargues's theorem implies

$$p^{1/(p-1)}(W^\circ)' \subset W^\circ \subset (W^\circ)'\ ;$$

however, none of the two inclusions is an equality in general.

(vi) Coming back to (iv), we note that an exact sequence of p-divisible groups gives an exact sequence on Λ's and W°'s. Is the converse true? It seems not unreasonable to hope that the answer is yes.[6]

(vii) Let K be a discretely valued complete extension of $\mathbb{Q}_p$ with perfect residue field. Then one can reprove the following theorem of Breuil, [4], (in case $p \neq 2$), and Kisin, [21], (in general) 'by descent':

THEOREM 4.18. *The category of p-divisible groups over $\mathcal{O}_K$ is equivalent to the category of lattices in crystalline representations of $\mathrm{Gal}(\bar{K}/K)$ with Hodge-Tate weights 0, 1.*

4.4. A p-adic analogue: The geometric perspective.

DEFINITION 4.19. Let R be a ring which is p-torsion, and let G/R be a p-divisible group (considered as an fpqc sheaf on schemes over R). Define the universal cover of G as $\tilde{G} = \varprojlim_{\times p} G$, as an fpqc sheaf on schemes over R. Moreover, let $TG = \ker(\tilde{G} \to G)$; one has a short exact sequence of fpqc sheaves

$$0 \to TG \to \tilde{G} \to G \to 0\ .$$

We note that TG is a sheafified version of the Tate module: For any R-algebra R',

$$TG(R') = \varprojlim G[p^n](R') = \mathrm{Hom}_{R'}(\mathbb{Q}_p/\mathbb{Z}_p, G)\ .$$

Also, $\tilde{G} = TG \otimes_{\mathbb{Z}_p} \mathbb{Q}_p$.

PROPOSITION 4.20. (i) *The functor $G \mapsto \tilde{G}$ turns isogenies into isomorphisms.*

---

[5] Using the Fargues-Fontaine curve, [13], one can describe the functor up to isogeny.

[6] V. Pilloni has communicated to us a simple proof of this statement, using Fargues's results on degrees for finite locally free group schemes, [11].

(ii) *Let $S \to R$ be a surjection with nilpotent kernel, and G_S a lift of G to S. Then for any S-algebra S',*

$$\tilde{G}_S(S') \xrightarrow{\cong} \tilde{G}(S' \otimes_S R) \ .$$

In particular, $\tilde{G}_S$ depends only on $\tilde{G}$: One can consider $\tilde{G}$ as a crystal on the infinitesimal site of R.

PROOF. Part (i) is clear. For part (ii) we may assume $S' = S$. Recall that by a result of Illusie, [**18**], the categories of p-divisible groups up to isogeny over R and S are equivalent. Thus,

$$\tilde{G}_S(S) = \mathrm{Hom}_S(\mathbb{Q}_p/\mathbb{Z}_p, G_S)[p^{-1}] = \mathrm{Hom}_R(\mathbb{Q}_p/\mathbb{Z}_p, G)[p^{-1}] = \tilde{G}(R) \ .$$

$\square$

In some cases, one can write down $\tilde{G}$: One gets examples of perfectoid spaces, by taking generic fibres!

PROPOSITION 4.21 ([**31**, Corollary 3.1.5]). *Let G be a connected p-divisible group over $W(\kappa)$, for some perfect field κ. Then there is an isomorphism of fpqc sheaves,*

$$\tilde{H} \cong \mathrm{Spf}\, W(\kappa)[[T_1^{1/p^\infty}, \ldots, T_d^{1/p^\infty}]] \ .$$

Fix an embedding $\overline{\mathbb{F}}_p \hookrightarrow \mathcal{O}_C/p$.

THEOREM 4.22 ([**31**, Theorem 5.1.4 (i)]). *Let G be a p-divisible group over $\mathcal{O}_C$. Then there is a p-divisible group $H/\overline{\mathbb{F}}_p$ (unique up to isogeny) and a quasi-isogeny $\rho : G \otimes_{\mathcal{O}_C} \mathcal{O}_C/p \to H \otimes_{\overline{\mathbb{F}}_p} \mathcal{O}_C/p$.*

In particular, one finds that $\tilde{G} \cong \tilde{H}_{\mathcal{O}_C}$, where the latter denotes the evaluation of $\tilde{H}$ on $\mathcal{O}_C$, considered as a crystal on the infinitesimal site. Thus, in any dimension, there are only finitely many possibilities for the universal cover of G, and these are given by the Dieudonné-Manin classification of p-divisible groups up to isogeny.

Now fix a p-divisible group H over $\overline{\mathbb{F}}_p$, of height h and dimension d. Note that for any p-divisible group G over $\mathcal{O}_C$, we have the $\mathbb{Z}_p$-lattice $\Lambda \subset \tilde{G}(\mathcal{O}_C)$, where $\Lambda = TG(\mathcal{O}_C)$ denotes the Tate module. In particular, we get a fully faithful functor from the category of pairs (G, ρ), where G is a p-divisible group over $\mathcal{O}_C$, and $\rho : G \otimes_{\mathcal{O}_C} \mathcal{O}_C/p \to H \otimes_{\overline{\mathbb{F}}_p} \mathcal{O}_C/p$ is a quasi-isogeny, to the category of $\mathbb{Z}_p$-lattices $\Lambda \subset \tilde{H}(\mathcal{O}_C)$. Here, we use ρ to identify $\tilde{G}$ and $\tilde{H}$.

Thus, as in the case over $\mathbb{C}$, one can ask the question for which $\mathbb{Z}_p$-lattices $\Lambda \subset \tilde{H}(\mathcal{O}_C)$ one can form the quotient $\tilde{H}/\Lambda$ to get a p-divisible group. In order to state the answer, we need the following proposition.

PROPOSITION 4.23. *For any p-divisible group G over $\mathcal{O}_C$, there is a natural logarithm map $\log_G : G(\mathcal{O}_C) \to \mathrm{Lie}\, G \otimes C$. One gets a short exact sequence*

$$0 \to \Lambda[p^{-1}] \to \tilde{G}(\mathcal{O}_C) \to \mathrm{Lie}\, G \otimes C \to 0 \ ,$$

where $\Lambda = TG(\mathcal{O}_C)$ is the Tate module of G. Moreover, there is a natural 'quasi-logarithm' map $\mathrm{qlog} : \tilde{H}(\mathcal{O}_C) \to M(H) \otimes C$, where $M(H)$ is the (covariant)

Dieudonné module, such that for any (G, ρ), the diagram

$$
\begin{array}{ccc}
\tilde{H}(\mathcal{O}_C) & \xrightarrow{\mathrm{qlog}} & M(H) \otimes C \\
\downarrow{\scriptstyle \cong} & & \downarrow \\
\tilde{G}(\mathcal{O}_C) \longrightarrow G(\mathcal{O}_C) & \xrightarrow{\mathrm{log}} & \mathrm{Lie}\, G \otimes C
\end{array}
$$

commutes. Here, the map $M(H) \otimes C \to \mathrm{Lie}\, G \otimes C$ comes from Grothendieck-Messing theory.

THEOREM 4.24 ([**31**, Theorem D]). *The category of pairs (G, ρ) is equivalent to the category of $\mathbb{Z}_p$-lattices $\Lambda \subset \tilde{H}(\mathcal{O}_C)$ such that the cokernel $V = \mathrm{coker}(\Lambda \otimes C \to M(H) \otimes C)$ is of dimension d, and the sequence*

$$
0 \to \Lambda[p^{-1}] \to \tilde{H}(\mathcal{O}_C) \to V \to 0
$$

is exact.

5. Rapoport-Zink spaces

Using these results, we show in [**31**] that Rapoport-Zink spaces at infinite level carry a natural structure as a perfectoid space. More precisely, consider a p-divisible group H over $\bar{\mathbb{F}}_p$ of dimension d and height h. Rapoport-Zink, [**24**], define the following deformation space.

THEOREM 5.1 ([**24**]). *The functor sending a $W(\bar{\mathbb{F}}_p)$-algebra R on which p is nilpotent to the set of isomorphism classes of pairs (G, ρ), where G/R is a p-divisible group and*

$$
\rho : H \otimes_{\bar{\mathbb{F}}_p} R/p \to G \otimes_R R/p
$$

is a quasi-isogeny, is representable by a formal scheme $\mathcal{M}$.

Moreover, its generic fibre $\mathcal{M}_\eta$, considered as an adic space, has a natural system of coverings $\mathcal{M}_n \to \mathcal{M}_\eta$ parametrizing isomorphisms $(\mathbb{Z}/p^n\mathbb{Z})^h \to G[p^n]$. Each of them is a rigid-analytic variety. However, the inverse limit of the $\mathcal{M}_n$ does not make sense within rigid geometry. In [**31**], we prove however the following result. For simplicity, we work with the base-change $\mathcal{M}_{n,K}$ of $\mathcal{M}_n$ to $\mathrm{Spa}(K, \mathcal{O}_K)$, where $K/\mathbb{Q}_p$ is a perfectoid field.

THEOREM 5.2. *There is a unique (up to unique isomorphism) perfectoid space $\mathcal{M}_{\infty,K}$ over $\mathrm{Spa}(K, \mathcal{O}_K)$ such that $\mathcal{M}_{\infty,K} \sim \varprojlim \mathcal{M}_{n,K}$. In fact, $\mathcal{M}_{\infty,K}$ is a locally closed subspace of the generic fibre of $\tilde{H}^h_{\mathcal{O}_K}$, defined by certain explicit conditions.*

Here, we use $\sim$ as in Definition 2.20.

REMARK 5.3. We note that the generic fibre of $\tilde{H}^h_{\mathcal{O}_K}$ is a perfectoid space, see Proposition 4.21 in the connected case. The map

$$
\mathcal{M}_{\infty,K} \to (\tilde{H}_{\mathcal{O}_K})^h_\eta
$$

is precisely the map from Theorem 4.24 on C-valued points, sending a deformation to the corresponding lattice in the universal cover.

We remark that the description of $\mathcal{M}_{\infty,K}$ as an explicit locally closed subspace of the generic fibre of $\tilde{H}^h_{\mathcal{O}_K}$ allows us to prove duality isomorphisms between Rapoport-Zink spaces as isomorphisms of perfectoid spaces. We refer to [**31**] for details. The abstract setup is however as in the following proposition, which gives the expected cohomological consequences (which were not included in [**31**]).

PROPOSITION 5.4. *Let C be an algebraically closed and complete extension of $\mathbb{Q}_p$. Let G, $\check{G}$ be two p-adic reductive groups, and let $\mathcal{M}_U$, $\check{\mathcal{M}}_{\check{U}}$ be two towers of partially proper smooth adic spaces over C parametrized by compact open subgroups $U \subset G(\mathbb{Q}_p)$, resp. $\check{U} \subset \check{G}(\mathbb{Q}_p)$, and with an action of $G(\mathbb{Q}_p)$, resp. $\check{G}(\mathbb{Q}_p)$, on the tower, such that for $U' \subset U$ an open normal subgroup, $\mathcal{M}_{U'}$ is a finite étale Galois cover of $\mathcal{M}_U$ with Galois group U/U', respectively the similar statement for the tower $\check{\mathcal{M}}_{\check{U}}$. Moreover, assume $\check{G}(\mathbb{Q}_p)$ acts continuously on each $\mathcal{M}_U$, and $G(\mathbb{Q}_p)$ acts continuously on each $\check{\mathcal{M}}_{\check{U}}$, such that all group actions are compatible and commute.*

Finally, assume that there is a perfectoid space $\mathcal{M}$ over C with a continuous action of $G(\mathbb{Q}_p) \times \check{G}(\mathbb{Q}_p)$ such that

$$\mathcal{M} \sim \varprojlim_U \mathcal{M}_U \ , \ \mathcal{M} \sim \varprojlim_{\check{U}} \check{\mathcal{M}}_{\check{U}} \ ,$$

where both maps to the inverse limit are $G(\mathbb{Q}_p) \times \check{G}(\mathbb{Q}_p)$-equivariant. Let $\ell \neq p$ be a prime.

(i) *For any $m \geq 1$, there are $G(\mathbb{Q}_p) \times \check{G}(\mathbb{Q}_p)$-equivariant isomorphisms*

$$\varinjlim_U H^i_c(\mathcal{M}_U, \mathbb{Z}/\ell^m\mathbb{Z}) \cong H^i_c(\mathcal{M}, \mathbb{Z}/\ell^m\mathbb{Z}) \cong \varinjlim_{\check{U}} H^i_c(\check{\mathcal{M}}_{\check{U}}, \mathbb{Z}/\ell^m\mathbb{Z}) \ .$$

(ii) *The action of $\check{G}(\mathbb{Q}_p)$ on $H^i_c(\mathcal{M}_U, \mathbb{Z}_\ell)$ is smooth for any $U \subset G(\mathbb{Q}_p)$; similarly, the action of $G(\mathbb{Q}_p)$ on $H^i_c(\check{\mathcal{M}}_{\check{U}}, \mathbb{Z}_\ell)$ is smooth for any $\check{U} \subset \check{G}(\mathbb{Q}_p)$.*

(iii) *There is a $G(\mathbb{Q}_p) \times \check{G}(\mathbb{Q}_p)$-equivariant isomorphism*

$$\varinjlim_U H^i_c(\mathcal{M}_U, \mathbb{Z}_\ell) \cong \varinjlim_{\check{U}} H^i_c(\check{\mathcal{M}}_{\check{U}}, \mathbb{Z}_\ell) \ .$$

Moreover, both identify with the $G(\mathbb{Q}_p) \times \check{G}(\mathbb{Q}_p)$-smooth vectors in $H^i_c(\mathcal{M}, \mathbb{Z}_\ell)$.

REMARK 5.5. Some remarks about the definitions of the various objects involved. We recall that for partially proper spaces, cohomology with compact support is defined as the derived functor of the functor Γ_c of taking sections with quasicompact support. This leads to the following equivalent definition. Fix some U, and let $V_{0U} \subset V_{1U} \subset \ldots \mathcal{M}_U$ be a sequence of qcqs open subsets exhausting $\mathcal{M}_U$. Let $V'_{0U} \subset V'_{1U} \subset \ldots \mathcal{M}_U$ be a second such sequence, such that $V_{kU} \subset V'_{kU}$ is a strict inclusion for all k, i.e. the $\overline{V_{kU}} \subset V'_{kU}$. Let $j_{kU} : V_{kU} \to V'_{kU}$ be the open inclusion. Then

$$H^i_c(\mathcal{M}_U, \mathbb{Z}/\ell^m\mathbb{Z}) = \varinjlim_k H^i(V'_{kU}, j_{kU!}\mathbb{Z}/\ell^m\mathbb{Z}) \ .^7$$

The same applies for $\mathcal{M}$, i.e. at infinite level.

---

[7] The latter groups are also usually denoted $H^i_c(V_{kU}, \mathbb{Z}/\ell^m\mathbb{Z})$. In that case however, they are not the derived functors of global sections with compact support: V_{kU} is not partially proper.

For $\mathbb{Z}_\ell$-cohomology, one has to take inverse limits, and one has to take some care about the order. We define

$$H^i_c(\mathcal{M}_U, \mathbb{Z}_\ell) = \varinjlim_k H^i(V'_{kU}, j_{kU!}\mathbb{Z}_\ell) = \varinjlim_k \varprojlim_m H^i(V'_{kU}, j_{kU!}\mathbb{Z}/\ell^m\mathbb{Z}) \ .$$

We remark that a result of Huber in the book [**17**] ensures that the groups

$$H^i(V'_{kU}, j_{kU!}\mathbb{Z}/\ell^m\mathbb{Z}) = H^i_c(V_{kU}, \mathbb{Z}/\ell^m\mathbb{Z})$$

are finite. We define $H^i_c(\mathcal{M}, \mathbb{Z}_\ell)$ in the same way.

PROOF. (i) There are obvious $G(\mathbb{Q}_p) \times \check{G}(\mathbb{Q}_p)$-equivariant maps, and we have to prove that they are isomorphisms; thus we may restrict to one tower, $\mathcal{M}_U$.

Fix some U, and let $V_{0U} \subset V_{1U} \subset \ldots \mathcal{M}_U$ be a sequence of qcqs open subsets exhausting $\mathcal{M}_U$. Let $V'_{0U} \subset V'_{1U} \subset \ldots \mathcal{M}_U$ be a second such sequence, such that $V_{kU} \subset V'_{kU}$ is a strict inclusion for all k, i.e. the $\overline{V_{kU}} \subset V'_{kU}$. Let $j_{kU} : V_{kU} \to V'_{kU}$ be the open inclusion. Then

$$H^i_c(\mathcal{M}_U, \mathbb{Z}/\ell^m\mathbb{Z}) = \varinjlim_k H^i(V'_{kU}, j_{kU!}\mathbb{Z}/\ell^m\mathbb{Z}) \ .$$

For $U' \subset U$, let us denote by $V_{kU'}$ etc. the corresponding objects induced by base-change, as well as $V_k \subset \mathcal{M}$. It is enough to prove that

$$\varinjlim_{U' \subset U} H^i(V'_{kU'}, j_{kU'!}\mathbb{Z}/\ell^m\mathbb{Z}) \to H^i(V', j_{k!}\mathbb{Z}/\ell^m\mathbb{Z})$$

is an isomorphism. But this follows from [**30**, Corollary 7.18].

(ii) It suffices to check for U small enough (using a Hochschild-Serre spectral sequence), so we may assume U is pro-p. Now, the map

$$H^i_c(\mathcal{M}_U, \mathbb{Z}/\ell^m\mathbb{Z}) \to \varinjlim_{U' \subset U} H^i_c(\mathcal{M}_{U'}, \mathbb{Z}/\ell^m\mathbb{Z}) = H^i_c(\mathcal{M}, \mathbb{Z}/\ell^m\mathbb{Z})$$

is injective. (The transition map to $U' \subset U$ normal has an inverse, given by averaging over U/U'.) On the right-hand side, the action of $\check{G}(\mathbb{Q}_p)$ is continuous by the comparison to the other tower in part (i), thus it is continuous on $H^i_c(\mathcal{M}_U, \mathbb{Z}/\ell^m\mathbb{Z})$.

The result follows by noting that for actions of pro-p-groups on finitely generated ℓ-adic modules, smoothness is equivalent to continuity. (By definition, one may write the cohomology as a direct limit of finitely generated $\mathbb{Z}_\ell$-modules.)

(iii) It suffices to identify the $G(\mathbb{Q}_p) \times \check{G}(\mathbb{Q}_p)$-smooth vectors in $H^i_c(\mathcal{M}, \mathbb{Z}_\ell)$ with

$$\varinjlim_U H^i_c(\mathcal{M}_U, \mathbb{Z}_\ell) \ .$$

By parts (i) and (ii), the direct limit injects into the $G(\mathbb{Q}_p) \times \check{G}(\mathbb{Q}_p)$-smooth vectors in $H^i_c(\mathcal{M}, \mathbb{Z}_\ell)$. On the other hand, take a vector $v \in H^i_c(\mathcal{M}, \mathbb{Z}_\ell)$ which is invariant under a compact open subgroup $U \subset G(\mathbb{Q}_p)$, without loss of generality pro-p. By averaging over U (which is possible, as U is pro-p), we see that v comes from $H^i_c(\mathcal{M}_U, \mathbb{Z}_\ell)$, as desired.

$\square$

REMARK 5.6. We remark that the proof shows that in part (iii), the $G(\mathbb{Q}_p)$-smooth vectors are the same as the $\check{G}(\mathbb{Q}_p)$-vectors, which are then the $G(\mathbb{Q}_p) \times \check{G}(\mathbb{Q}_p)$-smooth vectors identified in part (iii).

In the equal characteristic case, Weinstein, [**35**], has considered the Lubin-Tate case explicitly. In that case, the theory of Drinfeld level structures give natural integral models of $\mathcal{M}_n$ for all n, showing that in fact

$$\mathcal{M}_n \cong (\operatorname{Spf} \bar{\mathbb{F}}_p[[X_{1,n}, \ldots, X_{h,n}]])_\eta \ ,$$

where the canonical coordinates $X_{1,n}, \ldots, X_{h,n}$ come from the level-n-structure. This permits one to show by explicit computation that

$$\mathcal{M}_\infty \cong (\operatorname{Spf} \bar{\mathbb{F}}_p[[X_1^{1/p^\infty}, \ldots, X_h^{1/p^\infty}]])_\eta \ ,$$

which obviously has the desired property of being perfectoid. It should be noted that these spaces live over the base field $\mathbb{F}_q[[t]]$, but it is rather hard to write down the element t as an element of $\bar{\mathbb{F}}_p[[X_1^{1/p^\infty}, \ldots, X_h^{1/p^\infty}]]$ (a formula appears at the very end of [**14**]).

The paper [**31**] also contains a result on Dieudonné theory.

THEOREM 5.7 ([**31**, Theorem 4.1.4]). *Let R be a ring of characteristic p which is the quotient of a perfect ring by a finitely generated ideal. Then the Dieudonné module functor on p-divisible groups is fully faithful up to isogeny.*

Our result is slightly more precise than that, and in the case that R is perfect, it recovers the fact that the Dieudonné module functor is fully faithful, not just up to isogeny. Interestingly, the proof of this fully faithfulness result requires the use of perfectoid spaces, and most notably the almost purity theorem! In fact, this result from Dieudonné theory bridges the gap between universal covers of p-divisible groups, and Fontaine's rings. Indeed, one gets the following corollary, where B_{cris}^+ is Fontaine's crystalline period ring (associated to C).

COROLLARY 5.8. *Let C be an algebraically closed complete extension of $\mathbb{Q}_p$. Let $H/\bar{\mathbb{F}}_p$ be a p-divisible group, and write $\tilde{H}$ for its universal cover (lifted canonically to $\mathcal{O}_C$). Then*

$$\tilde{H}(\mathcal{O}_C) = (M(H) \otimes B_{\mathrm{cris}}^+)^{\varphi=p} \ ,$$

where $M(H)$ is the Lie algebra of the universal vector extension of H. Under this identification, the quasi-logarithm map

$$\mathrm{qlog} : \tilde{H}(\mathcal{O}_C) \to M(H) \otimes C$$

gets identified with

$$1 \otimes \Theta : M(H) \otimes B_{\mathrm{cris}}^+ \to M(H) \otimes C \ ,$$

where $\Theta : B_{\mathrm{cris}}^+ \to C$ is Fontaine's map.

This translates also Theorem 4.24 into p-adic Hodge theory terms.

6. Shimura varieties, and completed cohomology

A result very similar to the result for Rapoport-Zink spaces holds for Shimura varieties. Let Sh_K, $K \subset G(\mathbb{A}_f)$ be a Shimura variety of Hodge type associated to some reductive group G over $\mathbb{Q}$, defined over the reflex field E. For convenience, let us assume that Sh_K is projective, so that we do not have to worry about compactifications. Let $\mathbb{C}_p$ be the completion of an algebraic closure of $E_\mathfrak{p}$, where $\mathfrak{p}|p$ is a chosen place of E, and denote by $\mathrm{Sh}_{K,\mathbb{C}_p}$ the adic space over $\mathrm{Spa}(\mathbb{C}_p, \mathcal{O}_{\mathbb{C}_p})$ associated to the base-change of Sh_K to $\mathbb{C}_p$. The following result is work in progress.

THEOREM 6.1 ([**28**]). *For any sufficiently small level $K^p \subset G(\mathbb{A}_f^p)$ away from p, there exists a perfectoid space $\mathrm{Sh}_{K^p,\mathbb{C}_p}$ over $\mathrm{Spa}(\mathbb{C}_p, \mathcal{O}_{\mathbb{C}_p})$ such that $\mathrm{Sh}_{K^p,\mathbb{C}_p} \sim \varprojlim_{K_p} \mathrm{Sh}_{K_p K^p, \mathbb{C}_p}$.*

Let us explain a consequence of this theorem, when combined with the first result on p-adic Hodge theory. Recall Emerton's definition of p-adically completed cohomology groups, in the torsion case (where no completion has to be taken):

$$H^i(K^p, \mathbb{F}_p) = \varinjlim_{K_p} H^i_{\text{ét}}(\mathrm{Sh}_{K_p K^p, \bar{\mathbb{Q}}}, \mathbb{F}_p) \ .$$

COROLLARY 6.2. *For $i > \dim \mathrm{Sh}_K$, we have $H^i(K^p, \mathbb{F}_p) = 0$.*

PROOF. First, we can rewrite

$$H^i(K^p, \mathbb{F}_p) = \varinjlim_{K_p} H^i_{\text{ét}}(\mathrm{Sh}_{K_p K^p, \bar{\mathbb{Q}}}, \mathbb{F}_p) = \varinjlim_{K_p} H^i_{\text{ét}}(\mathrm{Sh}_{K_p K^p, \mathbb{C}_p}, \mathbb{F}_p) \ .$$

It is enough to prove that $H^i(K^p, \mathbb{F}_p) \otimes_{\mathbb{F}_p} \mathcal{O}_{\mathbb{C}_p}/p$ is almost zero. But now

$$H^i(K^p, \mathbb{F}_p) \otimes_{\mathbb{F}_p} \mathcal{O}_{\mathbb{C}_p}/p = \varinjlim_{K_p}(H^i_{\text{ét}}(\mathrm{Sh}_{K_p K^p, \mathbb{C}_p}, \mathbb{F}_p) \otimes_{\mathbb{F}_p} \mathcal{O}_{\mathbb{C}_p}/p) \ ,$$

and the latter is almost equal to

$$\varinjlim_{K_p} H^i_{\text{ét}}(\mathrm{Sh}_{K_p K^p, \mathbb{C}_p}, \mathcal{O}^+_{\mathrm{Sh}_{K_p K^p, \mathbb{C}_p}}/p)$$

by Theorem 3.3. Now we use that $\mathrm{Sh}_{K^p,\mathbb{C}_p} \sim \varprojlim_{K_p} \mathrm{Sh}_{K_p K^p, \mathbb{C}_p}$ (which implies in particular a similar relation among étale topoi), giving

$$\varinjlim_{K_p} H^i_{\text{ét}}(\mathrm{Sh}_{K_p K^p, \mathbb{C}_p}, \mathcal{O}^+_{\mathrm{Sh}_{K_p K^p, \mathbb{C}_p}}/p) = H^i_{\text{ét}}(\mathrm{Sh}_{K^p, \mathbb{C}_p}, \mathcal{O}^+_{\mathrm{Sh}_{K^p, \mathbb{C}_p}}/p) \ .$$

But note that for affinoid perfectoid spaces X, $H^i_{\text{ét}}(X, \mathcal{O}^+_X/p)$ is almost zero for $i > 0$. It follows that $H^i_{\text{ét}}(\mathrm{Sh}_{K^p,\mathbb{C}_p}, \mathcal{O}^+_{\mathrm{Sh}_{K^p,\mathbb{C}_p}}/p)$ and $H^i_{\text{an}}(\mathrm{Sh}_{K^p,\mathbb{C}_p}, \mathcal{O}^+_{\mathrm{Sh}_{K^p,\mathbb{C}_p}}/p)$ are almost equal. But now standard bounds on the cohomological dimension of topological spaces give the desired vanishing result. □

The corollary implies Conjecture 1.5 of Calegari and Emerton, [**6**], in the case of (compact) Shimura varieties of Hodge type, except for nonstrict instead of strict inequalities on the codimensions. More applications of these ideas will appear in [**28**].

References

[1] F. Andreatta and A. Iovita. Comparison Isomorphisms for Formal Schemes. http://www.mathstat.concordia.ca/faculty/iovita/paper_14.pdf.

[2] E. Bierstone and P. D. Milman. Canonical desingularization in characteristic zero by blowing up the maximum strata of a local invariant. *Invent. Math.*, 128(2):207–302, 1997.

[3] S. Bosch, U. Güntzer, and R. Remmert. *Non-Archimedean analysis*, volume 261 of *Grundlehren der Mathematischen Wissenschaften [Fundamental Principles of Mathematical Sciences]*. Springer-Verlag, Berlin, 1984. A systematic approach to rigid analytic geometry.

[4] C. Breuil. Groupes p-divisibles, groupes finis et modules filtrés. *Ann. of Math. (2)*, 152(2):489–549, 2000.

[5] O. Brinon. Représentations p-adiques cristallines et de de Rham dans le cas relatif. *Mém. Soc. Math. Fr. (N.S.)*, (112):vi+159, 2008.

[6] F. Calegari and M. Emerton. Completed cohomology—a survey. In *Non-abelian fundamental groups and Iwasawa theory*, volume 393 of *London Math. Soc. Lecture Note Ser.*, pages 239–257. Cambridge Univ. Press, Cambridge, 2012.

[7] A. J. de Jong. Smoothness, semi-stability and alterations. *Inst. Hautes Études Sci. Publ. Math.*, (83):51–93, 1996.

[8] P. Deligne. Théorie de Hodge. I. In *Actes du Congrès International des Mathématiciens (Nice, 1970), Tome 1*, pages 425–430. Gauthier-Villars, Paris, 1971.

[9] P. Deligne. La conjecture de Weil. II. *Inst. Hautes Études Sci. Publ. Math.*, (52):137–252, 1980.

[10] G. Faltings. Almost étale extensions. *Astérisque*, (279):185–270, 2002. Cohomologies p-adiques et applications arithmétiques, II.

[11] L. Fargues. La filtration de Harder-Narasimhan des schémas en groupes finis et plats. *J. Reine Angew. Math.*, 645:1–39, 2010.

[12] L. Fargues. Groupes analytiques rigides p-divisibles. 2012. preprint.

[13] L. Fargues and J.-M. Fontaine. Vector bundles and p-adic Hodge theory. 2012. preprint.

[14] L. Fargues, A. Genestier, and V. Lafforgue. *L'isomorphisme entre les tours de Lubin-Tate et de Drinfeld*, volume 262 of *Progress in Mathematics*. Birkhäuser Verlag, Basel, 2008.

[15] J.-M. Fontaine and J.-P. Wintenberger. Extensions algébrique et corps des normes des extensions APF des corps locaux. *C. R. Acad. Sci. Paris Sér. A-B*, 288(8):A441–A444, 1979.

[16] O. Gabber and L. Ramero. *Almost ring theory*, volume 1800 of *Lecture Notes in Mathematics*. Springer-Verlag, Berlin, 2003.

[17] R. Huber. *Étale cohomology of rigid analytic varieties and adic spaces*. Aspects of Mathematics, E30. Friedr. Vieweg & Sohn, Braunschweig, 1996.

[18] L. Illusie. Déformations de groupes de Barsotti-Tate (d'après A. Grothendieck). *Astérisque*, (127):151–198, 1985. Seminar on arithmetic bundles: the Mordell conjecture (Paris, 1983/84).

[19] K. Kedlaya and R. Liu. Relative p-adic Hodge theory, I: Foundations. http://math.mit.edu/~kedlaya/papers/relative-padic-Hodge1.pdf.

[20] R. Kiehl. Der Endlichkeitssatz für eigentliche Abbildungen in der nichtarchimedischen Funktionentheorie. *Invent. Math.*, 2:191–214, 1967.

[21] M. Kisin. Crystalline representations and F-crystals. In *Algebraic geometry and number theory*, volume 253 of *Progr. Math.*, pages 459–496. Birkhäuser Boston, Boston, MA, 2006.

[22] W. Lütkebohmert. From Tate's elliptic curve to abeloid varieties. *Pure Appl. Math. Q.*, 5(4, Special Issue: In honor of John Tate. Part 1):1385–1427, 2009.

[23] M. Rapoport and T. Zink. Über die lokale Zetafunktion von Shimuravarietäten. Monodromiefiltration und verschwindende Zyklen in ungleicher Charakteristik. *Invent. Math.*, 68(1):21–101, 1982.

[24] M. Rapoport and T. Zink. *Period spaces for p-divisible groups*, volume 141 of *Annals of Mathematics Studies*. Princeton University Press, Princeton, NJ, 1996.

[25] W. Schmid. Variation of Hodge structure: the singularities of the period mapping. *Invent. Math.*, 22:211–319, 1973.

[26] P. Schneider. The cohomology of local systems on p-adically uniformized varieties. *Math. Ann.*, 293(4):623–650, 1992.

[27] P. Scholze. http://mathoverflow.net/questions/65729/what-are-perfectoid-spaces.

[28] P. Scholze. On torsion in the cohomology of locally symmetric varieties. in preparation.

[29] P. Scholze. p-adic Hodge theory for rigid-analytic varieties. 2012. arXiv:1205.3463, to appear in Forum of Mathematics, Pi.

[30] P. Scholze. Perfectoid Spaces. *Publ. Math. de l'IHÉS*, 116(1):245 – 313, 2012.

[31] P. Scholze and J. Weinstein. Moduli of p-divisible groups. 2012. arXiv:1211.6357.

[32] J. Steenbrink. Limits of Hodge structures. *Invent. Math.*, 31(3):229–257, 1975/76.

[33] J. T. Tate. p − $divisible$ $groups$.. In *Proc. Conf. Local Fields (Driebergen, 1966)*, pages 158–183. Springer, Berlin, 1967.

[34] M. Temkin. On local properties of non-Archimedean analytic spaces. *Math. Ann.*, 318(3):585–607, 2000.

[35] J. Weinstein. Formal vector spaces over a local field of positive characteristic. 2012. arXiv:1207.6424.

MATHEMATISCHES INSTITUT DER UNIVERSITÄT BONN, ENDENICHER ALLEE 60, 53115 BONN, GERMANY

E-mail address: scholze@math.uni-bonn.de

Duality, statistical mechanics, and random matrices

Thomas Spencer

ABSTRACT. This article will present an informal review of some results and conjectures about the spectral theory of large random matrices and related spin systems in statistical mechanics. A class of lattice spin models provides a dual representation for spectral problems in random matrix theory. Ordered and disordered phases of the spins correspond to different spectral types and quantum time evolutions. In three dimensions, we describe a phase transition for a supersymmetric statistical mechanics system inspired by random matrix theory. This transition has a classical interpretation in terms of a history dependent walk on the lattice. In the ordered phase the walk is diffusive while in the disordered phase it is localized near its starting point.

Contents

1. Overview

In this section we shall give a brief overview of conjectures and theorems about random matrices and related models of interacting lattice spins. We begin by defining various classes of large random matrices including Wigner matrices, random band matrices and the Schrödinger equation with a random potential. Spectral properties of these matrices such as density of states, eigenvalue spacing, time evolution and notions of localization and quantum diffusion are described in §1.1, §1.2 and §1.4. Some simple examples of duality appear in §1.3. In this case the dual statistical mechanics consists of a single "spin". Standard saddle point analysis yields precise asymptotic information. An introduction to supersymmetric spin models and their relation to the spectral theory of random matrices is explained in §2.

In §1.6 we review the ordered and disorded phases for some classical spin models of statistical mechanics. Symmetry is conjectured to describe some universal features of the ordered spin phase. In §1.5 and §3 we discuss a three dimensional supersymmetric spin model which has a phase transition. This supersymmetric model is equivalent to a history dependent walk called the vertex reinforced jump process. In the ordered phase the walk is diffusive while in the disordered phase it is localized near its starting point. A similar transition is believed to occur for the random Schrödinger equation.

1.1. Wigner Matrices. There has been a great deal of interest and activity in the spectral theory of random matrices arising from various branches of mathematics and physics. These areas include probability, statistics, number theory, combinatorics, communication theory, quantum chaos and the spectral theory of Schrödinger operators with a random potential. See [Bre] for a review of some of these applications.

The study of large random matrices in physics originated with the work of Eugene Wigner in the 1950's who used them to predict the energy level statistics of a large nucleus. He argued that because of the complex interactions in the nucleus there should be a random matrix model with appropriate symmetries, whose eigenvalues would describe the energy level spacing statistics. He introduced a class of $N \times N$ random Hermitian $H = H^*$ and real symmetric matrices, $H = H^t$, called Wigner matrices. These matrices have a *mean field* structure: their matrix elements are independent and have equal variance subject to symmetry constraints. In particular for Hermitian matrices

$$\langle H_{ij} \rangle = 0 \quad and \quad \langle H_{ij} \, H_{i'j'} \rangle = \langle H_{ij} \, \bar{H}_{j'i'} \rangle = \frac{1}{N} \, \delta_{ij'} \, \delta_{ji'}. \tag{1.1}$$

In the special case where the matrix elements have a Gaussian distribution these conditions uniquely specify the probability distribution and the ensemble is referred to as the Gaussian Unitary Ensemble (GUE). Its probability distribution is proportional to $e^{-N \, Tr \, H^2/2}$. The Gaussian Orthogonal ensemble (GOE) has the same weight except that we integrate over real symmetric matrices.

For a large class of matrices satisfying (1.1), most of the eigenvalues lie in the interval $[-2, 2]$ as $N \to \infty$. Wigner [Wig1] used the method of moments to show that as $N \to \infty$ the fraction of eigenvalues of H less than E is given by $\int_{-2}^{E} \rho(E') \, dE'$ where the density of states (eigenvalues) $\rho(E)$ satisfies the semicircle law:

$$\rho(E) = \frac{1}{2\pi} \sqrt{4 - E^2}, \qquad |E| < 2. \tag{1.2}$$

The density of states can be defined for a broad class of random matrices. It is not universal because it depends on the details of the random matrix distribution.

Universality: Wigner and Dyson conjectured that the statistics of suitably scaled eigenvalue spacings such as $(E_{k+1} - E_k)/\delta(E)$ is universal for many large complex quantum systems. The scaling factor $\delta(E) = (\rho(E)N)^{-1}$ is the average eigenvalue spacing about E. Universality means that such spacing statistics should only depend on symmetry (Hermitian GUE or real Symmetric GOE or Symplectic GSE). These statistics are often referred to as Wigner-Dyson statistics and are observed in many physical systems. See [Dys] and [Wig2] for Wigner's historical perspective. Universality implies that the simpler mean field Gaussian models should accurately predict local spacing laws for a broad class of quantum ensembles. For large N GUE matrices, the local eigenvalue correlation ρ_2 is given by

$$\frac{1}{\rho^2(E)}\rho_2(E, E + \xi\delta(E)) = 1 - sin^2(\pi\xi)/(\pi\xi)^2, \quad |E| < 2. \tag{1.3}$$

For small ξ we see that there is quadratic repulsion between nearby eigenvalues. The spacing distribution is more complicated and is expressed in terms of a Fredholm determinant.

Saddle manifold approximation. One of the goals of this article is to explain how Wigner-Dyson universality has a natural interpretation in terms of interacting spins or fields. We shall see that these spin systems are associated to a symmetric space related to the symmetry of the random matrix. Many lattice spin systems have a phase transition as temperature is varied. At high temperature, the spins are disordered whereas at low temperature they tend to be aligned or ordered in 3 or more dimensions. This is much like an iron magnet which when heated loses its ordered magnetic properties. In §1.6 the ordered (low temperature) phases of a wide class of spin models on $\mathbb{Z}^d$, $d \geq 3$ are conjectured to be described by the saddle manifold or 0-mode approximation. This is a mean field model determined by symmetry. To investigate a large random matrix near energy E, a dual spin system is studied at a temperature T(E). In the framework of supersymmetric statistical mechanics W-D statistics arise from the contribution of finite dimensional saddle manifold of spins in an ordered state, [Ef1, Ef2].

Let us remark that universality of mean field theory and the saddle manifold discussed in this article is *different* from the more subtle notion of universality in critical phenomena. Critical universality occurs at temperatures very near the transition temperature which marks the *borderline* between order and disorder.

Number Theory: Wigner-Dyson statistics seem to extend far beyond spectral theory of random matrices. The high zeros of the Riemann zeta function appear to satisfy Wigner-Dyson statistics for GUE after suitable scaling. This striking conjecture is due to H. Montgomery and F. Dyson followed by numerical work of A. Odlyzko. See [RS, KS] for further developments relating number theory to random matrices.

Mathematical Results on universality: Recent work by Erdős et al. [Er2] and by Tao and Vu [TV] established universality of spacing statistics for a wide class of Wigner matrices. This may be thought of as a noncommutatitve central limit theorem. Note that Wigner-Dyson describes the law of highly correlated random variables namely the eigenvalues of a random matrix. A more detailed review of these developments as well as recent results on sparse and band matrices is covered in the lectures of L. Erdős [Er5] and in [Er4].

There is a related family of matrices called *invariant* ensembles. In this case the matrix elements are given by a probability density of the form $e^{-N\,tr\,V(H)}$ where $V(x) \geq 0$ is a function with suitable growth for large $|x|$. Unless $V(x) = x^2$, the matrix elements are no longer independent and the density of states is not given by the semi-circle law. Universality has also been established for invariant ensembles by Deift et al [Dei] and by Pastur and Shcherbina [PS1, PS2]. This means that the eigenvalue spacing distribution does not depend on V but it will depend on whether the matrices are complex Hermitian or real symmetric. This has lead to progress in areas such as Riemann-Hilbert theory, orthogonal polynomials, integrable systems, matrix-large deviation theory. See also recent work of Bourgade, Erdős, and Yau [BEY] which generalizes this work to a more general class of statistical mechanics models with a logarithmic potential.

1.2. The Anderson model and Random Band matrices. A few years after Wigner's work on energy levels of a large nucleus, Philip Anderson [And] introduced another class of random matrices to study a quantum particle on a d dimensional crystal which is scattered by random defects or impurities. The Anderson model is given by a discrete space Schrödinger operator on $\mathbb{Z}^d$ with a random potential V:

$$H = -\Delta + \lambda V \qquad (Hw)(j) = \sum_{j':|j'-j|=1} (w(j) - w(j')) + \lambda V(j)\, w(j) \quad j \in \mathbb{Z}^d. \quad (1.4)$$

The V_j are usually assumed to be independent, identically distributed random variables with mean 0 and unit variance. The coupling λ measures the strength of the disorder. For $\lambda \gg 1$ Anderson predicted *localization* with probability one for all energies E. An eigenstate ϕ is said to be localized if it decays exponentially fast about some lattice point b,

$$|\phi(j)| \approx C e^{-|j-b|/\ell(E)} \qquad\qquad (1.5)$$

where $\ell(E)$ is called the localization length. Mathematically, localization at all energies means that with probablity one, the spectrum of H is *dense, pure point spectrum*, and eigenstates decay exponentially fast. Note that if we drop the discrete Laplacian in (1.4) each eigenstate is concentrated at a single lattice vertex. In physics, localization corresponds to an insulating phase - no conduction. The Anderson model is also be studied on $\mathbb{R}^d$ and its low energy spectral behavior will be much the same as on the lattice.

Goldsheid, Molchanov and Pastur [GMP] proved localization for all energies in one dimension for any $\lambda > 0$. Thus even small amounts of disorder changes the character of the eigenstates in 1D from plane wave to localized. For $d \geq 2$ localization was proved using the estimates of [FS2] for strong disorder $\lambda \gg 1$ or for small $\lambda > 0$ and $E \leq -C\lambda^2$. See [AM] for an elegant proof using fractional moments and [Sp1, Sto] for mathematical reviews of Anderson localization.

The Anderson transition is expected to occur in three dimensions for small disorder (eg. $|\lambda| \leq 1$). There is conjectured to be an interval of absolutely continuous spectrum (E_m, E'_m) corresponding to energies where conduction or quantum diffusion occurs. For E outside the interval, $\ell(E) < \infty$, and $\ell(E) = \infty$ inside the interval. The energies E_m *and* E'_m are well defined with probability one and depend on λ. They are called mobility edges. Although the existence of such an interval is not questioned in theoretical physics, a mathematical proof of its existence has not been established and is considered to be a major open problem. It will be rephrased more precisely in terms of Green's functions (1.16) below.

In statistical mechanics the role of E is replaced by temperature T(E). Localization corresponds to a high temperature T(E) phase where spins are disordered whereas quantum diffusion corresponds to an ordered phase, low temperature. In two dimensions, Abrahams, Anderson, Liciardello and Ramakrishnan [Abr] predicted that for any $\lambda > 0$ all states are localized with a localization length $\ell \approx e^{c\lambda^{-2}}$. See [EM] for a recent review of the predicted behavior of eigenfunctions with $E \approx E_m$ using supersymmetric statistical mechanics.

Random Band Matrices: This article will focus on spectral properties of large Hermitian random band matrices, **RBM**. These matrices, $H_{ij} = \bar{H}_{ji}$ are indexed by i, j in $\mathbb{Z}^d$ and concentrated in a band of width W about the diagonal, $|i - j| \leq W$. They interpolate between the mean field Wigner type matrices (when $W \approx N$) and the Anderson model. The Anderson model and the random band model with fixed W are expected to have very similar behavior. There is an approximate correspondence $\lambda \approx 1/W$. Note however that the Anderson model is real symmetric (time reversal invariant) whereas RBM we study are complex Hermitian and thus they belong to distinct Wigner-Dyson classes.

Definition of RBM: We shall assume that H_{ij} are independent random variables of zero mean subject to the Hermitian constraint. In addition we assume that H_{jk} are *Gaussian* and

$$\langle H_{ij} \rangle = 0 \quad and \quad \langle H_{ij} H_{i'j'} \rangle = J_{ij}\delta_{ij'}\delta_{ji'} \quad i,j \in \mathbb{Z}^d \cap [1,N]^d \equiv \Lambda_N \qquad (1.6)$$

with $J_{ij} \approx 0$ when $|i - j| \geq W$ and normalized so that $\sum_j J_{ij} = 1$. W is called the width of the band. For fixed W and large N, RBM reflects the geometry of the lattice. The special case where $W = N$ and $J_{ij} = 1/N$, defines the Gaussian unitary ensemble (GUE).

Topics of particular interest are the density of states, the statistical properties of eigenvalues and eigenvectors of H when H is restricted to a large box of side N. As in the Anderson model a closely related topic is a description of the long time evolution of the Schrödinger equation on $\mathbb{Z}^d$

$$i\partial\phi(t,j)/\partial t = H\phi(t,j) \qquad (1.7)$$

given an initial state $\phi(0,j)$. The time evolution (§1.4) will reflect localized and conducting states depending upon whether the initial state spreads.

Although there is a wealth of information for mean field models, the study of RBM is much less developed. For one dimensional RBM, Schenker [Sch] proved that the localization length $\ell(E) \leq W^8$ for all E. Using SUSY statistical mechanics, Fyodorov and Mirlin [FM] predicted that the length is $\approx W^2$. Sodin [Sod] proved that the low eigenvalues of a random band matrix form an Airy process which coincides with that of GUE or GOE provided $W \gg N^{5/6}$. His result is sharp and in particular it implies that the lowest eigenvalue fluctuates following the Tracy-Widom distribution. Recent work of Erdős, Knowles, Yau and Yin [Er3] proves results about eigenstates and diffusive time evolution for $W^{5/4} \gg N$. Wigner-Dyson statistics for local eigenvalue correlations are expected to occur for $E \in (-2,2)$ provided that $W^2 \gg N \gg 1$, [FM]. In three dimensions Wigner-Dyson statistics should hold for $W \gg 1$ fixed as N gets large. See [Er4, Sp2] for recent mathematical reviews of sparse matrices and RBM.

1.3. Duality and Green's functions. The statistical mechanics approach to the spectral theory of random matrices described in these lectures was pioneered by physicists F. Wegner [Weg], L. Schäffer [SW] and K. Efetov [Efe1] in the early 1980's. It extracts spectral information of RBM in terms of certain supersymmetric **(SUSY)** statistical lattice spin models. The word supersymmetry refers to the fact that the spins have both Grassmann (anticommuting) and real components which appear in a symmetric fashion. §2 gives a brief introduction to Grassmann variables and supersymmetric models. In theoretical physics there are a number of reviews of SUSY and random matrices, [Efe2, Ef4, Fyo, Mir, VWZ, Zir1]. This mathematical review has considerable overlap with [Sp3].

We will see that there is a precise duality between SUSY statistical mechanics and RBM. The mathematical challenge is to analyze the corresponding SUSY statistical mechanics model and to prove that spins are aligned at long distances. There are similar mathematical problems for the quantum ferromagnet, superfluid and Bose-Einstein condensation.

We begin with a simple example of our notion of duality given by Stirling's formula:

$$N! = \int_0^\infty e^{-t} t^N \, dt = N^{N+1} \int_0^\infty e^{-N[s-\ln s]} \, ds \approx N^N e^{-N} \sqrt{2\pi N}\,. \qquad (1.8)$$

To obtain the right side we set $t = Ns$, and note that the integrand is maximized for $s^* = 1$. Fluctuations about the maximum of $-[s - \ln s] = -1 - (s-1)^2/2 + \ldots$ are Gaussian as indicated by the second term of the expansion. The right side of (1.8) follows from the identity $\int e^{-Ns^2/2} ds = \sqrt{2\pi/N}$. Higher order terms produce corrections of order $1/N$.

A more sophisticated version of duality is the asymptotic formula for counting the number of partitions of a large integer N given by Hardy and Ramanujan [HR]:

$$P(N) = \frac{1}{2\pi i} \oint \frac{1}{\prod_m (1 - z^m)} \, z^{-N-1} dz \approx \frac{1}{4\sqrt{3}N} e^{\pi \sqrt{2N/3}}\,. \qquad (1.9)$$

In this case, the complex variable z is on a contour about 0 lying inside $|z| = 1$. The saddle points depend upon N and are close to the roots of unity. A detailed analysis of this integral uses the theory of modular forms. See [Apo] for an exposition of Rademacher's work, [Rad].

To illustrate a simple form of duality for random matrices, let H be an N by N Hermitian matrix with the GUE (Gaussian Unitary Ensemble) distribution proportional to $e^{-NTrH^2/2}$. The Gaussian average of the determinant of the resolvent, $(E_\varepsilon - H)^{-1}$ can be expressed as

$$\langle \det(E_\varepsilon - H)^{-1} \rangle = \frac{(-iN)^N}{(N-1)!} \int_0^\infty e^{-N(s^2/2 + iE_\varepsilon s)} s^N \, ds/s\,, \quad E_\varepsilon \equiv E - i\varepsilon,\ \varepsilon > 0\,.$$

$$(1.10)$$

For large N and $|E| < 2$ this integral is dominated by the contributions near its saddle point $s^* = -iE/2 + \sqrt{1 - (E/2)^2}$. To analyze the right side we deform the contour of integration so that it passes through s^* and the integrand achieves it maximum modulus there. Then expand to second order about s^* as in Stirling's formula.

Key Identities: The derivation of (1.10) will be given in §(2.1). It relies on the unitary invariance of $e^{-NTrH^2/2}$ and the identity

$$\pi^N \det M^{-1} = \int e^{-z^* \cdot Mz} \prod_{i=1}^{N} dx_i \, dy_i\,. \qquad (1.11)$$

Above we have set $z_j = x_j + iy_j$, $1 \leq j \leq N$ and assumed that M is an N by N matrix with $Re\, M > 0$ so that the integral is well defined. There is a similar expression for the determinant of M using Grassmann variables $\bar\psi, \psi$

$$\det M = \int e^{-\bar\psi \cdot M\psi} \prod_{i=1}^{N} d\bar\psi_i d\psi_i\,. \qquad (1.12)$$

See §(2.1) for properties of the anticommuting variables $\bar\psi_i, \psi_i$. These formulas will play a key role in the SUSY formulation of random matrices by setting $M = i(E_\varepsilon - H)$. If H is Gaussian we can calculate the expectation of (1.11) and (1.12) explicitly in terms of $\bar z, z$ and $\bar\psi, \psi$.

Statistical Mechanics: Duality for more general Gaussian N by N random matrices, such as band matrices, involves integrals of *many* spin variables rather than the single spin s above. This is the domain of statistical mechanics which naturally suggests the use of nonperturbative techniques such as steepest descent, collective coordinates and renormalization to analyze spectral properties of random matrices. We shall see that unexpected symmetries also emerge in statistical mechanics formulation of RBM and these will help explain Wigner-Dyson universality of local eigenvalue statistics.

Green's Functions: In order to investigate the spectral properties of a random matrix H one studies averages of the resolvent or Green's function

$$G(E_\varepsilon; j, k) = (E_\varepsilon - H)^{-1}(j, k) \qquad E_\varepsilon = E - i\varepsilon, \quad \varepsilon > 0 \qquad (1.13)$$

where E is real and $j, k \in Z^d \cap [-N, N]^d$ label the matrix elements. Note $Im[(E - x) - i\varepsilon]^{-1} \approx \pi\delta_\varepsilon(E - x)$, where δ_ε is a regularized Dirac delta function. Thus the imaginary part of the Green's function gives spectral information about H in an ε neighborhood of E. The ultimate goal is to study G for $\varepsilon \approx N^{-d}$ which is the typical separation of eigenvalues. In order to understand evolution of the quantum dynamics up to time T, we need good estimates on G for $\varepsilon^{-1} \approx T$. The identity

$$\pi\rho_\varepsilon(E) = Im\langle G(E_\varepsilon; 0, 0)\rangle = \varepsilon \sum_j \langle |G(E_\varepsilon; 0, j)|^2\rangle \qquad (1.14)$$

is easily proved by applying the resolvent identity to $G - \bar{G}$. It reflects conservation of probability or unitarity and is some times referred to as a Ward identity.

Note that by Cramer's rule, Green's functions may be expressed as a ratio of determinants. This is why identities (1.11) and (1.12) are useful. We shall see that the average Green's function may be identified with the correlation of a spin S_j and S_k, at temperature T(E). The Gibbs weight for the corresponding supersymmetric statistical mechanics is generally rather complicated and will later be described in special cases.

In §2.1 and §2.2 we shall explain how to use statistical mechanics to get estimates on the density of states $\rho(E)$ for GUE and random band matrices. The density of states can be expressed as

$$\pi\rho(E) = \lim \varepsilon \downarrow 0 \; Im\langle G(E_\varepsilon; 0, 0)\rangle. \qquad (1.15)$$

To obtain information about the eigenstates or time evolution we must analyze the average of the modulus squared of the Green's function which distinguishes localization and diffusion. This is considerably more difficult to analyze than (1.15). These are two basic conjectured scenarios which will depend on the dimension of the lattice, W and the energy E:

Extended States and quantum-diffusion in 3D:

$$\langle |G(E_\varepsilon; j, k)|^2\rangle \cong \frac{\rho(E)}{-D\Delta + \varepsilon}(j, k) \approx C\,(|j - k| + 1)^{-1}. \qquad (1.16)$$

for $j, k \in \mathbb{Z}^3$ and $E \in (E_m, E'_m)$. For RBM in 3D we expect $E_m \approx -2$ and $E'_m \approx 2$ for large W. Note the right side is proportional to the Laplace transform of the heat kernel. The quantum particles of energy E are conducting states and thus correspond to the metallic phase. In a large finite box the eigenstates are uniformly spread out. GUE Wigner-Dyson statistics should govern the eigenvalue spacings near E. For the real symmetric Anderson model (1.4), these statistics should be given in terms of GOE matrices. In infinite volume, energies for which $\langle |G(E_\varepsilon; j, j)|^2 \rangle$ is bounded as $\varepsilon \downarrow 0$ belong to the absolutely continuous spectrum of H with probability one.

Extended states and quantum diffusion are expected to hold for the Anderson model for small $\lambda > 0$ and for RBM with fixed large W in 3D.

Localization:

$$\langle |G(E_\varepsilon; j, k)|^2 \rangle \leq C_\ell \, \varepsilon^{-1} \, e^{-|j-k|/\ell(E)} \, . \tag{1.17}$$

Note that the factor of $1/\varepsilon$ on right hand side must be present because of (1.14) whenever $\langle |G(E_\varepsilon; 0, k)|^2 \rangle$ is summable. If (1.17) holds as $\varepsilon \to 0$ then there is an interval of dense pure point spectrum in a neighborhood of E. An alternate way to establish localization due to Aizenman and Molchanov [AM] is to prove that for some α, $0 < \alpha < 1$

$$\langle |G(E_\varepsilon; j, k)|^\alpha \rangle \leq C e^{-|j-k|/\ell(E, \alpha)} \tag{1.18}$$

uniformly in ε.

1.4. Quantum Dynamics. There is a very natural dynamical interpretation of localization and extended states.

Define: $P(t, x) \equiv \langle |\delta_x \cdot e^{itH} \delta_0|^2 \rangle_V$, $x \in Z^d$, where H is the Anderson model and $\langle \cdot \rangle_V$ denotes the expectation over V. We could also define P for a RBM in infinite volume with W fixed. P(t,x) is the probability of finding a particle at x at time t assuming it started at 0. Note that by unitarity $\sum_x P(t, x) = 1$ and the mean square displacement is defined by

$$R^2(t) \equiv \sum_x P(t, x) |x|^2 \, . \tag{1.19}$$

Roughly speaking we expect that time evolution has three basic forms:

Dynamical Localization: $R^2(t) \leq Const.$

This holds if $|\lambda|$ is large or if the localization length $\ell(E)$ is uniformly bounded for all energies. For dynamical localization to hold more generally, the initial condition must be projected onto a region of spectrum where there is a uniform bound on $\ell(E)$.

Quantum Diffusion: $R^2(t) \approx Dt$

Quantum diffusion is closely related to (1.16) and the existence absolutely continuous spectrum in 3D, for $|\lambda| \leq 1$. The intuition for diffusion comes from thinking of a classical particle being weakly scattered by the potential or impurities as it moves on $\mathbb{Z}^3$. If we ignore memory effects, one might imagine that over long time scales the particle behaves like a random walk and diffuses. Although this classical picture is appealing it is probably harder to justify than its quantum counterpart. In the case of random Schrödinger, the best rigorous result about time evolution is is due to Erdős, Salmhofer and Yau [Er1] who establish diffusion for times $\lambda^{-2} \leq t \leq \lambda^{-(2+\varepsilon)}$ for some positive ε. At earlier times the motion is ballistic. For some results on quantum diffusion for RBM, see [Er3].

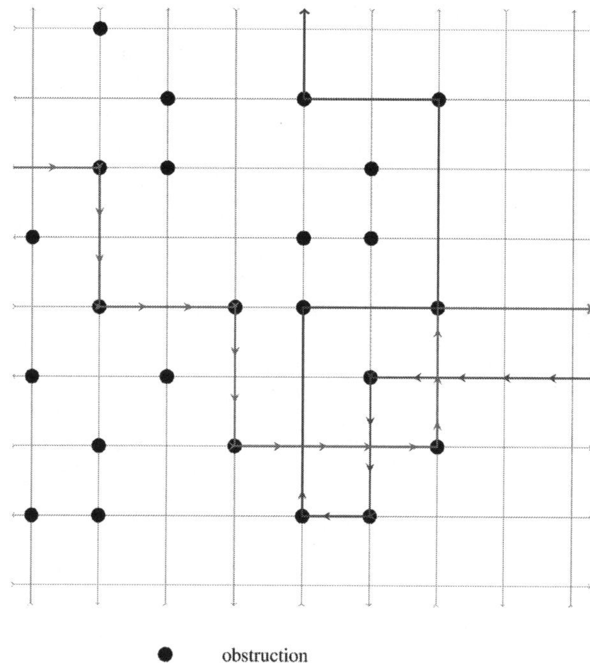

FIGURE 1. Orbits on Manhattan Lattice with obstructions

Ballistic motion: $R^2(t) \approx D\,t^2$

This can easily be established for $\lambda = 0$ by Fourier analysis. In this case there is no scattering and classically the particle moves in a straight line. It should also hold for periodic potentials using a Bloch wave analysis. Note that small λ is a *singular pertubation* since the time evolution is expected to change character from ballistic $\lambda = 0$ to diffusive in 3D when $\lambda > 0$ or localized in 1D and 2D.

Manhattan pinball: Next we describe a classical model of dynamics on the oriented Manhattan lattice with alternating orientations going north-south and east-west. See Figure 1. There are obstructions which appear independently on the vertices of the lattice with probability p. We imagine a particle moving along the edges at unit speed following the orientations of the lattice. The particle may not pass through the obstruction and turns only when it encounters an obstruction. Thus the dynamics are determined by an initial edge and the location of the fixed randomly located obstructions.

This model was proposed and analyzed by Beamond, Owczarek and Cardy [BOC] following related work of Gruzberg, Ludwig and Read [GLR]. Note that distinct orbits may pass through the same vertex but not the same edge.

Remarks: The Manhattan model looks classical but it is equivalent to a model of *unitary quantum dynamics* with $SU(2)$ edge disorder at a special energy. The phases arising from quantum mechanics are integrated out and in this case produce a positive measure. This is a quantum network model which arose from attempts to understand the Quantum Hall effect at the mobility edge. It belongs to the class C models in which time reversal symmetry is broken but Zeeman splitting is ignored.

Theorem (Chalker) If $p > 1/2$ all paths are closed with probability 1.

The proof is based on results on bond percolation in 2D. To see this draw an edge passing through each obstruction making $\pm 45°$ with the axes and connecting the midpoints of the squares of the dual lattice. The edge orientation $\pm 45°$ is selected to be plus if the vertical arrow passing through the obstruction is up and the horizontal arrow points right and minus for the down, right arrow. Such edges act as reflectors for an incoming particle. If $p > 1/2$ these edges percolate and thus trap the particle.

It is also interesting to study this model on an infinite 1D cylinder of circumference L. It is obvious that for fixed $p > 0$ the localization length is bounded by an exponential in L, p^{-L}. However, this estimate is very crude and it is natural to conjecture that the localization length is less than $C_p L$. Such an estimate is also expected for the Anderson model. The linear dependence on L is known for spin systems such as the XY and Heisenberg model (see §1.6), using a simple deformation argument in [MS]. For the edge reinforced random walk a closely related estimate was proved in [MR2].

In 2D, a renormalization group argument by Beamond, Owczarek and Cardy ([BOC]) suggests that for any $0 < p < 1$, all paths are closed. Moreover, the average diameter of an orbit $\approx e^{c/p^2}$. This assertion is compatible with the prediction by Abrahams et al. [Abr] that all states are localized for the 2D Random Schrödinger model. At the moment there is no proof of localization for any value of p with $0 < p < 1/2$.

There are related network models in 3D, [OSC]. In this case one expects that there are diffusive orbits, and $\langle (X(t) - X(0))^2 \rangle \approx Dt$. For a recent review of the relation between classical and quantum models see Cardy [Car].

1.5. SUSY models and a History Dependent Random walk. In [Weg, SW] F. Wegner used the replica trick to study spectral properties of the Anderson model in terms of spin systems with hyperbolic symmetry. Following Wegner's work, K. Efetov introduced a class of supersymmetric "spin" models on the lattice which provided a mathematically precise *dual* representation for Green's functions. Efetov expresses expectations such as (1.16) as products of ratios of determinants of $E_\varepsilon - H$ using identities closely related to (1.11) and (1.12). For Hermitian RBM the spins are 4 by 4 matrices indexed by a lattice. They have both complex components in $\mathbb{C}$ and anti-commuting or Grassmann components which appear symmetrically. Efetov identified a non compact $SU(1; 1|2)$ symmetry which contains a hyperbolic $SU(1; 1)$ part as well as compact $SU(2)$ part arising from the Grassmann variables. The statistical weight of Efetov's model is invariant under this symmetry. Real symmetric matrices such as the Anderson model are described by another supersymmetric space parametrized by 8×8 matrices.

The 3D Efetov spin model is expected to have a phase transition from ordered to disordered phases of the spins which is equivalent to the Anderson transition described above. In statistical mechanics this transition is much like that of a magnetic bar whose spins are aligned at low temperature but when heated the spins are disordered and the bar no longer exhibits magnetization. More precisely if the spins in the supersymmetric Efetov model are ordered (aligned at low temperature), the corresponding normalized eigenvectors v in a box of side N are extended: $max_j |v(j)| \approx N^{-d/2}$ as N gets large. In this case one expects quantum diffusion and the scaled eigenvalue spacing $(E_{k+1} - E_k)/\delta(E)$ should obey Wigner-Dyson statistics. On the other hand, when the spins are disordered, corresponding to high temperature, there is localization: the eigenstates decay exponentially fast about some lattice site. In this case, Poisson statistics describe the eigenvalue spacings. In §1.6 we review results about phase transitions for some classical models.

In practice, most theoretical physicists work with the nonlinear sigma model approximation to the Efetov model. In this case the matrix spins s_j are constrained to lie on a manifold and in particular $s^2 = I_4$. See §3.3 for a brief description. Thus it generalizes the well known Ising, XY and Heisenberg spins described later. Although this approximation is no longer the exact dual of a RBM, it preserves locality and symmetry and for most problems it is expected to give accurate and detailed information about the spectral theory of random matrices.

In 3 dimensions a proof of the ordered phase for the Efetov models has not been established. The key difficulty is due to the fact that weights on the spin configurations are not positive (oscillations and Grassmann variables) and there are gapless excitations responsible for diffusion. In the theoretical physics literature calculations are carried out perturbatively using spin wave and renormalization group analysis. However, in §2.3 we describe the Efetov model in one dimension where formulas are much simpler and correlations can be rigorously analyzed.

The $H^{2|2}$ model: In 1991 Zirnbauer [Zir2] introduced a simpler, supersymmetric spin system, which should have many features in common with of RBM and the Efetov model. We shall refer to it as the $\mathbf{H^{2|2}}$ **model** because the spins are vectors with values in the hyperbolic two plane augmented by two Grassmann fields. Zirnbauer established localization in one dimension. In three dimensions, [DSZ, DS] proved there is a phase transition from disorder to order. This transition is *analogous* to the conjectured Anderson transition. The $H^{2|2}$ model is defined in §3 and a sketch of the proof of the ordered phase is presented.

The Vertex Reinforced Jump Process: Although the $H^{2|2}$ model reflects many features expected to appear in quantum dynamics, it does not directly give spectral information about the Anderson model or RBM. However, recent results of Sabot and Tarres [ST] show that the $H^{2|2}$ model in d dimensions is equivalent to vertex reinforced jump process (VRJP) on $\mathbb{Z}^d$. This is a probabilistic, history dependent process in which a particle jumps to an adjacent vertex favoring those vertices at which it has spent more time. More precisely the VRJP is a continuous time process $X(t)$ where X jumps from the lattice site j to a neighbor j' with rate $\beta(1 + L_{j'}(t))$ where $L_{j'}(t)$ is the amount of time it has spent at j' up to time t. The phase transition in 3D may be interpreted as follows: For weak reinforcement (β large) the particle has diffusive behavior, whereas for strong reinforcement (β small) the particle is exponentially localized near its starting point, $Prob\{|X(t) - X(0)| \geq r\} \leq Ce^{-r/\ell}$. The parameter β is proportional to the inverse temperature in the $H^{2|2}$ model. In two dimensions it is natural to expect that the long time dynamics is localized even for weak reinforcement. We have seen that quantum dynamics in two and three dimensions is expected to have similar properties. Sabot and Tarres [ST] also show that the discrete time linearly edge reinforced random walk is given by a modified $H^{2|2}$ model.

1.6. Statistical Mechanics and Mean Field Universality. We now describe some classical spin systems on a lattice in which the spins take values in a compact symmetric space $\mathcal{S}$. Well studied examples are $s_j \in \mathcal{S} = Z_2$, $\mathbb{S}^1$, $\mathbb{S}^2$ which are referred to as the Ising model, the XY model and the Heisenberg model. Here $\mathbb{S}^m$

denotes the m-dimensional sphere. In later sections we shall consider noncompact spaces

$$\mathcal{S} = H^2 = SU(1;1)/U(1), \quad H^{2|2}, \quad U(1;1|2)/(U(1|1) \times U(1|1)) \qquad (1.20)$$

which parametrize the hyperbolic sigma model, the supersymmetric hyperbolic model and the (Hermitian) Efetov sigma model respectively. Spin systems whose targets are symmetric spaces or manifolds are often referred to as nonlinear sigma models. We shall see that the noncompact models are closely related to RBM.

The energy of a spin configuration s_j, $j \in Z^d \cap [-N, N]^d \equiv \Lambda_N$ in a periodic box of side 2N is given by the nearest neighbor ferromagnetic interaction $\sum_{|j-j'|=1}(s_j - s_{j'})^2$ with a rotationally invariant expectation given by

$$\langle A \rangle_N(\beta) = Z_{\Lambda_N}^{-1} \int A(\{s_j\}) e^{-\beta \sum_{|j-j'|=1}(s_j-s_{j'})^2} \prod_{j \in \Lambda_N} d\mu(s_j) . \qquad (1.21)$$

Here $\beta > 0$ is the inverse temperature and $d\mu(s_j)$ is the uniform measure on the sphere. The partition function Z is a normalization constant chosen so that $\langle 1 \rangle_N(\beta) = 1$.

For high temperature, (β small), the spins are disordered and correlations decay exponentially fast:

$$0 \le \langle s_k \cdot s_j \rangle_N(\beta) \le Ce^{-|k-j|/\ell(\beta)}, \qquad |j-k| \le N/2 \qquad (1.22)$$

where ℓ is the correlation length which is independent of N. Such estimates are relatively easy to prove for compact target spaces $\mathcal{S}$. This phase is analogous to localization for strong disorder.

Note that for large β, the spins tend to be locally aligned. In three dimensions the spins are ordered at long distances (uniformly in N) [FSS] and hence the magnetization, M, is non zero:

$$|\Lambda_N|^{-2} \sum_{j,k} \langle s_j \cdot s_k \rangle_N(\beta) = M_N^2(\beta) \ge [1 - \frac{m+1}{\beta} G(0,0)] > 0. \qquad (1.23)$$

Here G is the Green's function of the 3-dimensional lattice Laplacian. Note that $G(0,0)$ is finite in three dimensions but diverges in two dimensions. This corresponds to the fact that a random walk in two dimensions is recurrent whereas in 3D it is transient. If we then add a small external magnetic field in the 1 direction then the energy is defined to be

$$\sum_{|j-j'|=1} (s_j - s_{j'})^2 - \varepsilon \sum_j s_j^1 \quad and \quad M = \lim \varepsilon \downarrow 0 \langle s_0^1 \rangle(\beta, \varepsilon) \qquad (1.24)$$

where the expectation is defined in the infinite volume limit. If M in (1.24) does not vanish in the limit $\varepsilon \to 0$ then we say that there is *symmetry breaking* since $\langle s_0^1 \rangle(\beta, 0) = 0$ by symmetry. The Mermin-Wagner theorem states that components of the spin s^a, $a \ne 1$ perpendicular to the external field satisfy the lower bound

$$\sum_j e^{ij \cdot p} \langle s_0^a s_j^a \rangle(\beta, \varepsilon) \ge M^2(\beta p^2 + \varepsilon M)^{-1} \qquad (1.25)$$

for $|p_i| \le \pi$. This relation is valid in any dimension for systems with continuous symmetry and is proved using by integration by parts. In two dimensions this relation shows that M=0 as $\varepsilon \to 0$. If not, the integral over p on the right side is divergent in 2D but the integral over the left side equals $(s_0^a)^2 \le 1$. The $1/p^2$

indicates the presence of gapless modes called Goldstone modes which appear when there is continuous symmetry breaking as in (1.24) for $d \geq 3$. Note that in the noncompact case $\varepsilon > 0$ is needed to make integrations well defined and is related to the imaginary part of the energy. There is some progress in extending Mermin-Wagner type arguments to noncompact systems in [MR1, DMR].

In two dimensions, the XY model has the Kosterlitz-Thouless [KT] phase: slow power decay for large β, $< s_0 \cdot s_x > (\beta) \approx |x|^{1/(2\pi\beta)}$. An upper bound of this type for the XY was proved in [MS] and the lower bound in [FS1]. On the other hand, Polyakov [Pol] gave a renormalization argument which suggests that in the 2D Heisenberg model, spin correlations always decay exponentially fast with a correlation length $\ell(\beta) \approx e^{C\beta}$. Note the similarity to the localization length predicted for the two dimensional Anderson model, and RBM, [Abr] . In §3.1 we shall present arguments which suggest that the 2D $H^{2|2}$ model also has exponential decay for all β. A slightly more detailed discussion of spin systems appears in [Sp3].

The proof of the ordered phase (1.23) is based on reflection positivity, [FSS]. Unfortunately, this technique is not very flexible and it does not apply to the Efetov or $H^{2|2}$ models in 3D. The ordered phase for the $H^{2|2}$ model was proved in [DSZ] by using a family of identities which come from supersymmetry and by applying induction on length scales. A sketch of the proof is given in §3. T. Balaban [Bal] has developed another way to prove an ordered phase for systems with continuous symmetry via rigorous renormalization group techniques. The ordered phase for the Efetov sigma model in 3D is still open and is very closely related to the Anderson transition in 3D.

Next we formulate a mean field universality conjecture for the XY and Heisenberg models. Let Λ_N be a periodic cube of side length N and define

$$G_N(h,\beta) = \langle e^{h \cdot \sum_\Lambda S_j / |\Lambda|} \rangle_N(\beta) \quad where \ h \in R^2 \ or \ R^3. \tag{1.26}$$

Mean Field Universality Conjecture: In 3D,

$$G_N(h,\beta) \to \int e^{Mh \cdot S_0} d\mu(S_0)[1 + O(\frac{1}{\beta N})] \tag{1.27}$$

as $N \to \infty$. The right side encodes the collective behavior of the spins by a single mean field spin, S_0. The spatial lattice structure has disappeared. The set of directions is determined by the orbit of a spin under the symmetry group of the interaction. Thus the leading term of (1.27) is governed by the 0 - mode. Note that β only enters through the magnetization $M(\beta)$. Intuitively the right side is roughly obtained for large β by supposing that all the spins are aligned in some direction S_0. To second order in h (1.27) is an identity but higher order terms in h require proof.

Remark: The leading term on the right side is roughly like the law of large numbers. We expect corrections of central limit type arising from the free Gaussian field fluctuations. These are crudely accounted for in error term $[1 + O(\frac{1}{\beta N})]$. Universality of means that the details of the interaction on the left are unimportant. In particular the spins need not have nearest neighbor interaction nor need they be restricted to the sphere. Once order is established, the right side is governed by the group invariance of the spin distribution.

SUSY Motivation: The motivation for the above conjecture comes from SUSY statistical mechanics. The leading term is analogous to Wigner-Dyson local eigenvalue correlation (1.3) identified by [Ef1]. The finite volume correction formally was derived by Kravtsov and Mirlin [KM, Mir] by calculating the Gaussian fluctuations via spin wave theory. The leading term of the scaled local energy correlation (1.3) is corrected by the term $cg_N^{-2}sin^2(\pi\xi)$ where g_N is the conductance $\approx \beta N^{d-2}$. In SUSY statistical mechanics, the role of $M(\beta)$ is played by the density of states $\rho(E)$. Note however there is an important difference between the SUSY and XY or Heisenberg case. M is an order parameter for classical spin systems but $\rho(E)$ is not an order parameter for SUSY models since it does not vanish in either the localized phase or diffusive phase. Moreover, in a large box $\rho(E)$ has very weak dependence on N (see §2.2) but M_N (1.23) has a power law dependence on 1/N.

Jürg Fröhlich has recently explained a conceptual way to understand (1.27) by expressing the infinite volume expectation as a uniform superposition of pure states $<>_u (\beta)$ indexed by u in the sphere $\mathbb{S}^m$. Thus for $m = 2$ we have

$$G(h) = \int_{S^2} \langle e^{h \cdot \Sigma_\Lambda S_j/|\Lambda|} \rangle_u (\beta) d\mu(u) = \int_{S^2} e^{\langle h \cdot \Sigma_\Lambda S_j/|\Lambda|\rangle_u} d\mu(u)(1 + O(\frac{1}{\beta N}))$$

$$= \sinh(M|h|)/M|h| \,[1 + O(\frac{1}{\beta N})]. \tag{1.28}$$

The exponent of the right hand side of (1.28) is given by $Mh \cdot u$. The terms arising from the second cumulant go to zero because by infra-red bounds [FSS] the spin correlations $\langle S_j S_k \rangle_u - \langle S_j \rangle_u \langle S_k \rangle_u$ have a Fourier transform bounded by $(\beta p^2)^{-1}$ for $p \neq 0$ hence there is effectively a $1/r$ decay in the spins producing an error term as above. For the XY model, the analog of (1.28) can be proved using a theorem of Fröhlich and Pfister [FP] which states that for almost all β all states are superpositions of the form above. For more general symmetries such as those arising in SUSY with $U(1;1|2)$ symmetry the expected formulas are similar. However, rigorous spin wave analysis is needed to control fluctuations about the 0 mode.

Acknowledgments. I would like thank the organizers for inviting me talk at the conference on Current Developments in Mathematics. I thank M. Disertori, M. Shamis, T. Shcherbina and W-M. Wang for helpful comments on an earlier draft of this review.

2. SUSY statistical mechanics

In this section we shall give an introduction to SUSY statistical mechanics and apply it to obtain detailed information about the density of states for GUE and also 3D band matrices. A recent theorem of T. Shcherbina will illustrate a form of mean field universality for 1D Gaussian RBM. In one dimension, Efetov's sigma model can be analyzed and localization is established with a localization length proportional to $\beta \approx W^2\rho(E)^2$. The origins of hyperbolic symmetry and a theorem about hyperbolic σ-model in 3D are explained in §2.3.

2.1. Gaussian and Grassmann integrals.

The *dual* representation of the average of Green's function relies on averages of ratios of determinants. We first consider the inverse of the determinant expressed by a Gaussian integral identity.

Let $z = (z_1, z_2, ..., z_N)$ with $z_j = x_j + iy_j$ denote an element of $\mathbb{C}^N$ and define the quadratic form

$$[z \,;\, H\,z] = \sum_{k,j} \bar{z}_k \, H_{kj} \, z_j \,. \tag{2.1}$$

Recall $E_\varepsilon = E - i\varepsilon$. Then we can calculate the following Gaussian integrals:

$$\int e^{-i\,[z;(E_\varepsilon - H)z]} \, Dz = (-i)^N \det(E_\varepsilon - H)^{-1} \quad where \quad Dz \equiv \prod_j^N dx_j \, dy_j / \pi \tag{2.2}$$

and

$$\int e^{-i\,[z;(E_\varepsilon - H)z]} z_k \bar{z}_j \, Dz = (-i)^{N+1} \det(E_\varepsilon - H)^{-1} \, G(E_\varepsilon; k, j) \,. \tag{2.3}$$

$G(E_\varepsilon)$ is the Green's function (1.13). It is important to note that the integrals above are *convergent* provided that $\varepsilon > 0$. The quadratic form $[z; (E - H)z]$ is real so its contribution only oscillates. The factor of $i = \sqrt{-1}$ in the exponent is needed because the matrix $E - H$ has an indefinite signature when E is in the spectrum of H.

There is a similar identity in which the complex commuting variables z are replaced by anticommuting Grassmann variables ψ_j, $\bar{\psi}_j$, j = 1, 2 ... N which satisfy

$$\psi_j \psi_k = -\psi_k \psi_j, \quad \bar{\psi}_j \psi_k = -\psi_k \bar{\psi}_j, \quad \bar{\psi}_j \bar{\psi}_k = -\bar{\psi}_k \bar{\psi}_j, \quad \psi_j^2 = \bar{\psi}_j^2 = 0.$$

Let A be an $N \times N$ matrix and define

$$[\psi; A\,\psi] = \sum \bar{\psi}_k \, A_{kj} \, \psi_j \qquad D\psi \equiv \prod_j^N d\bar{\psi}_j d\psi_j \,.$$

The integral of any polynomial in $\bar{\psi}, \psi$ with respect to $D\psi$ is algebraically defined to be equal to the coefficient of the top degree monomial written as $\prod^N \psi_j \bar{\psi}_j$. For example

$$\int e^{-a[\psi;\psi]} D\psi = \int \prod^N (1 - a\bar{\psi}_j \psi_j) D\psi = \int \prod^N (1 + a\psi_j \bar{\psi}_j) D\psi = a^N \,.$$

The next two formulas are similar to those for the complex Gaussian integral

$$\int e^{-[\psi\,;\,A\psi]} D\psi = \det A, \qquad \int \psi_j \bar{\psi}_k e^{-[\psi\,;\,A\psi]} D\psi = (A^{-1})_{jk} \det A \,. \tag{2.4}$$

The bar superscript is just a label to denote a separate family of Grassmann variables. See [Ber, Mir, Sal, Sp3] for more details about Grassmann integration and its applications.

Main SUSY formulas: We shall see that Grassmann integration is extremely useful. It is used to cancel the unwanted determinant in (2.3) and enables us perform the average over the randomness in H. The basic SUSY (supersymmetric) representation for the Green's function follows from (2.3) and (2.4)

$$G(E_\varepsilon; k, j) = i \int e^{-i[z;(E_\varepsilon - H)z]} e^{-i[\psi;(E_\varepsilon - H)\psi]} z_k \bar{z}_j \, Dz \, D\psi \,. \tag{2.5}$$

This equation is the starting point for all SUSY formulas. Notice that if H has a Gaussian distribution, the expectation of (2.5) can be explicitly performed since H appears linearly. We obtain:

$$\langle tr\, G(E_\varepsilon)\rangle = i\int [z;z]\, e^{-iE_\varepsilon([z;z]+[\psi;\psi])} e^{-\frac{1}{2}\langle\{[z;Hz]+[\psi;H\psi]\}^2\rangle}\, Dz\, D\psi. \qquad (2.6)$$

The resulting lattice field model will be quartic in the z and ψ fields. Notice that if the observable $i\,[z;z]$ were absent from (2.6), then the determinants would cancel and the integral would be equal to 1 for all parameters. Thus in SUSY systems, the *partition function is identically 1*. In a similar fashion we can obtain more complicated formulas for $< G(E_\varepsilon;0,j)\bar{G}(E_\varepsilon;0,j) >$. To do this we must introduce additional variables $w \in \mathbb{C}^N$ and independent Grassmann variables $\chi, \bar\chi$ to obtain the second factor, $\bar{G}$, which is the complex conjugate of G. These spin systems are much harder to analyze.

Now let H be an $N \times N$, GUE matrix. The average in (2.6) is calculated with (1.1)

$$< \{[z;Hz]+[\psi;H\psi]\}^2 >= \{[z;z]^2 + 2[\psi;z][z;\psi] - [\psi;\psi]^2\}/N. \qquad (2.7)$$

Proof of (1.10): If we omit the Grassmann variables

$$(i)^{-N} < \det(E_\varepsilon - H)^{-1} >=< \int e^{-i[z;\,(E_\varepsilon-H)z]}\, Dz >= \int e^{-[z;E_\varepsilon z]-[z;z]^2/2N}\, Dz\,.$$

Let $r = [z,z] = \sum |z_j|^2$. Then we have

$$< \det(E_\varepsilon - H)^{-1} >= C_N \int_0^\infty e^{-\frac{1}{2N}r^2 - iE_\varepsilon r} r^{N-1}\, dr$$

After scaling $r \to Ns$ we obtain an integral of the form

$$\int_0^\infty e^{-N(s^2/2-\ln s + iE_\varepsilon s)} ds/s\,. \qquad \square$$

To calculate the average Green's function for GUE we shall apply the *Hubbard-Stratonovich* identity which converts quartic expressions in z and ψ in (2.7) into *quadratic* expressions we can explicitly calculate. Let $s_1, s_2 \in \mathbb{R}$ and note that

$$e^{-\frac{1}{2N}[z;z]^2} = c_N \int e^{-Ns_1^2/2} e^{is_1[z;z]} ds_1, \qquad e^{\frac{1}{2N}[\psi;\psi]^2} = c_N \int e^{-Ns_2^2/2} e^{-s_2[\psi;\psi]} ds_2$$

$$(2.8)$$

where $c_N = \sqrt{N/2\pi}$. Using (2.7) and (2.8) we calculate the N dimensional integral (2.6) over ψ, z in terms an integral over s_1, s_2. After integration by parts in s_1 we have

$$\frac{1}{N}tr\langle G(E_\varepsilon)\rangle = \frac{i}{N}\int [z;z]\, e^{-iE_\varepsilon([z;z]+[\psi;\psi])}\, e^{-\{[z;z]^2+2[\psi;z][z;\psi]-[\psi;\psi]^2\}/2N} DzD\psi = \langle s_1\rangle$$

$$\langle s_1\rangle \equiv c_N^2 \int s_1 e^{-N(s_1^2+s_2^2)/2}\, e^{-i(E_\varepsilon-s_1)[z;z]} e^{-i(E_\varepsilon-is_2)[\psi;\psi]}\, [1-[\psi;z][z;\psi]/N]\, ds_1 ds_2 DzD\psi$$

$$= c_N^2 \int s_1 e^{-N(s_1^2+s_2^2)/2} \frac{(E_\varepsilon - is_2)^N}{(E_\varepsilon - s_1)^N}\, R(s_1,s_2)\, ds_1\, ds_2 \qquad (2.9)$$

where

$$R = 1 - (E_\varepsilon - s_1)^{-1}(E_\varepsilon - is_2)^{-1}\,.$$

In (2.9) we have expanded $e^{-[\psi,z][z,\psi]/N} = 1 - [\psi,z][z,\psi]/N$. It produces R and is called the Fermion-Boson interaction. In this case it has a very simple form.

There are 4 saddle points

$$s_1^{*\pm} = E_\varepsilon/2 \pm i\sqrt{1 - (E_\varepsilon/2)^2}\,, \quad s_2^{*\pm} = -iE_\varepsilon/2 \pm \sqrt{1 - (E_\varepsilon/2)^2}\,. \qquad (2.10)$$

For s_1' we choose the saddle with positive imaginary part so that the pole of $(E - i\varepsilon - s_1)^{-N}$ is not crossed. Note that $s^* = (s_1^{*+}, s_2^{*+})$ is the dominant saddle because $R(s_1^{*+}, s_2^{*-}) = 0$. We shift our contour of integration so that it passes through the two saddles $(s_1^{*+}, s_2^{*\pm})$. Along this contour one checks that for E satisfying $|E| \leq 1.8$, the maximum modulus of the integrand occurs at the saddle s^*. In particular, this deformation of contour avoids the small denominator $E_\varepsilon - s_1$ occurring when $s_1 \approx E$. The Wigner semicircle law is given by

$$\rho_N(E) = \frac{Im}{N\pi}\, tr\, G(E_\varepsilon) = \frac{1}{\pi}Im < s_1 > \approx \frac{1}{\pi}Im\, s_1^*\,. \qquad (2.11)$$

A more detailed exposition of the calculations presented here can be found in [Dis]. See also [Sha] where the GOE case is calculated without using Hubbard-Stratonovich. The main result for GUE is that for $|E| \leq 2 - \delta,\, \delta > 0$

$$\rho_N(E) = \frac{1}{\pi}\sqrt{1 - (E/2)^2} + \frac{1}{N}Osc_N(E) + O(N^{-3/2}) \qquad (2.12)$$

where $Osc_N(E)$ is a bounded but highly oscillatory contribution which can be explicitly computed. This contribution arises from a careful analysis of the second saddle. At the edges of the spectrum $E = \pm 2$ these saddle point analysis must be refined because the Hessian vanishes at the saddle so that the leading contribution is cubic. This gives rise to an Airy function instead of Wigner's semicircle. See [Dis].

2.2. Density of states for Gaussian RBM. The above results for GUE can also be carried out by using the classical methods of orthogonal polynomials. In fact, asymptotics of Hermite polynomials are obtained by integral representations much like those above. However, for Gaussian RBM orthogonal polynomial techniques do not seem to apply and pertubative or moment methods can only be controlled when $\varepsilon \geq W^{-1}$. The SUSY statistical mechanics techniques for Gaussian RBM are natural and estimates have little dependence on $\varepsilon \to 0$. On the other hand, they are not easy to apply for non-Gaussian distributions.

Let $\Lambda \subset \mathbb{Z}^3$ be a cube of side N with periodic boundary conditions. We shall assume that J in (1.6) is

$$J_{jk} = (-W^2\Delta_\Lambda + 1)^{-1}(j,k) \approx C\,\frac{e^{-|j-k|/W}}{W^2|j-k|} \qquad (2.13)$$

where Δ_Λ is the lattice Laplacian with periodic boundary conditions. For a random band or GUE matrix the density of states is given by the limit as $N \to \infty$

$$\rho_N(E) = \lim \varepsilon \downarrow 0\, \frac{Im}{\pi(N)^d}\langle tr(E_\varepsilon - H)^{-1}\rangle\,. \qquad (2.14)$$

The average Green's function for Gaussian *band* matrices in three dimensions is a supersymmetric statistical mechanics model.

The analysis of random band matrices in [DPS] is in the same spirit as above except that we have many variables. Let $S = (S_1(j), S_2(j)) \in \mathbb{R}^2$, $j \in \Lambda \cap \mathbb{Z}^d$,

$$\rho_\Lambda(E, \varepsilon) \equiv \frac{1}{|\Lambda|}\langle tr(H - E - i\varepsilon)^{-1}\rangle_{RBM} = \langle S_1(0)\rangle_{SUSY}$$

$$= C_\Lambda \int S_1(0) e^{-\Sigma_j [\mathbf{W}^2 (\nabla S)(j)^2 + S(j)^2]/2} \cdot \mathbf{R} \cdot \prod_{j \in \Lambda} \frac{(i\, S_2(j) - E_\varepsilon)}{(S_1(j) - E_\varepsilon)} dS_1(j) dS_2(j) \ . \quad (2.15)$$

The band width W is large but fixed and $\Lambda \uparrow \mathbb{Z}^d$ as in statistical mechanics. The Grassmann variables have all been integrated out and produce the Fermion-Boson factor coupling S_1 and S_2 given by

$$\mathbf{R} = \det\{-W^2 \Delta_\Lambda + 1 - \delta_{ij}(S_1(j) - E_\varepsilon)^{-1}(i\, S_2(j) - E_\varepsilon)^{-1}\} \ . \quad (2.16)$$

For large W the spins are nearly equal and thus (2.15) is similar to the formula for GUE.

Although this formula looks rather complicated it can be analyzed rigorously as $\varepsilon \to 0$. We deform the contour of integration as we did for the GUE case. An essential feature of this model which makes it accessible to analysis is that the Hessian of the total action at the saddle point has a bounded inverse which decays exponentially. The cluster expansion [Sal] takes advantage of this fact and shows that spins at distances longer than W are approximately independent. Thus estimates in a large box Λ can essentially be reduced to a box of side W. Earlier work by Constantinescu et al. [Con] on Wegner's N orbital model obtained similar results also using SUSY statistical mechanics.

The main results of [DPS] are that for fixed $W \gg 1$, $\rho(E)$ is smooth and it has an asymptotic expansion in W^{-1} .

Theorem Let d=3, $|E| \leq 1.8$ and J be given by (2.13). For $W \geq W_0$, $\rho(E)$ is smooth and the average Green's function for RBM $< G(E_\varepsilon, j, j) >$ is uniformly bounded in ε *and* Λ. It is approximately given by the semicircle distribution with corrections of order $1/W^2$. Moreover, we have

$$| \langle G(E_\varepsilon; 0, x)\, G(E_\varepsilon; x, 0) \rangle | \leq C e^{-m|x|} \quad (2.17)$$

for $m \propto W^{-1}$.

Remark: One expects smoothness of the density of states for the Anderson model. However, this is a more difficult problem to analyze due to oscillatory contributions in SUSY produced by Δ.

Remark: If we consider $< |G(E_\varepsilon; j, k)|^2 >$ given by (1.13) or averages such as $< \det^2(H - E) >$, we shall see that the Hessian has eigenvalues near 0 and there is slow decay of correlations which is related to quantum diffusion. This is also more difficult to control rigorously because of long range correlations.

2.3. Hyperbolic symmetry and $\langle |\det(E_\varepsilon - H_\Lambda)|^{-2} \rangle$**.** In this section we describe how hyperbolic symmetry arises and briefly describe results about hyperbolic spins systems on the lattice. We shall see that there are interesting new features due to the continuous symmetries that arise as $\varepsilon \to 0$. These symmetries can produce gapless modes in 3D called Goldstone modes. In this case, rather than a saddle point we have a *saddle manifold*.

In section 2.1 we derived a simple formula for the average of $\det(E_\varepsilon - H)^{-1}$ for the GUE distribution. We now study the expressions of the form $\langle |\det(E_\varepsilon - H)|^{-2} \rangle$ which are associated with hyperbolic symmetry. To make this symmetry clear let us consider the very simple case when $N = 1$ and $H = h$ is a real Gaussian variable of unit variance. To simplify notation set E=0, and use two complex variables z, w to get the identity

$$\langle |\det(E_\varepsilon - H)|^{-2} \rangle = \langle |(E_\varepsilon - h)|^{-2} \rangle = \frac{1}{\pi^2} \int \langle \exp[ih(z^*z - w^*w) - \varepsilon(z^*z + w^*w)] \rangle \, dz \, dw$$

$$= \frac{1}{\pi^2} \int \exp[-\frac{1}{2}(z^*z - w^*w)^2 - \varepsilon(z^*z + w^*w)] \, dz \, dw \approx C\varepsilon^{-1} . \qquad (2.18)$$

Note that ε breaks the hyperbolic symmetry so that the integral is well defined. If we had no absolute value, we would get $(z^*z + w^*w)^2$ and hence (2.18) would be convergent even as $\varepsilon \to 0$.

In three dimensions, the hyperbolic sigma model was analyzed in [SZ]. The spins h_j take values in a hyperboloid $h_j = (x_j, y_j, z_j)$ which satisfies the constraint $z_j^2 - x_j^2 - y_j^2 = 1$ with $z > 0$. The nearest neighbor Gibbs weight is proportional to

$$e^{-\beta \sum_{j,j' \in \Lambda} h_j \cdot h_{j'} - \varepsilon \sum_j z_j} \qquad (2.19)$$

where $h \cdot h' = zz' - xx' - yy'$ and the sum above ranges over nearest neighbor j, j' in a 3D periodic box Λ. This model describes $\langle |\det(E_\varepsilon - H_\Lambda)|^{-2} \rangle$ where H is a 3D RBM in the sigma model approximation. It is a special case of Wegner's hyperbolic sigma model.

The analysis of the hyperbolic sigma model relies on the horospherical parametrization of the hyperboloid:

$$z = \cosh t + s^2 e^t / 2, \quad x = \sinh t - s^2 e^t / 2, \quad y = s e^t \qquad (2.20)$$

where $s, t \in \mathbb{R}$. This parametrization is a consequence of the Iwasawa decomposition. It plays a crucial role here and in the $H^{2|2}$ model in §3. The Gibbs weight now has the form $e^{-A(s,t)} \prod e^{t_j} ds_j dt_j$ where

$$A(s,t) = \beta \sum_{j \sim j' \in \Lambda} [cosh(t_j - t_{j'}) + \frac{1}{2}(s_j - s_{j'})^2 e^{(t_j + t_{j'})}] + \varepsilon \sum_{j \in \Lambda} [cosh(t_j) + \frac{1}{2}s^2 e^{t_j}] . \qquad (2.21)$$

The advantage of these coordinates is that the action is *quadratic* in the s_j variables. In fact we shall see in §3.1 that the quadratic form in s is the generator of a random walk in a random environment with local conductance $e^{t_j + t_{j'}}$ across each edge j, j' of the lattice. After integrating out the s_j we get an effective action of t_j which is convex for all β, ε, thus there is *no phase transition* for this sigma-model. D. Brydges has given a simple argument for convexity based on the matrix tree theorem. Hence the Brascamp-Lieb [BL] inequalities can be applied. The Hessian is bounded below as a quadratic form by $-\beta\Delta$. Let $G_0 = (-\beta\Delta + \varepsilon)^{-1}$.

Theorem ([SZ]) In the three dimensional hyperbolic sigma model all moments of the form $< cosh^p(t_0) >$ are bounded for all β. The estimates are uniform in ε provided we first take the limit $\Lambda \to \mathbb{Z}^3$. Moreover there is a constant C such that the spin correlations satisfy

$$\frac{1}{C}[f; G_0 f] \leq \sum_{j,k} f(j)f(k)\langle y_j \, y_k \rangle(\beta, \varepsilon) \leq C[f; G_0 f] \qquad (2.22)$$

where $f(j) \geq 0$ is any function with rapid decay. This estimate indicates that correlations $\langle y_j \, y_k \rangle(\beta, \varepsilon)$ have a slow decay $\approx |j - k|^{-1}$ matching the Green's function of $-\Delta$ in 3D. See [Sp3] for a slight improvement of the results of [SZ].

2.4. Efetov's SUSY σ-model in one dimension. Efetov's supersymmetric sigma model with $SU(1,1|2)$ is used to describe many quantum systems with disorder and is widely studied in theoretical physics. The sigma model is an approximation to the exact SUSY dual statistical mechanics model which describes RBM. The sigma approximation preserves locality and the symmetry of the exact model. It is expected to describe all the qualitative features of RBM in 3D and as well as 1D and 2D systems. In 3D this model seems to be well beyond the reach of present rigorous mathematical techniques.

However, in one dimension the Efetov sigma model has a very appealing form which we describe below. The Grassmann variables can be explicitly integrated out and the resulting model is a classical Heisenberg coupled to a hyperbolic sigma model which can be simply analyzed. In [Ef3] Efetov derived the following expression for a 1D sigma model describing the conductance in a chain of length L. It is a nearest neighbor spin model with *positive* weights given as follows. Let h_j and σ_j take values in a hyperboloid and the sphere S^2 respectively, see §1.6, §2.3. The Gibbs weight is then proportional to

$$\prod_{j=0}^{L} \left(h_j \cdot h_{j+1} + \sigma_j \cdot \sigma_{j+1}\right) e^{\beta(\sigma_j \cdot \sigma_{j+1} - h_j \cdot h_{j+1})} \ . \tag{2.23}$$

To make the integral well defined we fix each of the spins $h_0, \sigma_0, h_L, \sigma_L$ equal to $(0,0,1)$. Physically this corresponds to placing leads at the end of the chain to measure its conductance. As in classical statistical mechanics, the parameter $\beta > 0$ is referred to as the inverse temperature and $\beta \approx W^2 \rho(E)^2$ where $\rho(E)$ is the density of states, and W is the band width. Both the compact and noncompact symmetries are apparent. The Fermion-Boson coupling is just given by the first factor in the product above. In higher dimensions, integrating out the Grassmann variables leads to much more complicated expressions.

In 1D when $\beta \gg 1$, the spins are aligned provided $L \ll \beta \approx \rho^2 W^2$. At longer lengths L the spins are disordered and the conductance of the chain goes to 0, as $e^{-cL/\beta}$. Once the explicit integration over the Grassmann variables is performed, the proof of these statements is straightforward and will appear in joint work with Disertori following ideas of M. Zirnbauer.

Remark: We believe that similar results hold without the sigma approximation. This would imply localization for 1D RBM with $\ell(E) \approx W^2 \rho(E)^2$.

2.5. Average of Determinants and the Heisenberg model. In this section we consider Gaussian random band matrices H with covariance $J_{j,k}$ given by (2.13) in a periodic box of side N. This is often referred to as the Fermion-Fermion sector. Define

$$F_N(E, E') = \langle \det(H - E) \det(H - E') \rangle . \tag{2.24}$$

We shall use Grassmann integration and duality to show that this expression is closely related to the Heisenberg model described in §1.6. The sphere parametrizes the zero modes. Let $\bar{\psi}_j, \psi_j, \bar{\chi}_j, \chi_j$ denote Grassmann variables indexed by $j \in \Lambda_N$. Then using (2.4) we have

$$\det(H - E) \det(H - E') = \int e^{-[\bar{\psi};(H-E)\psi] - [\bar{\chi}(H-E')\chi]} D\psi D\chi .$$

Let M_j be defined by

$$M_j = \begin{pmatrix} \bar\psi_j\psi_j & \bar\psi_j\chi_j \\ \bar\chi_j\psi_j & \bar\chi_j\chi_j \end{pmatrix}$$

and set $\tilde E = diag(E, E')$. Next we average over the H. Since H is Gaussian and has covariance J_{jk} we get

$$\langle \det(H - E)\det(H - E')\rangle = \int e^{-\sum J_{j,k} Tr\,(M_j M_k) + \sum_j Tr\,(M_j \tilde E)}\, D\psi D\chi.$$

Now we use Hubbard-Stratonovich and let X_j be 2×2 Gaussian Hermitian matrices such that

$$\langle e^{+i\,Tr(Y_j X_j)}\rangle_X = e^{-\sum J_{jk} Tr(Y_j Y_k)}$$

where the subscript is the average over X. Then we have

$$F_N(E, E') = \int \langle e^{\sum_j tr(iX_j + \tilde E)M_j}\rangle_X D\psi D\chi. \tag{2.25}$$

The integral over the M in (2.25) can be computed since the Grassmann variables appear quadratically and factor. Applying (2.4) we get

$$F_N(E, E') = \langle \prod_j \det(iX_j + \tilde E)\rangle_X = \langle (-1)^N \prod_j \det(X_j - i\tilde E)\rangle_X. \tag{2.26}$$

By (2.13) the Gibbs weight for the Gaussian measure in X is proportional to

$$e^{-\frac{1}{2}Tr \sum_j [W^2(\nabla X)_j^2 + X_j^2]} \prod DX_j \tag{2.27}$$

where DX is the Lebesgue measure on two by two Hermitian matrices.

Since W is assumed to be large, the X_j typically depend slowly on j. Hence it is natural to look for a constant saddle point (manifold) of the action and consider fluctuations about it. The saddle point satisfies the following equation:

$$\partial/\partial X \, \{Tr[-X^2/2 + \log(X - i\tilde E)]\} = 0 \quad thus \quad X^2 - i\tilde E X - I_2 = 0 \tag{2.28}$$

where $I_2 = diag(1,1)$. Let E=E'. Then the saddle point X_s has the following form:

$$\check X_s = \{iE/2 \pm \bar\rho(E)\}\, I_2 \quad or \quad \hat X_s = iE/2\, I_2 \pm \bar\rho(E)\, \sigma_3 \tag{2.29}$$

where

$$\bar\rho(E) = \pi\rho(E) = \sqrt{1 - (E/2)^2}$$

and $\sigma_3 = diag(1, -1)$. Note that $U^* \hat X_s U$ is also a critical point for any unitary U and solves (2.28). The orbit forms a saddle manifold isomorphic to the sphere.

We claim that the dominant contribution is near $\hat X_s$. To see this, consider fluctuations about a saddle X_s, $X_j = X_s + Y_j$ where Y_j is a small 2 by 2 Hermitian matrix. In terms of Y, the action is

$$\frac{1}{2}Tr \sum_j [W^2(\nabla Y)_j^2 + (Y_j + X_s)^2] - \sum_j \ln \det(Y_j + X_s - iEI_2).$$

Note that if $\overline X$ is the complex conjugate of X,

$$X_s - iEI_2 = \bar X_s \quad and \quad \bar X_s^{-1} = X_s.$$

The linear terms cancel since we are at a saddle point and the contribution of the constant terms have equal modulus. The quadratic contribution in Y is

$$\frac{1}{2}Tr\left[W^2(\nabla Y)_j^2 + (Y_j^2) - (X_s Y_j)^2\right].$$

If $X_s = \check{X}_s$ then

$$Tr[(Y_j^2) - (X_s Y_j)]^2 = TrY^2[1 - \{iE/2 \pm \bar{\rho}(E)\}^2]$$

Note that real part of the right side is definite for $|E| < 2$. However, if $X_s = \hat{X}_s$ then the off diagonal elements of Y cancel:

$$Tr[(Y_j^2) - (\hat{X}_s Y_j)^2] = y_{11}^2(1 - \{iE/2 + \bar{\rho}(E)\}^2) + y_{22}^2(1 - \{iE/2 - \bar{\rho}(E)\}^2)$$

In conclusion, although the two saddles have equal contributions in modulus, the Hessian about the second saddle produces the dominant contribution because of the coefficient of the variables $|y_{12}|^2$ vanishes. This means that the saddle manifold, $\mathbb{S}^2$, should govern the large W behavior. Since the coefficient $|y_{12}|^2$ vanishes of our Hessian has two (zero) Goldstone modes corresponding to the dimension of $\mathbb{S}^2$.

In the sigma model approximation we fix the eigenvalues of X. Thus for $U_j \in SU(2)$ we have

$$X_j = iE/2\, I_2 + \bar{\rho}(E)\, U_j^* \sigma_3 U_j \equiv iE/2\, I_2 + \bar{\rho}(E)\, S_j. \tag{2.30}$$

In this approximation we get the Heisenberg model described in §1.6 at inverse temperature $\beta = W^2\bar{\rho}^2$. If the Heisenberg model is ordered then one might expect that the saddle manifold produces the main contribution to F_N in (2.26). If we substitute (2.30) into (2.26) and let $\tilde{E} = diag(E, E') = EI_2 + \xi/\rho N^{-d}\sigma_3$, $e_3 = (0, 0, 1)$ we get

$$\frac{F_N(E + \xi/\rho N^d,\ E - \xi/\rho N^d)}{F_N(E, E)} \approx \langle e^{2i\pi\xi N^{-d}\sum_j e_3 \cdot S_j}\rangle_N(\beta).$$

The expression on the right should be compared to (1.28).

The informal discussion above is made precise in 1D for RBM by a theorem of T. Shchberina.

Theorem ([Shc2]). Let H denote a Gaussian RBM with covariance given by (2.13). Let F_N be defined by (2.24). In one dimension, for $W^2 \gg N$ and $|E| < 2$

$$\frac{F_N(E + \xi/\rho N,\ E - \xi/\rho N)}{F_N(E, E)} \to \frac{\sin(2\pi\xi)}{2\pi\xi} \quad as \quad N \to \infty. \tag{2.31}$$

Remarks: The theorem also holds for GUE matrices, hence it illustrates the validity of mean field theory discussed in §1.6. In one dimension the Heisenberg model is ordered at distances less than $\beta \approx \rho^2 W^2$. However, if $N \gg W^2$ the spins are independent at long distances in 1D so the left side should go to 1. In three dimensions, it is natural to conjecture that (2.31) holds for fixed large W as N gets large since the 3D Heisenberg model is ordered at all distances, [FSS]. More general products of 2m determinants of Wigner matrices were studied in [Shc1]. For $m \geq 2$ the expectation of product is close to that for GUE except there are simple corrections arising from the fourth cumulant of the distribution.

Remark: In [Shc3], Shcherbina has recently applied related arguments to a more complicated SUSY model to prove universality of the pair correlation ρ_2 for RBM with any finite number of blocks.

3. A phase transition for the $H^{2|2}$ model

In this section we describe results for a simpler version of the supersymmetric Efetov models which was introduced by Zirnbauer in (1991), [Zir2]. This is a system of interacting spins indexed by $\mathbb{Z}^d$. We call this model the $H^{2|2}$ model since the spins have two real components in a hyperboloid and two Grassmann partners to make it supersymmetric. It is expected to qualitatively reflect the phenomenology of Anderson's model described in §1.2 .

The $H^{2|2}$ is related to the hyperbolic sigma model discussed in §2.3 but the additional Grassmann variables change the character of the model. A key feature of the $H^{2|2}$ model is that in horospherical coordinates the Grassmann degrees of freedom can be explicitly integrated out to produce a real effective action with positive weights. Thus probabilistic methods can be applied. In fact we will show that correlations in the model are expressed as a random walk in a correlated random environment.

In three or more dimensions, [DSZ, DS] proved that the $H^{2|2}$ model has a phase transition which is *analogous* to the *Anderson transition*. In the ordered phase $\beta \gg 1$, there are gapless modes and diffusion. For $0 < \beta \ll 1$ there is exponential localization. The analysis of the phase transition relies heavily on identities arising from symmetries and on the study of a random walk in a strongly correlated random environment, see (3.5) below. The generator of this walk is not uniformly elliptic. If it were, we would always be in a diffusive phase. At low temperature in 3D we must control fluctuations of the environment. These are controlled by proving inductively that fluctuations of the local conductance, are bounded at successively longer length scales. Once strong estimates on these fluctuations are proved, diffusion follows since then our generator is effectively elliptic. When $0 < \beta \ll 1$ the effective conductance goes to 0.

As mentioned in §1.5, Sabot and Tarres proved that the $H^{2|2}$ model is equivalent to a history dependent process on $\mathbb{Z}^d$ called the vertex reinforced jump process (VRJP). In this process the particle jumps to an adjacent site favoring those where it has spent more time. The transition in 3D can be described as a transition from localization and recurrence (strong reinforcement, high temperature) to diffusion and transience for weak reinforcement. In two dimensions we conjecture that the walk does not diffuse even when the reinforcement is weak. As in the Anderson model or Heisenberg model we conjecture that the walk localizes about its initial point with a localization length of $e^{c\beta}$ where β is inversely proportional to the magnitude of the reinforcement. See §3.1 for additional comments.

3.1. Definition of the model and Theorems. In order to define the $H^{2|2}$ sigma model, let u_j be a vector at each lattice point $j \in \Lambda \subset \mathbb{Z}^d$ with three real components and two Grassmann components $u_j = (z_j, x_j, y_j, \xi_j, \eta_j)$, where ξ, η are odd elements and z, x, y are even elements of a real Grassmann algebra. The scalar product is defined by

$$(u, u') = -zz' + xx' + yy' + \xi\eta' - \eta\xi' , \qquad (u, u) = -z^2 + x^2 + y^2 + 2\xi\eta \quad (3.1)$$

and the action is obtained by summing over nearest neighbors j, j'

$$\mathcal{A}[u] = \frac{1}{2} \sum_{(j,j')\in\Lambda} \beta(u_j - u_{j'}, u_j - u_{j'}) + \sum_{j\in\Lambda} \varepsilon_j(z_j - 1). \quad (3.2)$$

The sigma model constraint, $(u_j, u_j) = -1$, is imposed so that the field lies on the SUSY hyperboloid, $H^{2|2}$.

We choose the branch of the hyperboloid so that $z_j \geq 1$ for each j. It is very useful to parametrize this manifold in horospherical coordinates:

$$x = \sinh t - e^t \left(\tfrac{1}{2}s^2 + \bar{\psi}\psi\right) , \quad y = se^t, \quad \xi = \bar{\psi}e^t , \quad \eta = \psi e^t,$$

and

$$z = \cosh t + e^t \left(\tfrac{1}{2}s^2 + \bar{\psi}\psi\right) \tag{3.3}$$

where t and s are real and $\bar{\psi}$, ψ are odd elements of a real Grassmann algebra.

In these coordinates, the weight of a field or spin configulation is $e^{-\mathcal{A}}$ where the sigma model action $\mathcal{A}$ is given by

$$\mathcal{A}[t, s, \psi, \bar{\psi}] = \sum_{(ij)\in\Lambda} \beta(\cosh(t_i - t_j) - 1)$$

$$+\tfrac{1}{2}[s; D_{\beta,\varepsilon}s] + [\bar{\psi}; D_{\beta,\varepsilon}\psi] + \sum_{j\in\Lambda} \varepsilon_j(\cosh t_j - 1). \tag{3.4}$$

Here $D_{\beta,\varepsilon} = D_{\beta,\varepsilon}(t)$ is the generator of a *random walk in random environment*, given by the quadratic form

$$[v \,;\, D_{\beta,\varepsilon}(t)\, v]_\Lambda \equiv \beta \sum_{(j\sim j')} e^{t_j + t_{j'}} \, (v_j - v_{j'})^2 + \sum_{k\in\Lambda} \varepsilon_k\, e^{t_k} v_k^2 . \tag{3.5}$$

If $t_j = 0$ then D is the lattice Laplacian. Note that the action is quadratic in the Grassmann and s variables. We define the corresponding expectation by $< \cdot > = < \cdot >_{\Lambda,\beta,\varepsilon}$.

The weights, $e^{t_j + t_{j'}}$ in (3.5), are the *local conductances* across a nearest neighbor edge j, j'. The $\varepsilon_j\, e^{t_j}$ term is a killing rate for the walk at j. For the random walk starting at 0 without killing, we take $\varepsilon_0 = 1$ and $\varepsilon_j = 0$ otherwise. If we set $\varepsilon_j = \varepsilon$ then ε is analogous to the imaginary part of the energy in RBM.

After integrating over the Grassmann variables ψ, $\bar{\psi}$ and the variables $s_j \in \mathbb{R}$ we get the effective field theory with action $\mathcal{E}_{\beta,\varepsilon}(t)$ and partition function

$$Z_\Lambda(\beta, \varepsilon) = \int e^{-\mathcal{E}_{\beta,\varepsilon}(t)} \prod e^{-t_j}\, dt_j \equiv \int e^{-\beta\mathcal{L}(t)} \cdot [\det D_{\beta,\varepsilon}(t)]^{1/2} \prod_j e^{-t_j} \frac{dt_j}{\sqrt{2\pi}}. \tag{3.6}$$

where

$$\mathcal{L}(t) = \sum_{j\sim j'} [\cosh(t_j - t_{j'}) - 1] + \sum_j \frac{\varepsilon_j}{\beta}[(\cosh(t_j) - 1].$$

Note that the determinant is a positive but nonlocal functional of the t_j hence the effective action, $\mathcal{E} = \mathcal{L} - 1/2 \ln Det D_{\beta,\varepsilon}$, is also nonlocal. By the matrix tree theorem $\ln Det D_{\beta,\varepsilon}(t)$ is convex as a function of t. Thus $\mathcal{L}$ and $\ln Det D$ compete. The additional factor of e^{-t_j} arises from a Jacobian. Because of the internal supersymmetry, we know that for all values of β, ε the partition function $Z(\beta, \varepsilon) \equiv 1$. This identity holds even if β is edge dependent, β_e. Some of the Ward identities needed for our analysis can be obtained by taking derivatives in β_e.

For large β the value of $\{t_j\}$ which maximizes the integrand is a constant t^* given by appendix in [DSZ]

$$1 - e^{2t^*} = (-\beta\Delta + \varepsilon e^{-t^*})^{-1}(0,0),$$

$$\text{1D: } \varepsilon\,e^{-t^*} \simeq \beta^{-1}, \quad \text{2D: } \varepsilon\,e^{-t^*} \simeq e^{-\beta}, \quad \text{3D: } t^* \approx 0. \tag{3.7}$$

In one and two dimensions ($0 < \varepsilon \ll 1$) *large negative values of t are favored* and the sensitive dependence of t^* on ε suggests disorder and localization. On the other hand in 3D order should be present for large β. For small values of β, t^* has sensitive dependence on ε for all dimensions. Although it is reasonable to expect that t^* gives a reasonable approximation to the behavior of t, note that one must be careful about interpreting it due large deviations of the t field. For example, by a Ward identity we know that $< e^{t_j} >= 1$ for all β, ε. In 1D and 2D, $< e^{t_j + t_{j'}} >$ is expected to diverge, whereas $< e^{t_j/2} >$ should become small as $\varepsilon \to 0$. One way to adapt the saddle approximation so that it is sensitive to different observables is to include the observable when computing the saddle point. For example, when taking the expectation of $e^{p\,t_0}$, the saddle is only slightly changed when $p = 1/2$ but for $p = 2$ it will give a divergent contribution when there is localization.

The *analog* of the Green's function $< |G(E_\varepsilon; 0, x)|^2 >$ of the Anderson model is given by

$$< y_0 y_x > (\beta, \varepsilon) =< s_0 e^{t_0} s_x e^{t_x} > (\beta, \varepsilon) =< e^{(t_0 + t_x)} D_{\beta,\varepsilon}(t)^{-1}(0, x) > (\beta, \varepsilon) \tag{3.8}$$

where we have used (3.3) and (3.4) to obtain these identities. This formula expresses correlations for the $H^{2|2}$ model as a random walk in a random environment. If $\{t_j = 0\}$ then we see from (3.5), that (3.8) will have diffusive behavior. The expectation also has a Schrödinger representation since

$$e^{-t} D_{\beta,\varepsilon}(t)\, e^{-t} = -\beta\Delta + \beta V(t) + \varepsilon e^{-t_j}, \quad V_j(t) = \sum_{i:|i-j|=1}(e^{t_i - t_j} - 1). \tag{3.9}$$

In one and two dimensions the value of t^* suggests a mass $\approx \varepsilon e^{-t^*} = 1/\beta$, $e^{-\beta}$ even as ε becomes small.

There are two basic Ward identities which hold for all $\beta, \varepsilon > 0$, $q \in \mathbb{R}$,

$$< e^{q\,t_j} >=< e^{(1-q)\,t_j} > \quad and \quad \sum_x < y_0 y_x >= \varepsilon^{-1}.$$

Note the similarity with the Ward identity (1.14) for the Green's function.

The following theorem gives a partial description of the ordered state of the $H^{2|2}$ model in 3D for large β. Let $G_0 = (-\beta\Delta + \varepsilon)^{-1}$.

Theorem 3.1 ([DSZ]) In 3 or more dimensions for $\beta \geq \bar{\beta} \gg 1$ the fluctuations of the field t are uniformly bounded in x, y, ε and Λ : $\langle \cosh^m(t_x - t_y)\rangle_{\Lambda,\beta,\varepsilon} \leq 2$ and the spin correlations satisfy

$$\frac{1}{C}[f; G_0 f] \leq \sum_{j,k} f(j)f(k)\langle y_j\, y_k\rangle(\beta, \varepsilon) \leq C[f; G_0 f] \tag{3.10}$$

for $f(j) \geq 0$.

Remarks: Note that (3.10) is similar to (2.22) for the hyperbolic sigma model. The lower bound in [SZ] and [DSZ] was improved by Y. Capdebosq, see [Sp3]. The proof Theorem (3.1) is different from that for the hyperbolic sigma model because the effective action is not convex.

The next theorem describes the disordered or localized phase of $H^{2|2}$ model for small β in any dimension d .

Theorem 3.2 ([DS]) If $0 < \beta < \bar{\beta}$ and $\bar{\beta}^{1/2}\log(\bar{\beta}^{-1}) < 1/(2d-1)$, there is a finite localization length ℓ such that

$$0 \leq\, < y_0 y_x > (\beta,\varepsilon) \leq \frac{C_\ell}{\varepsilon} e^{-|x|/\ell(\beta)} \,. \tag{3.11}$$

For a one dimensional chain $\ell(\beta) \leq C/\beta$ for all β.

Remarks: The estimate on $\bar{\beta}$ in the disordered phase in 3D is probably reasonably sharp. However, in the 3D ordered phase $\bar{\beta}$ is much larger than it is expected to be. Recent work of Disertori, Merkl and Rolles [DMR] establishes exponential localization for one dimensional strips of arbitrary width for any $\beta < \infty$. In [MR2] similar results were proved for the linearly edge reinforced random walk (ERRW) which is closely related to a variant of the $H^{2|2}$ model. In this case, the localization length is shown to be proportional to the width of the strip. In [MR1] a slow power law decay for the conductance was proved in 2D for the ERRW even for weak reinforcement. These results are similar to the power law obtained in [MS] for the 2D XY model. The proofs in [MR1, MR2, DMR, MS] use a deformation of the statistical mechanics measure and is related to Mermin-Wagner.

Hessian suggests localization in 2D? The Hessian of the action for $H^{2|2}$ model about $t_j = 0$ is approximately given by the quadratic form in $\{v_j\}$ as follows:

$$\beta \sum_j (\nabla v)_j^2 - \sum_{j,k} (\nabla v)_j (\nabla G_0)^2 (j-k)(\nabla v)_k + \frac{\varepsilon^2}{\beta^2} \sum_{j,k} v_j v_k G_0^2(j-k) \tag{3.12}$$

where $G_0 = (-\Delta + \varepsilon/\beta)^{-1}$. In 2D note that $\sum_j (\nabla G_0)^2(j)$ has a log divergence as $\varepsilon \to 0$. A similar expression holds about t = const. Since $\ln DetD(t)$ is convex, the second term of (3.12) is negative for any t. The Fourier transform of (3.12) is $(\beta - g_\varepsilon(p))p^2$. Thus the effective spin stiffness (coefficient of p^2) goes to 0 as p and $\varepsilon \to 0$. In 3D, $g_\varepsilon(p)$ is uniformly bounded thus there is only a minor correction to β. For the hyperbolic sigma model the first minus sign in (3.12) is plus so the effective spin stiffness grows in 2D.

3.2. Ward identities and sketch of proof. The proof of Theorems 3.1 and 3.2 above relies on Ward identities which are a consequence of internal supersymmetry. We focus on the proof of Theorem 3.1 since it is technically more complicated. The main goal is to show that fluctuations of the form $\langle \cosh^m(t_j - t_k)\rangle_{\Lambda,\beta,\varepsilon}$ are bounded using induction on $|j - k|$.

We state some Ward identities following [DSZ]. Let S be an integrable function of the variables x, y, z, ξ, η which is supersymmetric, i.e., it is invariant under transformations preserving

$$x_i x_j + y_i y_j + \xi_i \eta_j - \eta_i \xi_j$$

then $\int S = S(0)$. In horospherical coordinates the function S_{ij} given by

$$S_{ij} = B_{ij} + e^{t_i + t_j}(\bar{\psi}_i - \bar{\psi}_j)(\psi_i - \psi_j), \quad B_{ij} = cosh(t_i - t_j) + \frac{1}{2}e^{t_i + t_j}(s_i - s_j)^2 \tag{3.13}$$

is supersymmetric. If i and j are nearest neighbors, $S_{ij} - 1$ is a term in the action $\mathcal{A}$ given in (3.4) and it follows that the partition function $Z_\Lambda(\beta, \varepsilon) \equiv 1$. More generally for each m we have

$$1 = <S_{ij}^m>_{\beta,\varepsilon} = <B_{ij}^m \left[1 - mB_{ij}^{-1} e^{t_i + t_j}(\bar\psi_i - \bar\psi_j)(\psi_i - \psi_j)\right]>_{\beta,\varepsilon} . \qquad (3.14)$$

Here we have used that $S_{ij}^m e^{-\mathcal{A}_{\beta,\varepsilon}}$ is integrable for $\varepsilon > 0$. Since the action is quadratic in $\bar\psi, \psi$ the integration over the Grassmann variables is explicitly given using (2.4) by

$$G_{ij} = \frac{e^{t_i + t_j}}{B_{ij}} \left[(\delta_i - \delta_j); D_{\beta,\varepsilon}(t)^{-1}(\delta_i - \delta_j)\right]_\Lambda \quad and \quad 1 \leq cosh^m(t_i - t_j) \leq B_{ij}^m .$$

$$(3.15)$$

Thus we have the identity

$$<B_{ij}^m(1 - mG_{ij})> = 1 . \qquad (3.16)$$

When $|i - j| = 1$ it is easy to show that $0 \leq G_{ij}(t) \leq \beta^{-1}$ for all t. Then by (3.16)

$$m\beta^{-1} \leq \frac{1}{2} \quad \Rightarrow \quad <cosh^m(t_i - t_j)> \leq <B_{ij}^m> \leq 2 . \qquad (3.17)$$

Remarks: The vector $(\delta_i - \delta_j)$ is orthogonal to the 0 mode of $D_{\beta,\varepsilon}(t)$ in finite volume containing i and j. If we set $t_k = 0$, $s_k = 0$ for all k, then it is clear that G_{ij} is uniformly bounded by $C\beta^{-1}$ for all $i, j \in \mathbb{Z}^3$. However, in two dimensions it diverges logarithmically in $|i - j|$. From the definition of $D_{\beta,\varepsilon}$ given in (3.5) we might expect that for large β, G in (3.15) is also of order $1/\beta$ in 3D. However, there are rare configurations of $t_k \ll -1$ with k on a surface separating i and j for which G_{ij} can diverge as $\varepsilon \to 0$. To see this consider G with Neumann boundary conditions separating i and j. Then in a finite box G_{jj} diverges as ε goes to 0.

For distances $|x - y| > 1$, there is no uniform bound on G_{xy}. In 3D sufficient conditions on the field t to get the estimate $G_{xy} \leq C/\beta$ are given by

$$1 \leq \cosh(t_j - t_x) \leq B_{jx} \leq a\,|j - x|^\alpha , \quad 0 < \alpha < 1/2 , \qquad (3.18)$$

and the same for $\cosh(t_j - t_y)$. The number a is a constant, say $a > 10$. It will turn out that these estimates are needed only for the sites j in a 3D diamond-type region, R_{xy}, containing x and y. Notice that since the exponent α is positive, we are allowing larger fluctuations at larger scales. The probability that such a condition is violated will be shown to be small by induction.

These conditions described above are initially expressed in terms of supersymmetric characteristic functions $\chi_{x,y}$ inserted in (3.16). It is important to show that the nilpotent (or Grassmann) part of $\chi_{x,y}$ is not important, so we may think of $\chi_{x,y}$ in the usual classical sense. The remaining problem is to obtain unconditional estimates on the fluctuations and thereby prove Theorem 3.1. This is first done for short scales. For larger scales we use induction.

Our induction hypothesis is that

$$\left\langle \prod_{i=1}^n B_{x_i y_i}^m \right\rangle \leq 2^n \qquad (3.19)$$

holds under the assumption that the diamond-type regions associated with $i = 1, \ldots, n$ have disjoint interiors. The induction is in ℓ, defined as the maximal separation $|x_i - y_i|$ in the product over $i = 1, \ldots, n$. For $\ell = 1$ this hypothesis is easily verified.

In order to prove unconditional estimates on the fluctuations, first consider a site b in R_{xy} closest to x or y such that condition (3.18) is violated for $j = b$. We shall then prove by induction that the probability for such an event to occur is small. The inequality $B_{xy}^m < 2^m B_{xc}^m B_{cy}^m$ is used for a point c near b. Since the distances $|x - c|$ and $|c - y|$ are less than $|x - y|$, induction can be applied. The factor 2^m is offset by the small probability of the event when β is large.

3.3. The Efetov sigma model. In this section we follow [Ef1, Fyo, Mir, Sp3, Zi1] to give a brief discussion of the Efetov sigma model. In order to obtain information about averages of the form $\langle |G(E_\varepsilon; 0, j)|^2 \rangle$ we introduce a field with four components

$$\Phi_j = (z_j, w_j, \psi_j, \chi_j)$$

where z, w are complex fields and ψ, χ are Grassmann fields. Let $L = diag(1, -1, 1, 1)$, and $\Lambda = diag(1, -1, 1, -1)$. For a Hermitian matrix H, define the action

$$A(E, \varepsilon) = \Phi^* \cdot L \left\{ i(H - E) + \varepsilon \Lambda \right\} \Phi \,. \tag{3.20}$$

Note that the signature of L is chosen so that the z and w variables appear as complex conjugates of each other. Then we have the identity:

$$|G(E_\varepsilon; 0, j)|^2 = \int z_0 \bar{z}_j w_0 \bar{w}_j \, e^{-A(E, \varepsilon)} \, D\Phi \tag{3.21}$$

where

$$D\Phi \equiv Dz \, Dw \, D\psi \, D\chi \,.$$

Without the observable $z_0 \bar{z}_j w_0 \bar{w}_j$, $\int e^{-A} = 1$. The integral over Gaussian H can be calculated as in (2.6) and will produce a quartic interaction in Φ. However, now the Hubbard-Stratonovich transformation, which is usually used in the Bosonic sector, involves a subtle analytic continuation first worked out in [SW], see also [Fyo, Zir1].

Now let us define a matrix of the form

$$M = \begin{pmatrix} [BB] & [BF] \\ [FB] & [FF] \end{pmatrix}$$

where each block is a 2×2 matrix. M will be called a *supermatrix* if the diagonal blocks, BB and FF are made of commuting (even elements) variables while the off diagonal blocks FB and BF consist of odd elements in the Grassmann algebra. Define the supertrace

$$Str(M) \equiv Tr([BB] - [FF]) \,.$$

Note that for supermatrices A and B we have $Str(AB) = Str(BA)$. We define the adjoint of a supermatrix M by

$$M^\dagger = \begin{pmatrix} [BB]^* & [FB]^* \\ -[BF]^* & [FF]^* \end{pmatrix} \,.$$

The symbol * denotes the usual transpose followed by conjugation. For Grassmann variables we have $\overline{\psi_a \psi_b} = \bar{\psi}_a \bar{\psi}_b$. But $\bar{\bar{\psi}} = -\psi$ so that $\dagger$ is an involution and

$$\Phi_1^\dagger(M\Phi_2) = (M^\dagger \Phi_1)^\dagger \Phi_2 \,.$$

For $\varepsilon = 0$ the action is invariant under the action of matrices T in $SU(1; 1|2)$ which by definition satisfy:

$$T^\dagger L T = L \,. \tag{3.22}$$

As in §2.5 the sigma approximation of the spin is given by matrices

$$S_j = T_j^{-1} \Lambda T_j \qquad (3.23)$$

as T ranges over $SU(1; 1|2)$. It is the orbit of a critical point, which is proportional to Λ under the action of $SU(1, 1|2)$. Thus the matrix S ranges over a supersymmetric space $U(1, 1|2)/(U(1|1) \times U(1|1))$. The SUSY sigma model has a Gibbs density defined by

$$\exp\{-\beta Str \sum_{j \sim j'} (S_j S_{j'}) - \varepsilon\, Str \sum_j \Lambda S_j\}. \qquad (3.24)$$

If we drop the FB contribution we get the hyperbolic sigma model, (BB), and the ferromagnetic Heisenberg model, (FF). The sign of the action is correct in (3.24) but incorrect in [Sp3]. It is the FB contribution which couples these sectors and makes mathematical analysis difficult. A general discussion of supersymmetric spaces appears in [Zir3].

In a one dimensional chain of length L with $\varepsilon = 0$ except at the end points, the Grassmann variables can be explicitly integrated over producing the formula (2.31). An explicit parametrization of the 4×4 matrices S_j and integration measure is given in [Ef1, Mir]. As in §2.5 and the discussion of the Heisenberg model, fluctuations about the saddle should produce massless modes - Goldstone Bosons in 3D for the Efetov model which are responsible for quantum diffusion.

4. Concluding Remarks

(1) Universality of local eigenvalue spacing is well understood for many mean field models, [ER5]. In these lectures we have tried to present a picture for why one should expect universality of mean field theory in RBM and the Anderson model provided that the dual spin system is *ordered*. T. Shcherbina has recently applied SUSY statistical mechanics (with no approximations) to prove universality of ρ_2 for certain RBM made of GUE blocks [Shc3]. However, much more needs to be done to analyze the ordered phases of classical spin models with continuous symmetry and SUSY models whose 0 modes give Wigner-Dyson statistics.

(2) The average Green's function for a 3D Gaussian band matrix has been successfully analyzed in detail by the SUSY formalism. For weak disorder, smoothness of the density of states and exponential decay of $\langle G(E_\varepsilon; j, k) \rangle$ have not been proved for the Anderson model in 3D due to strong oscillations produced by $-\Delta - E$ in the SUSY model.

(3) Although SUSY statistical mechanics provides a natural approach for the analysis of random band matrices and also for certain history dependent walks, results about eigenvectors and universality of eigenvalue spacing for random band matrices are still very limited. The most promising avenue of success is for one dimensional band matrices of width W. When $W^2 \gg N$ one expects that the eigenstates are extended, [FM], and that the eigenvalue spacing is given by GUE or GOE for $|E| < 2 - \delta, \delta > 0$.

(4) In 2 dimensions we expect that all eigenstates are localized in the infinite volume limit. In statistical mechanics this means that even at low temperature the corresponding spin correlations decay exponentially fast at long distances. For example the Manhattan model, the vertex reinforced jump process (VRJP) and the Anderson model are expected to have localized dynamics for all non zero strengths

of disorder. An exception to this localization conjecture are quantum systems with spin orbit coupling. Renormalization calculations predict these systems exhibit superdiffusion in 2D, [Ef1, Ef2].

(5) What happens at energies E_m, E'_m or temperature T_c which mark the borderline between localization and diffusion? There has been much analysis in theoretical physics devoted to the multi-fractal behavior of eigenfunctions with energy near E_m, see for example [EM]. The behavior of localization near E_m is poorly understood and there is no upper critical dimension. This is in contrast to the Ising or XY models which have Gaussian free field behavior at T_c when the dimension $d \geq 4$. Fyodorov and Zirnbauer have suggested that the transition for the $H^{2|2}$ model may have many interesting features in three or more dimensions. This transition has been explored [DFZ], in the Migdal approximation.

References

[Abr]   E. Abrahams, P. W. Anderson, D. C. Licciardello and T. V. Ramakrishnan, Scaling theory of localization: Absence of quantum diffusion in two dimensions, Phys. Rev. Lett. 42, 673–677 (1979).

[AM]   M. Aizenman and S. Molchanov, Localization at large disorder and at extreme energies: an elementary derivation, Commun. Math. Phys. **157**, 245–278 (1993).

[And]   P. W. Anderson, Absence of Diffusion in Certain Random Lattices, Phys. Rev. **109**, 1492–1505 (1958).

[Apo]   T. Apostol, Modular Functions and Dirichlet Series in Number Theory, **41**, Graduate Texts in Mathematics, Springer Verlag.

[Bal]   T. Balaban, A low temperature expansion for classical N-vector models. I. A renormalization group flow, Commun. Math. Phys. **167** 103–154 (1995).

[Ber]   F. A. Berezin, Introduction to Superanalysis, (Reidel, Dordrecht, The Netherlands, 1987).

[BEY]   P. Bourgade, L. Erdős, H.-T. Yau, Universality of general β-Ensembles, arXiv:1104.2272.

[Bre]   E. Brezin, V. Kazakov, D. Serban, P. Wiegman, A. Zabrodin, Applications of Random matrices to physics, *Nato Science Series* **221**, Springer 2006.

[BL]   H. Brascamp and E. Lieb, On extensions of the Brunn-Minkowski and Prekopa-Leindler theorems, including inequalities for log concave functions, and with an application to the diffusion equation, *J. Func. Anal.* **22**, 366–389 (1976).

[BOC]   E. J. Beamond, A. L. Owczarek, J. Cardy, Quantum and classical localisation and the Manhattan lattice, *J. Phys. A, Math. Gen.* **36** 10251 (2003), arXiv:cond-mat/0210359.

[Car]   J. Cardy, Quantum Network Models and Classical Localization Problems, Int. J. Mod. Phys. B, **24**, 1989–2014 (2010).

[Con]   F. Constantinescu, G. Felder, K. Gawedzki, A. Kupiainen, Analyticity of density of states in a gauge-invariant model for disordered electronic systems, Jour. Stat. Phys. **48**, 365–391 (1987).

[Dei]   P. Deift, T. Kriecherbauer, K. T.-K. McLaughlin, S. Venakides, X. Zhou, Uniform asymptotics for polynomials orthogonal with respect to varying exponential weights and applications to universality questions in random matrix theory, Comm. Pure Appl. Math. **52**, 1335–1425 (1999).

[Dis]   M. Disertori, Density of states for GUE through supersymmetric approach, *Rev. Math. Phys.* **16**, 1191–1225 (2004).

[DFZ]   W. Drunk, D. Fuchs, M.R. Zirnbauer, *Migdal-Kadanoff renormalization of a nonlinear supervector model with hyperbolic symmetry*, Ann. Physik **1** 134–150 (1992).

[DMR]   M. Disertori, F. Merkl, S. Rolles, Localization for a nonlinear sigma model in a strip related to vertex reinforced jump processes, preprint.

[DPS]   M. Disertori, H. Pinson, T. Spencer, Density of States for Random Band Matrices, *Commun. Math. Phys.* **232**, 83–124 (2002), arxiv.org/math-ph/0111047.

[DS]   M. Disertori, T. Spencer, Anderson Localization for a SUSY sigma model, *Commun. Math. Phys.* **300** 659–671 (2010), arXiv.org/abs/0910.3325.

[DSZ]   M. Disertori, T. Spencer, M. R. Zirnbauer, Quasi-diffusion in a 3D Supersymmetric Hyperbolic Sigma Model, *Commun. Math Phys.*, **300**, 435–486 (2010); arxiv.org/abs/0901.1652.

[Dys]   F.J. Dyson, Statistical Theory of the Energy Levels of Complex Systems. I, II, III, J. Math. Phys. **3**, 140–175 (1962).

[Ef1]   K.B. Efetov, Supersymmetry and theory of disordered metals, *Adv. Phys.***32**, 874 (1983).

[Ef2]   K.B. Efetov, Supersymmetry in Disorder and Chaos, (Cambridge, UK, Cambridge University Press, 1997).

[Ef3]   K.B. Efetov, Minimum Metalic conductivity in the theory of localization, JETP Lett, **40**, 738–741 (1984).

[Ef4]   K.B. Efetov, Anderson Localization and Supersymmetry, In: 50 years of Anderson Localization, ed. by E. Abrahams, *World Scientific Publishing Company 2010*, arXiv.org/abs/1002.2632.

[Er1]   L. Erdős, M Salmhofer, H.-T. Yau, Quantum diffusion of the random Schrödinger evolution in the scaling limit, Acta Mathematica **200**, 211–277 (2008).

[Er2]   L. Erdős S. Peche, J. Ramírez, B. Schlein, and H-T. Yau, Bulk universality for Wigner matrices. Commun. Pure Appl. Math. **63**, No. 7, 895—925 (2010).

[Er3]   L. Erdős, A. Knowles, H.-T. Yau, J. Yin Delocalization and Diffusion Profile for Random Band Matrices, preprint, arXiv:1205.5669v3.

[Er4]   L. Erdős, H.-T. Yau, Universality of local spectral statistics of random matrices. Bull. Amer. Math. Soc. **49**, 377–414 (2012).

[Er5]   L. Erdsős, Universality for random matrices and log-gases, preprint, arXiv:1212.0839.

[EM]   F. Evers, A.D. Mirlin, Anderson Transitions, Rev. Mod. Phys. **80**, 1355–1417 (2008).

[FP]   J. Fröhlich, C. Pfister, Spin Waves, Vortices, and the Structure of Equilibrium States in the Classical X Y Model, Commun. Math. Phys. **89**, 303–327 (1983).

[FS1]   J. Fröhlich, T. Spencer, The Kosterlitz-Thouless Transition, Commun. Math. Phys. **81**, 527–602 (1981).

[FS2]   J. Fröhlich and T. Spencer, Absence of diffusion in the Anderson tight binding model for large disorder or low energy, Comm. Math. Phys. **88** (1983), 151–184.

[FSS]   J. Fröhlich, B. Simon, T. Spencer, Infrared bounds, phase transitions and continuous symmetry breaking, Commun. Math. Phys. **50** 75–95 (1976).

[FM]   Y.V. Fyodorov and A.D. Mirlin, Scaling properties of localization in random band matrices: a σ-model approach, Phys. Rev. Lett. **67**, 2405–2409, (1991).

[Fyo]   Y.V. Fyodorov, Basic Features of Efetov's SUSY. In: Mesoscopic Quantum Physics, (Les Houches, France, 1994), ed. by E. Akkermans et al.

[GLR] I.A. Gruzberg, A.W.W. Ludwig, N. Read, Exact exponents for the spin Quantun Hall transition, *Phys. Rev. Lett.* **82**, 4524–4527 (1999).

[GMP]   I. Goldsheid, S. Molchanov and L. Pastur, One-dimensional random Schrödinger operator has a pure point spectrum, Funct. Anal. Appl. **11**, 1–10 (1977).

[HR]   G. H. Hardy and S. Ramanujan, Asymptotic Fomulae in Combinary Analysis, Proc. London Math. Soc. series 2, **17**, 75–115 (1918).

[KS]   J.P. Keating and N.C. Snaith, Random Matrix Theorey and L functions at $s = 1/2$, Commun. Math. Phys. **214**, 91–110 (2000).

[KM]   V.E. Kravtsov and A.D. Mirlin, Level statistics in a metallic sample: Corrections to the Wigner–Dyson distribution JETP Lett., **60**, 645–649 (1994).

[KT]   J.M. Kosterlitz and D.J. Thouless, Ordering, metastability and phase transitions in two-dimensional systems, *J. Phys. C* **6**, 1181 (1973).

[Mir]   A.D. Mirlin, Statistics of energy levels. In: New Directions in Quantum Chaos, (Proceedings of the International School of Physics "Enrico Fermi", Course CXLIII), ed. by G.Casati et al. (IOS Press, Amsterdam), 223–298 (2000), arXiv.cond-mat/0006421.

[MR1]   F. Merkl and S.W.W. Rolles, Bounding a Random Environment for Two-dimensional Edge-reinforced Random Walk, Elec. J. of Probability **13** 530–565 (2008).

[MR2]   F. Merkl and S.W.W. Rolles. Edge-reinforced random walk on one-dimensional periodic graphs. Probab. Theory Related Fields, **145**(3–4):323–349 (2009).

[MS]   O. McBryan, T. Spencer, On the decay of dorrelations in SO(n) symmetric ferromagnets, Commun. Math. Phys. **53**, 299–302 (1977).

[OSC]   M. Ortuño, A. M. Somoza and J. T. Chalker, Random Walks and Anderson Localization in a Three-Dimensional Class C Network Model, Phys. Rev. Lett. **102**, 070603–7 (2009).

[PS1]   L. Pastur, M. Shcherbina; Universality of the local eigenvalue statistics for a class of unitary invariant random matrix ensembles. J. Stat. Phys. **86**, 109–147 (1997).

[PS2]   L. Pastur and M. Shcherbina; Eigenvalue distribution of large Random Matrices; Mathematical Surveys and Monographs **171**, American Mathematical Society (2011).

[Pol]   A.M. Polyakov, Interaction of Goldstone particles in two dimensions. Phys. Lett. **59B** 79–81 (1975).

[Rad]   H. Rademacher, On the Expansion of the Partition Function in a Series, Ann. of Math. **44** 416–422 (1943).

[RS]   Z. Rudnick and P. Sarnak, Zeros of principal L-functions and random-matrix theory, *Duke Math. J.* **81**, 269–322 (1996).

[Sal]   M. Salmhofer, Renormalization: An Introduction, (Springer, 1999).

[Sch]   J. Schenker, Eigenvector Localization for Random Band Matrices with Power Law Band Width, *Commun. Math. Phys.* **290**, 1065–1097 (2009).

[Sha]   M. Shamis, Density of States for GUE and GOE through a supersymmetric approach, preprint.

[Shc1] T. Shcherbina, On the correlation function of the characteristic polynomials of the hermitian Wigner ensemble, *Commun. Math. Phys.*, **308**, 1–21 (2011); http://arxiv.org/abs/1006.2536v1.

[Shc2] T. Shcherbina On the second mixed moment of the characteristic polynomials of the 1D band matrices, http://arxiv.org/abs/1209.3385v3.

[Shc3] T. Shcherbina, Universality of the local regime for the block band matrices with a finite number of blocks, preprint.

[Sod]   A. Sodin, The spectral edge of some random band matrices, *Annals of Mathematics*, **172**, 2223 (2010).

[Sp1]   T. Spencer, Mathematical aspects of Anderson Localization. In: *50 years of Anderson Localization*, ed. by E. Abrahams (World Scientific Publishing Company, 2010).

[Sp2]   T. Spencer, Random Band and Sparse Matrices, ed. G. Akemann, J. Baik, P. Di Francesco, in *The Oxford Handbook of Random Matrix Theory* (Oxford University Press, 2010).

[Sp3]   T. Spencer, SUSY Statistical Mechanics and Random Band Matrices, In *Quantum Theory from Small to Large Scales* edited by J. Fröhlich et al.; Oxford University Press, 2012.

[ST]   C. Sabot and P. Tarres, Edge-reinforced random walk, Vertex-Reinforced Jump Process and the supersymmetric hyperbolic sigma model, http://arxiv.org/abs/1111.3991.

[Sto]   G. Stolz, An Introduction to the Mathematics of Anderson Localization, Contemporary Mathematics **552**, 71–108 AMS (2011), arXiv:1104.2317.

[SW]   L. Schäfer and F. Wegner, Disordered system with n orbitals per site: Lagrange formulation, hyperbolic symmetry, and Goldstone modes *Z. Phys.* B **38**, 113–126 (1980).

[SZ]   T. Spencer and M. R. Zirnbauer, Spontaneous symmetry breaking of a hyperbolic sigma model in three dimensions, *Comm. Math. Phys.***252**, 167–187 (2004).

[TV]   Tao, T. and Vu, V.: Random matrices: Universality of the local eigenvalue statistics. Acta Math., **206**, 127–204 (2011).

[VWZ]   J.J.M. Verbaarschot, H.A. Weidenmüller, and M.R. Zirnbauer. *Phys. Reports*, **129**, 367–438 (1985).

[Weg]   F. Wegner, The Mobility edge problem: Continuous symmetry and a Conjecture, *Z. Phys.* B **35**, 207–210 (1979).

[Wig1] E. P. Wigner, Characteristic vectors of bordered matrices with infinite dimensions. Ann. of Math. **62**, 548–564 (1955).

[Wig2] E. P. Wigner, Random Matrices in Physics Siam Review **9** 1–23, (1967).

[Zir1] M. R. Zirnbauer, The Supersymmetry method of random matrix theory, http://arXiv.org/abs/math-ph/0404057 (2004).

[Zir2] M. R. Zirnbauer, Fourier analysis on a hyperbolic supermanifold with constant curvature, *Commun. Math. Phys.* **141**, 503–522 (1991).

[Zir3] M. R. Zirnbauer, Riemannian symmetric superspaces and their origin in random-matrix theory, *J. Math. Phys.* **37**, 4986–5018 (1996).

INSTITUTE FOR ADVANCED STUDY, PRINCETON, NEW JERSEY, U.S.A.
E-mail address: spencer@math.ias.edu

Made in the USA
Monee, IL
07 July 2026